D1756991

WITHDRAWN
FROM
UNIVERSITIES
AT
MEDWAY
LIBRARY

DRILL HALL LIBRARY
MEDWAY

HMSOP
1600
43^^

Frontiers in Probability and the Statistical Sciences

3065355

Susmita Datta • Bart J.A. Mertens

Editors

Statistical Analysis of Proteomics, Metabolomics, and Lipidomics Data Using Mass Spectrometry

Springer

Editors
Susmita Datta
Professor
Department of Biostatistics
University of Florida
Gainesville, FL, USA

Bart J.A. Mertens
Department of Medical Statistics
and Bioinformatics
Leiden University Medical Centre
RC Leiden, The Netherlands

Frontiers in Probability and the Statistical Sciences
ISBN 978-3-319-45807-6   ISBN 978-3-319-45809-0   (eBook)
DOI 10.1007/978-3-319-45809-0

Library of Congress Control Number: 2016960566

© Springer International Publishing Switzerland 2017
This work is subject to copyright. All rights are reserved by the Publisher, whether the whole or part of the material is concerned, specifically the rights of translation, reprinting, reuse of illustrations, recitation, broadcasting, reproduction on microfilms or in any other physical way, and transmission or information storage and retrieval, electronic adaptation, computer software, or by similar or dissimilar methodology now known or hereafter developed.
The use of general descriptive names, registered names, trademarks, service marks, etc. in this publication does not imply, even in the absence of a specific statement, that such names are exempt from the relevant protective laws and regulations and therefore free for general use.
The publisher, the authors and the editors are safe to assume that the advice and information in this book are believed to be true and accurate at the date of publication. Neither the publisher nor the authors or the editors give a warranty, express or implied, with respect to the material contained herein or for any errors or omissions that may have been made.

Printed on acid-free paper

This Springer imprint is published by Springer Nature
The registered company is Springer International Publishing AG
The registered company address is: Gewerbestrasse 11, 6330 Cham, Switzerland

Preface

Following the genomics revolution, proteomics, metabolomics, and lipidomics studies have emerged, among others, as a natural follow-up either in the investigation of human biology or, similarly, in animal or plant studies. The combined use of these omics fields may lead to a more comprehensive understanding of system biology. In medical research, it may generate novel biomarkers which may be used in early detection of disease and for the development of new screening programs. Examples are in practical clinical patient monitoring, and in the development of new patient disease management rules for complex diseases such as cancer and cardiovascular or inflammatory diseases, among others. Another key objective of such comprehensive system-level study is in the elucidation of the molecular biochemical process associated with these biomarkers. In plant biology, relevant applications may be in the prediction of desirable properties of novel foods or crops or in assisting genetic manipulation-based breeding programs for new varieties, for example.

Within the omics revolution, proteins play a key role in the study of living organisms as they provide the essential link with the genome on the one hand, while they are also key components of the physiological metabolic pathways of cells. Proteomics is therefore a fundamental research field which investigates the structure or function of protein complexes consisting of multiple proteins simultaneously. Metabolites, on the other hand, are small molecules formed from the breakdown products of larger molecules, such as proteins, after undergoing a metabolic process within an organism. They are involved in (cell) signaling processes through stimulatory and inhibitory effects on enzymes, among others. Comprehensive metabolic profiling or metabolomics can give an instantaneous snapshot of the physiology of the cell. Lipidomics is closely related to metabolomics but studies a specific set of non-water-soluble molecules consisting of glycerol and fatty acids. The collection of all lipids in an organism is referred to as *lipidome*, in analogy to *genome*, *proteome*, *metabolome*, and so on. Although all these molecules have differences in both structure and functions, they can all be studied experimentally using modern spectrometric technology—specifically mass spectrometry—to assess the required omics expression of interest.

Unlike measurement procedures and methodology in genomics research, which is reasonably standardized at the time of writing, mass spectrometry is itself a vast field with many forms and variants. Typical vehicles are time-of-flight (MALDI) mass spectrometry, liquid chromatography-mass spectrometry (LC-MS), and Fourier transform ion cyclotron resonance mass spectrometry (FTICRMS or FTMS), among others. The field is therefore still very much in flux, with many distinct mass spectrometric measurement technologies and hence also different study designs in use. The data types generated in these studies are also very different. They tend to have complex structures, while no consensus data analytical approaches have yet been agreed upon. For these reasons, expert knowledge gained with one specific measurement platform does not easily carry over to other mass spectrometric systems. Writing a comprehensive overview text on statistical data analytic methodology in the new mass spectrometry-based omics field would therefore not be realistic. Instead, we have chosen to bring together a group of established researchers to present their expert knowledge in their specific application area within this emerging field. With this book, we want to provide an overview of the current status of such mass spectrometry-based omics data analysis and give impetus to the emergence of a common view on the design and analysis of such data and experiments. In this way, the book could support the development of more standardized templates, research practices, and references for any data analyst such as statisticians, computer scientists, computational biologists, analytical chemists, and data scientists, both in the omics application fields we discuss and in related omics fields such as glycomics.

Materials presented in this book cover a broad range of topics. First, we discuss the preprocessing of mass spectrometry data such as data normalization, alignment, denoising, and peak detection including monoisotopic peaks. Second, it provides methods for identification of proteins from tandem mass spectrometry data. There is also a chapter on a software package for the analysis of such omics data. Additionally, it has chapters on downstream data analysis using Bayesian and frequentist statistical predictive modeling and classification techniques. Last but not the least, there are chapters on specific examples of biomarker detection using proteomics, metabolomics, and lipidomics data. We hope that this book will be suitable for the scientists, practitioners, software developers, as well as advanced students of statistics, computer science, computational biology, and biomedical sciences.

Gainesville, FL, USA   Susmita Datta
Leiden, The Netherlands   Bart J.A. Mertens

Contents

Transformation, Normalization, and Batch Effect in the Analysis of Mass Spectrometry Data for Omics Studies

Bart J.A. Mertens

1 Spectrometry and Data Transformation

There is a long-standing literature on the application of transformation in spectrometry data, predating the advent of mass spectrometry. A good example may be found in the literature on infrared (IR) and near-infrared (NIR) spectrometry. An excellent recent introduction to statistical methods in this field was written by Naes et al. [13]. The log-transform has a special significance in traditional spectrometry. A crucial component of the appeal of the log-transform in (near) infrared spectrometry is due to Beer's law (1852) [2], which states that absorbance of light in a medium is logarithmically related to the concentration of the material through which the light must pass before reaching the detector. In other words,

$$\text{Absorbance} = \log\left(\frac{I_0}{I}\right) = kLC,$$

where I_0 is the incident intensity of the light and I the measured intensity after passing through the medium, with C the concentration, L the path length the light travels through, and k the absorption constant. Another way to put this is that relative intensity is linearly related to concentration through the log-transform as

$$\log\left(\frac{1}{I}\right) = \alpha + \beta C,$$

B.J.A. Mertens (✉)
Department of Medical Statistics, Leiden University Medical Center,
PO Box 9600, 2300 RC Leiden, The Netherlands
e-mail: b.mertens@lumc.nl

© Springer International Publishing Switzerland 2017
S. Datta, B.J.A. Mertens (eds.), *Statistical Analysis of Proteomics, Metabolomics, and Lipidomics Data Using Mass Spectrometry*, Frontiers in Probability and the Statistical Sciences, DOI 10.1007/978-3-319-45809-0_1

where the path length, initial intensity, and absorption constants are subsumed in the parameters α, β. Similar formulae exist for light reflection spectrometry. The formula has been used to provide justification for the application of classical linear regression procedures with log-transformed univariate spectrometric intensity readings.

The above constitutes an argument in favor of the log-transform for spectrometry data based on non-linearity of spectral response. It is partly responsible for causing the early literature on statistical and chemometric approaches in the analysis of spectrometry data to be based on the log-transformed measures. Beer's law applies only to (univariate) IR or NIR spectrometric readings at a single wavelength. There is no multivariate extension of the law to cover full spectra consisting of the spectrometric readings across an entire wavelength range. Nevertheless, the log-transform would also be routinely applied once truly multivariate spectrometry became available, jointly recording the spectrometric response at several wavelengths, as shown in Fig. 1 which plots 50 NIR reflectance spectra across the range 1100–2500 nm.

Log-transforming was heavily embedded in the statistical spectrometry literature, when modern laser-based mass spectrometry measurement became routinely available by the end of the twentieth century. Although Beer's law does not apply to mass spectrometry, the log-transform continues to be key to mass spectrometry-based

Fig. 1 Log infrared reflectance measurements on 50 samples of mustard seeds within the 1100–2500 nm wavelength range

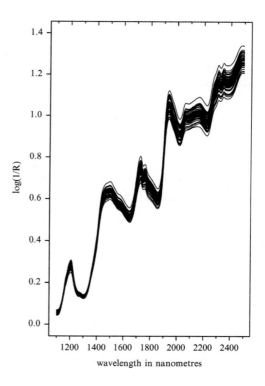

omics data analysis. This is because the statistical reasons and rational behind the log-transform are more enduring and powerful than the appeal to Beer's law might reveal. We discuss four distinct arguments in favor of the log-transform.

1.1   Scale and Order of Magnitude Effects

The first argument is associated with the objective of mass spectrometry to estimate protein composition in complex mixtures consisting of large numbers of proteins which may differ substantially **both** in the *masses* of the constituent proteins in the mixture and in the *abundances* in which these are present. It is precisely because of this objective that mass spectrometry has become a tool of choice in modern omics research. Indeed modern mass spectrometric instruments are specifically engineered to have the capability to measure protein concentrations across large ranges of abundance, typically spanning several orders of magnitude—known as the so-called dynamic range—and to simultaneously achieve this across a wide mass range.

Unfortunately, the spectrometry engineer's delight then becomes the statistical analyst's nightmare as this property renders data which spreads across several orders of magnitude in spectral response. A good example may be found in Fourier-transform spectrometry, which can easily display variation across 5–6 orders of magnitude. In extreme cases this may cause numeric overflow problems in analyses, although use of modern professional statistical software may reduce this problem. Taking logarithms removes the order of magnitude effects.

1.2   Skew and Influential Observations

A second issue related to the above is that spectral measures will tend to be extremely skewed, not only within an individual (across the within-sample spectrum responses), but also across samples or patients at a single m/z point. This may render the data unsuitable for standard analyses such as linear discriminant or similar when used on the original scale.

A related issue is that the skew may cause or be associated with a limited set of highly influential observations. This is particularly troubling in omics applications, as the spectra are typically very high-dimensional observations either when storing the response on a large grid of m/z values which can easily range in the thousands, or after reduction to an integrated peak list. Influential observations may affect the robustness of conclusions reported, as results may differ substantially after removal of a single—or isolated group of—spectra from the analysis. A good example may be found in the calculation of a principal component decomposition as a dimension reduction prior to application of some subsequent data-analytic procedure. Principal component analysis is known to be sensitive to extreme observations [7], particularly in high-dimensional applications with small sample size. The same phenomenon will, however, also tend to apply for other analysis approaches, such as regression methods and discriminant procedures. Transformation to log-scale may mitigate this problem.

1.3 Statistical Properties of Particle Counting

At some risk of oversimplification, mass spectrometers are in some sense nothing else but sophisticated particle counters, repeating a particle counting operation at each m/z position along the mass/charge range which is being investigated. As a consequence, spectrometry measures tend to have statistical properties reminiscent of those observed in Poisson (counting) processes. The variation of the spectral response tends to be related to the magnitude of the signal itself. This implies that multiplicative noise models are often a more faithfully description of the data. Multiplicative error data are, however, more difficult to analyze using standard software. Log-transforming can be used to bring the data closer to the additive error scenario.

1.4 Intrinsic Standardization

An interesting property of log-transforming is that it may lead to intrinsic standardization of the spectral response. Imagine, for example, that we are interested in calibrating the expected value $E(Y)$ of some outcome Y based on two spectral responses X_1 and X_2 and that we have a study available recording both the observed outcome and the two spectrometry measures across a collection of samples (such as patients). Let us also assume that we can use some generalized linear model to link the expected outcome to the spectrometry data via some link function f and that the true model may be written in terms of a linear combination of the log-transformed spectral measures

$$f(E(Y)) = \alpha + \log(X_1) - \log(X_2)$$

which reduces to

$$f(E(Y)) = \alpha + \log\left(\frac{X_1}{X_2}\right).$$

The result is that the linear dependence of the expected outcome via the link function on the log-transformed data actually implies regression on the log-ratio of the spectral responses X_1 and X_2, such that any multiplicative effect would cancel. This can be regarded as an implicit form of standardization through the log-transform. It is a general property which can be used in many statistical approaches such as (generalized) linear regression and discriminant analysis.

For spectrometry data generally and mass spectrometry particularly, generally good advice would be to replace the raw measurements with log-transformed values at an early stage of the analysis, by application of the transformation

$$\log(Y + a)$$

with a a suitably chosen constant. See also [14] for comments about transformation for mass spectrometry data and the log-transform in particular. Many statistical texts will also mention the Box-Cox transform when discussing the log-transform. For (mass) spectrometry data this approach is of limited value however, because the optimal transform may lack the multiple justifications given above—which might as well be used as a priori grounds for choosing the logarithm—but also because the approach is by definition univariate, while a modern mass spectrometry reading will consist of thousands of measures across m/z values and samples.

2   Normalization and Scaling

Normalization is an issue which is often encountered in omics data analyses, but is somewhat resistant to a precise definition. We can identify what are usually perceived as the main objectives of it and warn about the dangers associated with the topic, so that we may avoid the most common pitfalls.

The objective of normalization can be loosely described as removing any unwanted variation in the spectrometric signal which cannot be controlled for or removed in any other way, such as by modifying the experiment, for example. This sets it apart from batch effect, which we will discuss later. The latter can sometimes be adjusted for or accommodated by changing the experiment so that its effects can be either explicitly removed or adjusted for in subsequent analysis, by exploiting the structure of the experimental design. Not all effects can be accounted for in this manner, however.

Examples of such effects which may induce a need for normalization are variations in the amount of material analyzed, such as ionization changes, e.g., small changes in "spotting" sample material to plates, subtle fluctuations in temperature, small changes in sample preparation prior to measurement, such as bead-based processing to extract protein, differential sample degradation, sensor degradation, and so forth. An important feature of such variation is that, while we may speculate such variability sources are there and affect our experiment, they are difficult to either control or predict, which typically means all we can do is try to post hoc adjust for it, but prior to any subsequent analysis steps.

Important in devising an appropriate normalization strategy is that we should try to remove or reduce these effects on the measured spectral data, while retaining the relevant (biological) signal of interest. Unfortunately, this is typically problematic for spectrometry. This is because, as explained in the above paragraphs on transformation, the spectrometric signal and its variability are typically linked, often even after log-transform, while the unwanted sources of variation affect all measures derived from the spectrum.

Imagine we have a study recording a mass spectrum on a dense grid of finely spaced points along the mass range or alternatively storing the data as a sequence of integrated peaks representing protein or peptide abundances and this for a collection of samples (be it patients, animal, or other). We write the ordered sequence of spectrometry measures for each ith sample unit as $\mathbf{x}_i = (x_{i1}, \ldots, x_{ip})$, with p the number of grid points at which the spectrum is stored or the number of summary peaks. A transformation choice favored by some analysts is to apply early on in the analysis (possibly after first application of the log-transform) standardization to unit standard deviation of all spectral measures at each grid point separately. In other words, we replace the original data at each jth gridpoint with the measures $x_{ij}/\text{std}(\mathbf{x}^j)$ where $\text{std}(\mathbf{x}^j)$ is the standard deviation of the measures $\mathbf{x}^j = (x_{1j}, \ldots, x_{nj})^T$ across all n samples at that gridpoint. This procedure, sometimes also referred to as reduction to z-scores, is a form of scaling. It is identical to standardization to unit standard deviation of predictor variables in regression analysis ([16, pp. 124–125], [15, pp. 349, 357–358]) when predictor variables are measured at different measurement scales (different units, such as kg, cm, and mg/l). Indeed, in the early days of regression analysis, reporting standardized regression coefficients was an early attempt at assessing relative importance of effects.

For spectrometry data generally and mass spectrometry in particular, standardization is more complicated. Measurement units are by definition identical within a spectrum across the mass range. This would counsel against transforming to unit standard deviation as calibrated effects then remain directly comparable across the mass range on the original untransformed scale. There are, however, stronger arguments against this form of standardization in spectrometry. Figure 2 illustrates the issues. The plot shows mean (MALDI-TOF) spectra from a clinical case–control study, after suitable transformation. To ease comparison, we plot the negative control spectrum versus the mean case spectrum. The rectangular region highlights and enlarges a region between 1200 and 1900 Da where most of the discriminant effects are found between the cases and control groups, based on a discriminant analysis. Indicated are four key peaks at 1352.4, 1467.7, 1780, and 1867.2 Da which together summarize most of the between-group contrast between cases and controls. Figure 3 shows different statistics calculated on the same data within the same mass range. The top plot again shows the mean spectra for cases and controls within the 1200–1900 Da region as before, while the middle graph plots a graph of weighted discriminant coefficients obtained from a linear discriminant model calibrated from the data. It is obvious how the discriminant analysis identifies the peaks at 1354.2 and 1467.7 Da and contrasts these with the peaks at 1780 and 1867.2 Da. The below graph in Fig. 3 shows the first two principal components calculated on the same data and based on the pooled variance–covariance matrix.

There are several things to note in this picture. The first is how much the principal component and mean spectra curves resemble one another. The first component

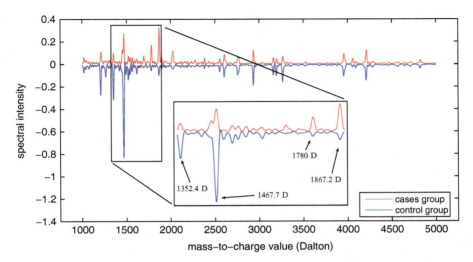

Fig. 2 Mean spectra for cases and controls separately in a case–control study. The negative of the mean control group spectrum is plotted to improve readability

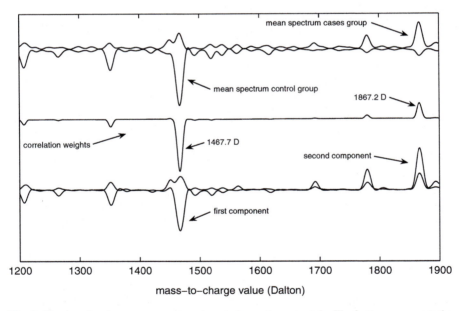

Fig. 3 The *top plot* shows mean cases and controls spectra separately. The *bottom curves* are the loadings of the first two principal components across the same mass/charge range. The *middle curve* shows the discriminant weights from a logistic regression model calibrated to distinguish cases from controls with the same data

closely approximates the mean control spectrum, while the second component does the same for the mean cases spectrum. At first sight, this might seem all the more remarkable, since the principal component decomposition is based on the pooled variance–covariance matrix, and hence on the "residual spectra" $x_i - \bar{x}_{g(i)}$, where $\bar{x}_{g(i)}$ denotes the mean spectrum of group $g(i)$ to which the ith observation belongs, with $g = 1, 2$ for the cases and control groups, respectively. So the figure shows two different aspects of the data. One is the systematic (mean) spectral response (top graphs), the other are the deviations relative to the mean spectral outcome (bottom graphs). From the figure, we can see that the component decomposition tells us that the peaks at 1352.4 and 1467.7 Da are highly correlated and account for much of the variation in the spectral data, as they weigh heavily in the first principal component. Similarly, the second component summarizes much of the expression in both peaks at 1780 and 1867.2, which are again highly correlated. Because of this, the classification might as well be summarized as a contrasting between the first and second principal component, since this would contrast peaks 1352.4 and 1467.7 with the expression at peaks 1780 and 1867.2 (see Mertens et al. [11] for the full analysis).

This feature of the data where the mean expression and deviations from the mean are closely linked as shown in the above example is typical of spectrometric variation. It is the consequence of the connection between mean expression and variance we mentioned above when discussing the log-transform and can be observed in almost all spectrometry data, often even after log-transforming. To put this differently, in spectrometry data, we will find the signal where the variance is (even if we correct the variance calculation for systematic differences in expression, as shown in our above example). It is for this reason that transforming to unit standardization should be avoided with spectrometry data, unless scale-invariant methods are explicitly used to counter this problem.

In addition to the above considerations, there are also other arguments for avoiding reduction to z-scores or transformation to unit standard deviation. An important argument here is that summary measures such as means and standard deviations are prone to outliers or influential observations, which can be a particular problem in high-dimensional statistics and with spectrometry in particular. A specific problem with such form of standardization is that it may cause problems when comparing results between studies. This is because systematic differences may be introduced between studies (or similarly, when executing separate analyses between batches—see further), due to distinct outliers which affect the estimates of the standard deviations for standardizing between repetitions of the experiment.

Our final comment on the above standardization approach is that medians and inter-quartile ranges (IQR) are sometimes used instead of means and standard deviations in an attempt to alleviate some of the robustness concerns. Other authors advocate use of some function of the standard deviation, such as the square root of the standard deviation instead of the standard deviation itself. This is sometimes referred to as Pareto scaling [17]. The rational for this amendment is that it upweights the median expressed features without excessively inflating the (spectral) baselines. An advantage of the approach may be that it does not completely remove the scale information from the data. Nevertheless, the choice of the square root would still appear to be an ad hoc decision in any practical data-analytic setting.

Some authors make a formal distinction between scaling and normalization methods and consider the first as operations on each feature across the available samples in the study [5]. Normalization is then specifically defined as manipulation of the observed spectral data measurements on the same sample unit (or collection of samples taken from the same individual) (within-spectrum or within-unit normalization). A potential issue with the above-described approaches to normalization via statistical transformation is that they are based on a borrowing of information across samples within an experiment. Another extreme form of such borrowing is a normalization approach which replaces the original set of spectral expression measures for a specific sample with the sample spectrum measures divided by the sample sum, such that the transformed set of measures adds to 1. An argument sometimes used in favor of such transformation is that it would account for systematic differences in abundance—possibly caused by varying degrees of ionization or similar effects from sample to sample—such that only the relative abundances within a sample are interpretable. Although this approach is unfortunately common, it has in fact no biological foundation [5]. Even if arguments based on either the physics or chemical properties of the measurement methodology could be found, these could not be used in favor of such data-analytic approach as described above, which we shall refer to as "closure normalization." The problem with the approach is that it actually *induces* spurious—and large—*biases* in the correlations between the spectral measures which mask the true population associations between the compounds we wish to investigate. Figure 4 shows the effect of closure normalization on uncorrelated normal data in three dimensions. The left plot shows scatterplots between each of the three normally distributed measures. The right shows scatterplots of the resulting transformed variables after closure normalization. The absolute correlations between each variable pair has increased from 0 (for the original uncorrelated data) to 0.5 (after transformation). This becomes particularly problematic should the subsequent objective be to perform some form of network or association analysis, in which case the prior closure normalization renders results meaningless. Similarly problems would, however, also apply to regression and discriminant analysis.

A variant of the above closure approach to normalization which is sometimes also used is to adjust to the maximum peak observed in a spectrum, where the maximum is either the spectrum-specific maximum, or the maximum at that spectral location which corresponds to the maximum mean spectrum across several samples. Adjustment is often carried out by dividing each spectrum by its maximum at the maximum location, such that the spectral response gets a constant expression at the maximum location in the transformed data. Just as for closure normalization, this procedure appears appealing on intuitive grounds at first sight, but suffers from similar problems, as the variation at the maximum induces severe correlations across the spectral range in the transformed data, which *cannot have biological interpretation*. Both approaches should be avoided.

Data normalized by the sum of the combined expression (closure normalization) can be viewed as an instance of compositional data [1]. Hence, instead of applying such normalization, one could therefore think of using special-purpose methods from the compositional data analysis literature, or to develop or adapt such methods for application in omics applications. This has not been attempted to our knowledge at the time of writing. As an alternative, it should be recommended to take a conservative approach and refrain from excessive transformation when the consequences are not well understood or accounted for in subsequent analysis. In such cases restricting to log-transformation as discussed earlier is safer. In any case, the original untransformed data should always be at hand and stored to allow verification of results through possible sensitivity analysis.

The above is only an introduction to some of the main forms of normalization in use at this time. Many other forms exist and will undoubtedly continue to emerge. An interesting one worth mentioning is the idea of "lagging" the spectra by taking differences between subsequent values within the spectral range. With log-transformed spectrometric data, this is another approach which induces ratios between subsequent spectral intensities which eliminate multiplicative change effects. An example is found in an interesting paper by Krzanowski et al. [10]. It has the drawback that results from subsequent statistical analysis can be more difficulty to interpret, but it might be of use in pure prediction problems. Other forms of standardization and normalization are also found in the literature, particularly methods which seem to inherit more from common approaches in microarray analysis, such as quantile normalization [6]. Ranking of spectral response including the extreme form of reducing to binary has also been investigated. The latter can be particularly useful as a simple approach when data are subject to a lower detection

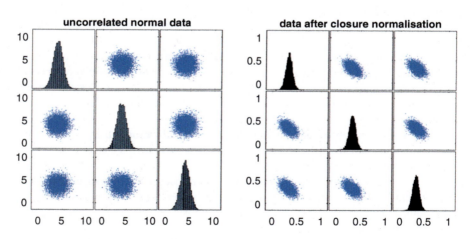

Fig. 4 Effect of closure normalization on uncorrelated data

limit [8]. Other forms of normalization and standardization worth mentioning at time of writing are scatter correction and orthogonal signal correction. We refer to Naes et al. [13] for a good introduction to these methods.

Which transformations should be applied first? What is a good order of applying distinct normalization or transformation steps? There is some difference of opinion between researchers on the precise sequence in which various normalization procedures are applied to the data. As a general rule it seems wise to apply logarithms early and calculate means and standard deviations only after log-transforming.

The issue of normalization is closely linked to the problem of standardization of mass spectra. Several definitions of standardization may be possible here. One option is to define the problem as "external" standardization, which would form part of the experiment itself (as opposed to the post-experimental data processing we describe before) where we somehow try to change the experiment so that part of the systematic experimental variation is either prevented from occurring or could be accounted for through post-processing of the data. Examples would be in the use of spike-in controls, on a sample plate, or even within the sample material itself, so that the spectral response can be adjusted for the expression of the known spike-in material which is added. Another example would be in the use of technical controls on a sample plate with known concentrations. Yet another example would be systematic equipment re-calibration to re-produce a (set of) known standards, so that sample-to-sample variation due to experimental drift is suppressed as much as possible. All these approaches to standardization are different from the above described methods in that they try to circumvent known sources of variation by changing the experiment itself, rather than post hoc attempting to adjust for it.

3   Batch Effect in Omics Studies

A batch effect is a source of unwanted variation within an experiment which is typically characterized by the property that the effect is due to a known structure within the experiment. Usually the structure, and thus also the existence of the associated effect, is known in advance of the experiment but cannot be avoided or eliminated from the measurement process. Furthermore, the effect of the experimental structure itself may not necessarily be predictable either, even when knowledge of the structure exists. A typical example would be distinct target plates which are used to collect sample materials prior to analysis in some mass analyzer. However, we can sometimes manipulate or modify the experiment to account for the known structure so that it is rendered innocuous and cannot unduly affect conclusions. Even better, we may be able to remove (part of) the effects due to batch structure, by taking suitable precautions at the experimental design stage.

In contrast to genetics, batch effect is much more problematic within proteomics and metabolomics, which is due to the measurement procedures involved. It should also be noted that batch effect and the accommodation of it are different from standardization and normalization issues in that we need to identify its presence and consider its potential effects both *before* and *after* the experiment. Before the experiment, because we may want to tweak or change the experimental structure to take the presence of the batch effect into account (this may involve discussion between both statistician and spectrometrist). The objective is to change the experimental design so as to avoid confounding of the batch structure with the effect or group structure of interest. After the experiment, because we may wish to apply some data-analytic approach to remove the effect (which may depend on the experiment having been properly designed in the first place). Accounting for batch effect in proteomic studies will hence involve two key steps.

1. To control the experimental design as much as possible in advance of the experiment to maximize our options to either remove or account for batch effect at the analysis stage.
2. To exploit the chosen structure afterwards to either remove or adjust for the batch variation in the analysis.

The objectives of these steps are at least threefold.

1. To ensure experimental validity (the experiment can deliver the required results)
2. To improve robustness of the experiment (conclusions will still be valid, even when experimental execution differs from experimental planning)
3. To improve (statistical) efficiency of effect estimates based on the data (statistical summaries will have lower variation).

In the following discussion on batch effect we will make a distinction between the following two types of batch effect which may occur in practical experimentation. The first are **time-fixed batch effects**. Examples of these are

- plates in mass spectrometry
- freezer used for sample storage
- change of cleaning or work-up fluid for sample processing
- batches of beads or cartridges for protein fractionation
- instrument re-calibration.

In contrast, **time-dependent batch effects** are due to experimental structures associated with time. Examples of these are found in the following situations:

- longitudinal experimentation, long observation tracks with repeated measurement per individual
- sample collection across extended time periods, patient accrual spread across an extended period of time
- time-indexed samples or instrument calibrations
- distinct days of measurement or sample processing.

Both types of batch effects may occur in proteomic experiments, but for different reasons and with distinct consequences. The treatment of both types of batch effect will also be different between the two.

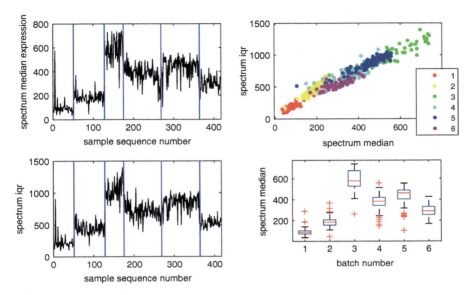

Fig. 5 Batch effect in a case–control study using six target plates. The *vertical lines* in the *left-side plots* indicate transition to a new plate (batch). The *distinct colors* in the scatter plot represent the different batches

3.1 Time-Fixed Batch Effects

We consider a case–control study as an example. The experiment contrasted 175 cases with 242 healthy controls. Due to the large sample size and multiple replicate measurements per sample, sample material needed to be assigned to six target plates prior to mass spectrometric measurement. The plates constitute a systematic—batch—structure within the experiment. On inspecting within-sample medians and inter-quartile ranges (IQR), systematic plate-to-plate differences were noted, shown in Fig. 5. Analysis of the data using a discriminant approach led to correct classification of 97 % of samples. Unfortunately, on closer investigation of the experimental design, it turned out that all case material had been assigned sequentially to the first three plates, after which the controls samples were thawed next and assigned to the subsequent plates. It is not possible to statistically adjust for such perfect confounding between the plate structure and the potential between-group effect which was the primary target of the research study. After discussion, the study was abandoned, leading to significant loss of both experimentation time and resources, among which the valuable sample materials.

A simple procedure exists to prevent such problems, called blocked randomization. It consists of assigning cases and controls in equal proportions and at random across the distinct plates. Table 1 shows such a design for a case–control study randomizing cancer cases and healthy controls to three target plates, as reported by Mertens et. al. [11], which gives more details about the study. In addition to

Table 1 Block-randomized case–control study in a study using three plates

	Plates													
Disease group	1				2				3				Total number	
Number of controls	17				17				16				50	
Number of cases	22				22				19				63	
Disease stage	1	2	3	4	1	2	3	4	1	2	3	4		
Cases	4	10	4	4	4	10	4	4	3	8	4	4		

Table 2 Case–control study assigning samples to 34 plates as they became available

ca	co	ca	co	ca	co	ca	co
4	0	1	3	0	15	1	6
3	0	0	9	1	0	1	9
11	0	4	0	3	0	0	7
5	0	1	0	0	4	0	4
12	0	1	0	1	0		
21	40	1	0	0	5		
1	3	2	0	1	9		
2	0	0	4	2	8		
16	13	0	3	0	4		
0	15	0	16	2	14		

The assignment order is from top-to-bottom, left-to-right

randomizing the cases and controls in roughly equal proportions across the plates, the study also tried to have cancer stages in roughly equal distributions from plate-to-plate. A recent overview of classical principles of statistical design for proteomics research can be found in a paper by Cairns [3] and the references therein.

Another example is a glycomics study which assigned cases and controls to target plates sequentially as the samples became available. By the end of the study period, 288 samples had become available, of which 97 were cases and 191 controls. The case–control assignment to plates is shown in Table 2. As can be seen, case–control assignment is perfectly confounded with the plate effect for 25 out of 34 experimental batches, which makes these measurements useless for between-group comparison. Note also how only two plates indicated in red typescript contain appreciable numbers of *both* cases and controls. After analysis of the data, it was found that the estimate of the batch effect (std) substantially exceeded the measurement error estimate (std). No clear evidence of differential expression of glycans between groups emerged, though it could be hypothesized that any differences might be small and at least smaller than the observed between-batch variation. This raised questions whether the "design" was to blame for the failure of the study to identify differential expression.

To investigate the consequences of using a "design" as used in the above glycomics study, we investigate a simulation study which contrasts several potential alternative designs. For each of these four designs we assume the same sample size of 288 samples with 97 cases and 191 controls, just as for the original study. We consider the following alternative scenarios:

1. Experiment 1 is a hypothetical experiment which is able to use only one plate (no plate effects present).
2. Experiment 2 is a randomized blocked version of the original experiment which uses three plates and distributes the cases and controls in approximately balanced ratios of 32/65, 32/63, and 33/63 to the first, second, and third plate, respectively.
3. Experiment 3 is a perfectly confounded experiment which again uses three plates, but assigns all 97 cases to the first plate and assigns 95 and 96 controls to the other two plates, respectively.
4. The last experiment uses the original glycomics experiment distribution of cases and controls to 34 plates, as shown above in Table 2.

We now simulate experimental data for each of the above four scenarios, generating effect sizes ranging from 0 to 1.5 for a *single* glycan (univariate simulation) and assuming between-batch effects with standard deviations σ_B taking values 3.6, 1.8, 0.9, and 0.45. The standard deviation of the error σ_E takes the value 1.8 throughout (these numbers inspired by results from the real data analysis).

In the analysis of the simulated data for the above experiments, we fit linear mixed effect models [12] to the simulated data of experiments 2, 3, and 4. The mixed effect models correct for the known batch structure using a random effect while estimating the between-group effect with a fixed effect term. For the first simulated experiment a simple linear regression model is used, which is equivalent to a two-sample pooled *t*-testing approach. Figures 6, 7, 8, 9, and 10 show the probabilities to detect the between-group effect (power) across the effect size range simulated and for the standard deviations of the batch effect indicated. As expected, we find that the probability to detect the effect increases as the effect size grows for the single-plate experiment [1]. The blocked experiment [3] matches the power of the single-plate experiment regardless of the size of the batch random effect. The power of the actually implemented glycomics experiment [4] depends on the size of the batch effect. As the batch effect gets smaller (associated std goes to zero) then we can eventually ignore the batch effect altogether and we obtain the same powers as if the batch effect was not present. This is of course a confirmation of what we would expect to find. If the batch effect is substantial however (associated std is large relative to the size of the between-group effect), then we pay a penalty in terms of seriously reduced powers of detecting the between-group effect. The perfectly confounded experiment [3] performs dramatically whatever the batch effect is, since we are forced to account for a known batch structure—irrespective of the true but unknown population batch effect—and thus loose all power of detecting the effect of interest. The excellent performance of the blocked version of the experiment again emphasizes the need and importance of pro-actively designing and implementing block-randomized experiments when batch structures are identified in advance of the experimentation.

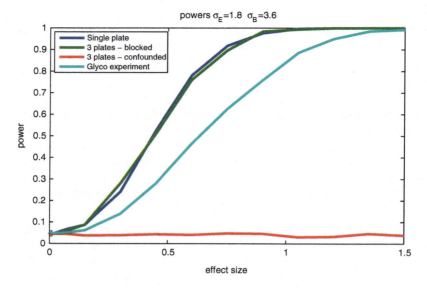

Fig. 6 Power curves for four distinct study designs, with error standard deviation $\sigma_E = 1.8$ and batch effect standard deviation $\sigma_B = 3.6$

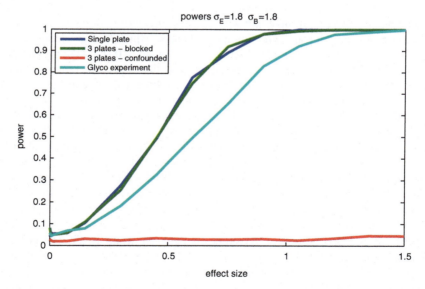

Fig. 7 Power curves for four distinct study designs, with error standard deviation $\sigma_E = 1.8$ and batch effect standard deviation $\sigma_B = 1.8$

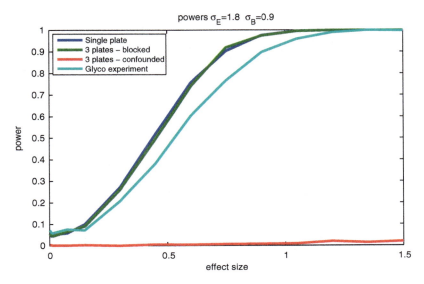

Fig. 8 Power curves for four distinct study designs, with error standard deviation $\sigma_E = 1.8$ and batch effect standard deviation $\sigma_B = 0.9$

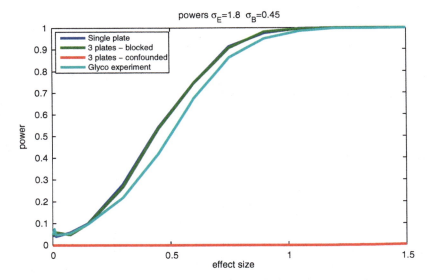

Fig. 9 Power curves for four distinct study designs, with error standard deviation $\sigma_E = 1.8$ and batch effect standard deviation $\sigma_B = 0.45$

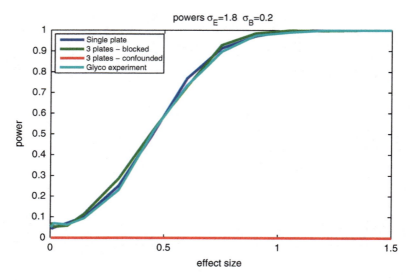

Fig. 10 Power curves for four distinct study designs, with error standard deviation $\sigma_E = 1.8$ and batch effect standard deviation $\sigma_B = 0.2$

3.2 Time-Dependent Batch Effects

Longitudinal experimentation with spectrometry-based omics is still in its infancy, but we should expect this area to mature. Experience with the analysis of such data must still develop. The same applies to pre-processing and methods for dealing with batch effects. We can, however, already point to some aspects which are likely to become a key concern in future applications for which omic science and technology will have to provide credible answers. A key aspect which is liable to be an issue in many omic studies is illustrated in Fig. 11. The figure displays a study design where samples are collected dispersed over time and kept in storage until a decision is taken to extract sample material and analyze the samples using some form of spectrometry at a common point in time. Patients themselves are followed up until end of the study—up to which point some outcome measure may be continually recorded during the follow-up—or some pre-defined endpoint (such as cancer occurrence or recurrence or similar outcome) prior to the study termination. A common and relatively simple form of such a design which is often used in clinical research could, for example, be a survival study [4, 9] with the so-called administrative censoring as depicted in Fig. 11, where end of the study represents the censoring time. The issues discussed in this section are, however, completely general to longitudinal study with staggered patient entry and not restricted to study with a specific (survival) outcome. The key issue is that such study designs are likely to suffer from potential ageing of the sample material, prior to spectrometric measurement, such that sample age and duration to observation of the outcome become confounded. This problem may

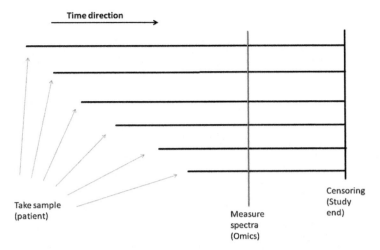

Fig. 11 Longitudinal omic study design with sampling-time confounded with follow-up

severely complicate future longitudinal studies. This point does not always seem to be realized in the present literature. This is probably also part due to the fact that genomic or gene expression data, as discussed before, is likely not to be affected by the same problem as would be the case for the more modern spectrometry-based omics data, as in proteomics, for example. An excellent text describing modern survival analysis methodology in novel high-dimensional data applications was recently provided by van Houwelingen and Putter [18, Part IV—Chaps. 11 and 12].

It is at time of writing not clear how this issue should be addressed in design and analysis of longitudinal and survival studies with spectrometry-based omic data generally. The problem is particularly important because it could affect all existing biobanks. One could propose that instead of—or in addition to—the development of *biobanks* which store sample materials, attention should be given to establishing *databanks* of (omic) spectra, which are measured at pre-specified and regular time points instead. Such an approach could break the confounding between the follow-up time of patients which is then de-coupled from the measurement times. The problem with the latter proposal is that the measurement devices (spectrometers) themselves may exhibit ageing, such that the ageing problem is replaced with a spectrometer calibration problem.

4   Discussion

We have critically discussed and contrasted the distinct issues of transformation, normalization, and management of batch effect in the analysis of omic data. Transformation, standardization, and normalization are typically dealt with "after" the experiment and usually by the data analyst or statistician involved. Batch effect and the presence thereof is an issue that should be considered both "before" and "after" the data-generating experiment. The objectives here should be to optimize the design for known batch effect such that these cannot unduly affect conclusions or invalidate the experiment. This task is usually carried out in collaboration and prior discussion between both statistician and spectrometrist when planning the experiment. Secondly, appropriate methods should be used after the experiment to either eliminate or otherwise accommodate the batch effect after the experiment when analyzing the data. The latter task will usually be carried out by the statistician solely. Discussion of such methods falls outside the scope of this chapter.

Methodological choice in pre-processing of spectrometry data should in practice depend on many aspects. Purpose of the study is a key consideration. If prediction or diagnostic models are to be calibrated, then pre-processing methods which can be applied "within-sample" without borrowing of information across distinct samples are more attractive because this can make subsequent calibration of any predictive rule easier. Other considerations may be ease of communication of results, established practice (insofar it is reasonable of course), variance stabilization, interpretability, and so forth. Our discussion has not been comprehensive but highlighted the main ideas and approaches instead, pointing to common pitfalls, opportunities, and future problems left to be solved.

Acknowledgements This work was supported by funding from the European Community's Seventh Framework Programme FP7/2011: Marie Curie Initial Training Network MEDIASRES ("Novel Statistical Methodology for Diagnostic/Prognostic and Therapeutic Studies and Systematic Reviews," www.mediasres-itn.eu) with the Grant Agreement Number 290025 and by funding from the European Union's Seventh Framework Programme FP7/ Health/F5/2012: MIMOmics ("Methods for Integrated Analysis of Multiple Omics Datasets," http://www.mimomics.eu) under the Grant Agreement Number 305280.
Thanks to Mar Rodríguez Girondo for critical comments on an early version of this text.

References

1. Aitchison, J. (1986). *The statistical analysis of compositional data*. Caldwell: Blackburn.
2. Beer, A. (1852). *Annalen der Physic und Chime, 86*, 78–88.
3. Cairns, D. A. (2011). Statistical issues in quality control of proteomic analyses: Good experimental design and planning. *Proteomics, 11*, 1037–1048.
4. Cox, D. R., & Oakes, D. (1984). *Analysis of survival data*. London: Chapman and Hall.
5. Craig, A., Cloarec, O., Holmes, E., Nicholson, J. K., & Lindon, J. C. (2006). Scaling and normalization effects in NMR spectroscopic metabonomic data sets. *Analytical Chemistry, 78*, 2262–2267.
6. Eidhammer, I., Barsnes, H., Eide, G. E., & Martens, L. (2013). *Computational and statistical methods for protein quantification by mass spectrometry*. Chichester: Wiley.
7. Jolliffe, I. T. (2002). *Principal component analysis*. New York: Springer.
8. Kakourou, A., Vach, W., Nicolardi, S., van der Burgt, Y., & Mertens, B. (2016). Accounting for isotopic clustering in Fourier transform mass spectrometry data analysis for clinical diagnostic studies. *Statistical Applications in Genetics and Molecular Biology, 15*(5), 415–430. doi:10.1515/sagmb-2016-0005.

9. Klein, J. P., van Houwelingen, H. C., Ibrahim, J. G., & Scheike, T. H. (2014). *Handbook of survival analysis*. Boca Raton: Chapman and Hall/CRC Press.
10. Krzanowski, W. J., Jonathan, P., McCarthy, W. V., & Thomas, M. R (1995). Discriminant analysis with singular covariance matrices: Methods and applications to spectroscopic data. *Applied Statistics, 44*, 101–115.
11. Mertens, B. J. A., De Noo, M. E., Tollenaar, R. A. E. M., & Deelder, A. M. (2006). Mass spectrometry proteomic diagnosis: Enacting the double cross-validatory paradigm. *Journal of Computational Biology, 13*(9), 1591–1605.
12. Molenberghs, G., & Verbeke, G. (2000). *Linear mixed models for longitudinal data*. New York: Springer.
13. Naes, T., Isaksson, T., Fearn, T., & Davies, T. (2002). *A user-friendly guide to multivariate classification and calibration*. Chichester: NIR Publications.
14. Sauve, A., & Speed, T. (2004). Normalization, baseline correction and alignment of high-throughput mass spectrometry data. In *Proceedings of the Genomic Signal Processing and Statistics Workshop*, Baltimore, MO.
15. Snedecor, G. W., & Cochran, W. G. (1980). *Statistical methods* (7th ed.). Ames: Iowa State University Press.
16. Vach, W. (2013). *Regression models as a tool in medical research*. Boca Raton: Chapman and Hall/CRC Press.
17. van den Berg, R. A., Hoefsloot, H. C. J., Westerhuis, J. A., Smilde, A. K., & van der Werf, M. J. (2006). Centering, scaling and transformations: Improving the biological information content of metabolomics data. *BMC Genomics, 7*, 142.
18. van Houwelingen, H. C., & Putter, H. (2012). *Dynamic prediction in clinical survival analysis*. Boca Raton: Chapman and Hall/CRC Press.

Automated Alignment of Mass Spectrometry Data Using Functional Geometry

Anuj Srivastava

1 Introduction

The use of mass spectrometry data in profiling metabolites present in a specimen is important in biomarker discovery, enzyme substrate assignment, drug development, and many other applications. Liquid chromatography-mass spectrometry (LC-MS) is a common data collection technique in this area and it provides information about retention times of different metabolites. These metabolites are identified by peaks (high y values) in observed chromatograms at the corresponding retention times (x axis). As stated in [17], it is difficult to reproduce run-to-run retention times across experiments/observations due to instrument instability. Different measurements can exhibit variability in peak locations when observing the exact same specimen. In liquid chromatography, the common causes of such shifts in peak locations are changes in column separation temperature, mobile phase composition, mobile phase flow rate, stationary phase age, etc. We illustrate this issue using an example taken from [7] involving proteomics data collected for patients having therapeutic treatments for Acute Myeloid Leukemia. This example studied earlier in [16] and shown here in Fig. 1 displays two chromatograms in blue and red, representing the same chemical specimen. Despite having the same chemical contents, the chromatograms show differences in peaks locations throughout the domain. Due to these shifts, a simple comparison of chromatograms using standard norms will result in unusually high differences despite representing the same material. Thus, any analysis of such data is faced with the challenge of random nonlinear shifts in the peaks, that needs to be reconciled before drawing statistical inferences.

A. Srivastava (✉)
Department of Statistics, Florida State University, Tallahassee, FL 32306, USA
e-mail: anuj@stat.fsu.edu

© Springer International Publishing Switzerland 2017
S. Datta, B.J.A. Mertens (eds.), *Statistical Analysis of Proteomics, Metabolomics, and Lipidomics Data Using Mass Spectrometry*, Frontiers in Probability and the Statistical Sciences, DOI 10.1007/978-3-319-45809-0_2

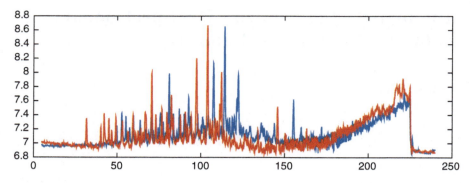

Fig. 1 Example of LC-MS data samples from a proteomics study. Taken from [16]

Consequently, an important goal in LC-MS data analysis is to align peaks in a principled way using some form of nonlinear time-warping.

The literature contains several ideas for handling the nonlinear alignment problem although they are neither fully automatic or nor completely successful in the alignment task. One prominent approach is to find dominant (taller) peaks across chromatograms, match them in a certain way, and then estimate warpings that facilitate that matching. However, the problems of detecting and matching prominent peaks are not straightforward themselves. They are often subjective and require human inputs. Wong et al. [18] developed a peak detection and alignment method that is based exactly on this idea. Bloemberg et al. [2] provide a survey of some current techniques in warping- based alignment of chromatograms, including some computer programs associated with these techniques. Several authors have also developed online tools for spectral alignment, see, e.g., [18]. A statistical approach for spectral alignment and modeling, using a Gaussian mixture model, is presented in [3].

In this paper, we present a fully automated method for alignment of chromatograms, by considering them as functional data and using the nonlinear time-warping of their domains for matching peak locations. The problem of functional data alignment has been studied by several authors, including [4, 6, 9, 14]. In our approach, the actual alignment becomes an optimization problem with a novel objective function which, in turn, is derived using ideas from functional information geometry. Any two chromatograms are aligned by minimizing a distance between them; this distance can be seen as an extension of the classical nonparametric *Fisher-Rao distance*, derived originally for comparing probability density functions, to more general class of functions. Although this distance has nice theoretical properties (invariance to time-warping, etc.), it is too complex to be useful in practical algorithms, especially for processing high-throughput data. This issue is resolved using a change of variable, i.e., replacing the original functions by their square-root slope functions (SRSFs) so that the required distance becomes simply the $\mathbb{L}^2$ norm between their SRSFs. Now, returning to the alignment of several chromatograms, the basic idea is to derive a mean of the given functions (SRSFs of chromatograms) and to align the individual SRSFs to this mean. This procedure is iterative as the mean itself is updated using the aligned SRSFs. Upon convergence, we obtain aligned SRSFs that can then be mapped back to aligned chromatograms using the inverse maps. This last step relies on the fact that the SRSF representation is invertible up to a constant. This framework, termed the *extended Fisher-Rao framework*, is a fully automated procedure and does not require any peak detection or matching. Instead, it utilizes the aforementioned metric (extended Fisher-Rao) and a corresponding representation (SRSF) to formulate alignment as an optimization problem.

The rest of this paper is as follows. We briefly describe the main mathematical challenges in solving the alignment problem and highlight the limitations of a commonly used approach based on the $\mathbb{L}^2$ norm. In Sect. 3, we lay out our mathematical framework, leading up to the presentation of the automated alignment algorithm. This is followed by experimental results on a number of simulated and real LC-MS datasets in Sect. 4, and the paper ends with a short conclusion in Sect. 5.

2 Fundamental Issues in Functional Alignment

As mentioned above, we view the observed chromatograms as real-valued functions on a fixed interval. Without any loss of generality, we will use $[0, 1]$ as domain of these functions. Let $\mathscr{F}$ momentarily denote the relevant set of real-valued functions on $[0, 1]$ although the precise definition of $\mathscr{F}$ comes later when we present more details. An important problem in statistical analysis of functional data (and specifically in LC-MS data analysis) is the alignment of functions using domain warping. The broad goal of an alignment process is to warp the retention-time (or parameter) axis in such a way that their peaks and valleys are better aligned. This alignment problem has also been referred to as the separation of *phase* and *amplitude* [10], or the *registration* [6], or the correspondence of functions in the given data.

Towards this goal, we need to specify the set of valid warping functions. We will use the set of positive diffeomorphisms of $[0, 1]$ as the set of allowed time-warping function; we will denote it by Γ. It is important to note that Γ is a *group* with composition as the binary operation. For any two elements $\gamma_1, \gamma_2 \in \Gamma$, their composition $\gamma_1 \circ \gamma_2 \in \Gamma$. The identity function $\gamma_{\mathrm{id}}(t) = t$ forms the identity element of Γ and, finally for every $\gamma \in \Gamma$, there exists an inverse element $\gamma^{-1} \in \Gamma$ such that $\gamma \circ \gamma^{-1} = \gamma_{\mathrm{id}}$.

In this context, we can pose two kinds of registration problems:

1. **Pairwise Alignment Problem**: Given any two functions f_1 and f_2 in $\mathscr{F}$, we define their pairwise alignment or registration to be the problem of finding a warping function γ such that a certain energy term $E[f_1, f_2 \circ \gamma]$ is minimized. Figure 2 shows an example of this idea where f_2 is time-warped to align it with f_1.

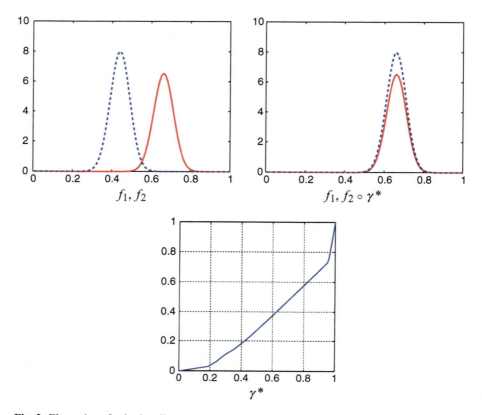

Fig. 2 Illustration of pairwise alignment

2. **Groupwise or Multiple Alignment**: In this case we are given a set of functions $\{f_i \in \mathscr{F} | i = 1, 2, \ldots, n\}$. The problem of finding a set of warping functions $\{\gamma_i | i = 1, 2, \ldots, n\}$ such that, for any $t \in [0, 1]$, the values $f_i(\gamma_i(t))$ are said to be registered with each other, is termed the problem of *joint* or *multiple alignment*. Figure 3 shows an example of this idea where the given functions $\{f_i\}$ (left panel) are aligned (middle panel) using the warping functions shown in the right panel.

We will start by considering the pairwise alignment problem and then later extend that solution to address multiple alignment.

The main question in solving pairwise registration is: What should be the optimization criterion E? In other words, what is a mathematical definition of a *good registration*? Visually one can evaluate an alignment by comparing the locations of peaks and valleys, but how should one do it in a formal, quantifiable and, most importantly, automated way? Before we present our solution, we look at one of the most popular ways of registering functional data in the current literature and highlight its limitations.

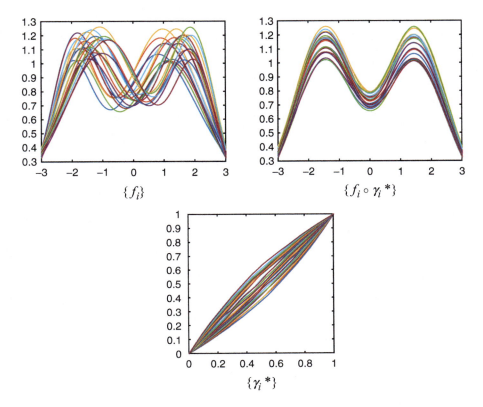

$\{f_i\}$ $\{f_i \circ \gamma_i {}^*\}$

$\{\gamma_i {}^*\}$

Fig. 3 Illustration of multiple function alignment

2.1 Problems in Using $\mathbb{L}^2$-Norm for Pairwise Registration

When searching for an objective function that measures alignment between f_1 and
$f_2 \circ \gamma$, a natural quantity that comes to mind is the $\mathbb{L}^2$ norm of their difference. That
is, we can define the optimal warping function to be:

$$\gamma^* = \arg \inf_{\gamma \in \Gamma} \|f_1 - f_2 \circ \gamma\|^2 . \tag{1}$$

It is well known that this formulation is problematic, as it leads to a phenomena
called the *pinching effect* [10]. What happens is that in matching of f_1 and f_2 one
can squeeze or pinch a large part of f_2 and make this cost function arbitrarily
close to zero. An illustration of this problem is presented in Fig. 4 using a simple
example. Here a part of f_2 is identical to f_1 over $[0, 0.6]$ and is completely different
over the remaining domain $[0.6, 1]$. Since f_1 is essentially zero and f_2 is strictly
positive in $[0.6, 1]$, there is no warping that can match f_1 with f_2 over that subinterval.

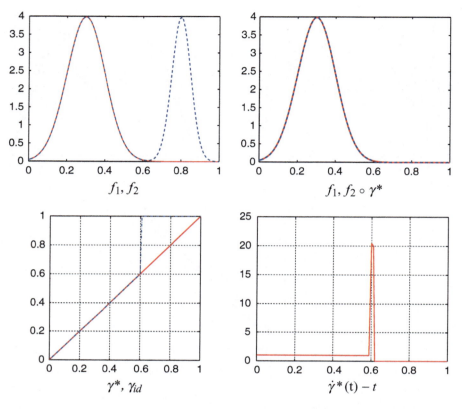

Fig. 4 Illustration of the pinching effect. The *top row* shows a degenerate analytical solution to matching under $\mathbb{L}^2$ norm for functions. The *bottom row* shows the corresponding warping function

The optimal solution is, therefore, to decimate that part of f_2 by using the following γ^*: it coincides with γ_{id} over $[0, 0.6]$, climbs rapidly to $1 - \epsilon$ around 0.6, and then goes slowly from $1 - \epsilon$ to 1 over the interval $[0.6, 1]$. It is easy to check that in the limit:

$$\lim_{\epsilon \to 0} \|f_1 - f_2 \circ \gamma_\epsilon\| = 0 .$$

The top-right panel of Fig. 4 shows the limiting case where f_1 and $f_2 \circ \gamma_0$ are identical. In the bottom row, we provide results from a numerical procedure for this optimization problem. Sometimes, in practice, we restrict γ to have positive bounded slope and do not obtain the same result as the theoretical limit. However, one can still see the pinching effect in numerical implementations.

To avoid the pinching problem, one frequently imposes an additional term to the optimization cost, a term that penalizes the roughness of γ. This term, also called a *regularization term*, results in the registration problem of the type:

$$\gamma^* = \arg\inf_{\gamma \in \Gamma} \left(\|f_1 - f_2 \circ \gamma\|^2 + \lambda \mathcal{R}(\gamma) \right) , \tag{2}$$

where $\mathcal{R}(\gamma)$ is the regularization term, e.g., $\mathcal{R}(\gamma) = \int \ddot{\gamma}(t)^2 dt$ and $\lambda > 0$ is a constant. While this solution avoids pinching, it has several other problems including the fact it does not satisfy inverse symmetry. That is, the registration of f_1 to f_2 may lead to a completely different result than that of f_2 to f_1. We illustrate this using an example.

Example 1. As a simple example, let $f_1(t) = t$ and $f_2(t) = 1 + (t - 0.5)^2$. In this case, for the minimization problem $\gamma_{12} = \min_{\gamma \in \Gamma} \|f_2 - f_1 \circ \gamma\|^2$, the optimal solution is as follows. Define a warping function that climbs quickly (linearly) from 0 to $1 - \epsilon$ on the interval $[0, \epsilon]$ and then climbs slowly (also linearly) from $1 - \epsilon$ to 1 in the remaining interval $[\epsilon, 1]$. The limiting function, when $\epsilon \to 0$, results in the optimal γ for this case.

For the inverse problem, $\gamma_{21} = \min_{\gamma \in \Gamma} \|f_1 - f_2 \circ \gamma\|^2$, the optimal warping is the following. Define a warping function that rises quickly from 0 to $0.5 - \epsilon$ in the interval $[0, \epsilon]$, climbs slowly from $0.5 - \epsilon$ to $0.5 + \epsilon$ in the interval $[\epsilon, 1 - \epsilon]$, and finally climbs quickly from $0.5 + \epsilon$ to 1 in the interval $[1 - \epsilon, 1]$. The optimal solution is obtained when $\epsilon \to 0$. A numerical implementation of these two solutions, based on the dynamic programming algorithm [1], are shown in Fig. 5. Since this implementation allows only a limited number of possible slopes for optimal γ, the results are not as accurate as the analytical solution. Still, it is clear to see a large difference in the solutions for the two cases.

2.2 Desired Properties in Alignment Framework

In view of these limitations of the popular $\mathbb{L}^2$-norm-based framework, we first enumerate a set of basic properties that any (alignment) objective function E should satisfy. Actually, only the first one is a fundamental property, the remaining two are simple consequences of the first one (and some additional structure). Still, we list all three of them to highlight different aspects of the registration problem.

1. **Invariance to Simultaneous Warping**: We start by noting that an identical warping of any two functions preserves their registration. That is, for any $\gamma \in \Gamma$ and $f_1, f_2 \in \mathcal{F}$, the function pair (f_1, f_2) has the same registration as the pair $(f_1 \circ \gamma, f_2 \circ \gamma)$. What we mean by that is the application of γ has not disturbed their point-to-point correspondence. This is easy to see since γ is a diffeomorphism. The two height values across the functions that were matched, say $f_1(t_0)$ and $f_2(t_0)$, for a parameter value t_0, remain matched. They are now labeled $f_1(\gamma(t_0))$ and $f_2(\gamma(t_0))$, and the parameter value has changed to $\gamma(t_0)$, but they still have the same parameter and, thus, are still matched to each other. This is illustrated using an example in Fig. 6 where the left panel shows a pair f_1 and f_2. Note that the two functions are nicely registered since their peaks and valleys are perfectly

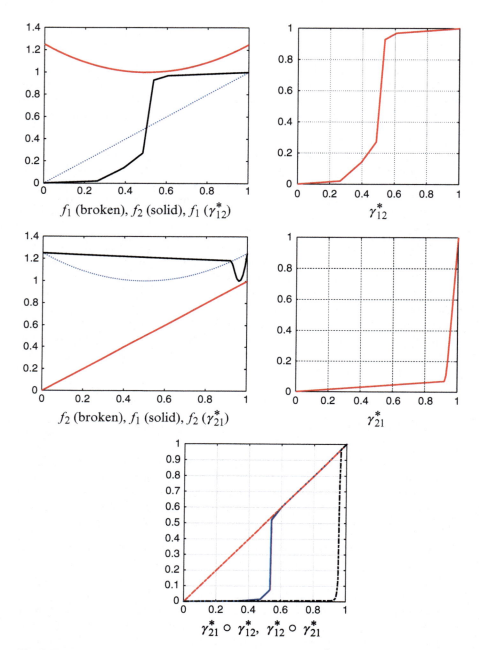

Fig. 5 Example of asymmetry in registration using penalized $\mathbb{L}^2$ norm. The *top row* shows solution of registering f_1 to f_2, while the *bottom two rows* shows the reverse case. *Bottom most* emphasizes that the two γ functions are not inverses of each other

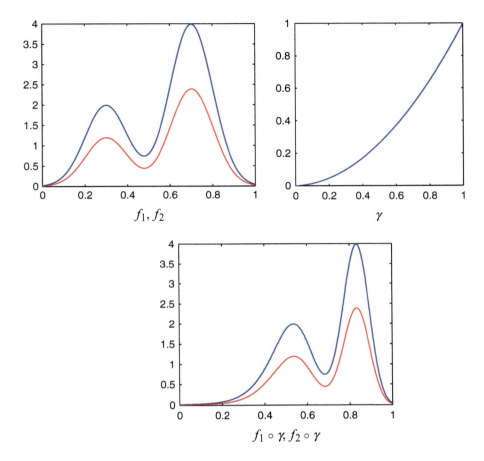

f_1, f_2

γ

$f_1 \circ \gamma, f_2 \circ \gamma$

Fig. 6 Identical warping of f_1, f_2 by γ preserves their registration

aligned. If we apply the warping function shown in the top right panel to both their domains, we get the functions shown in the bottom. The peaks and valleys, and in fact all the points, have the same matching as before. This motivates the following invariance property of E. Since E is expected to be a measure of registration of two functions, it should remain unchanged if the two functions are warped identically. That is, for all $f_1, f_2 \in \mathscr{F}$:

Invariance Property : $E[f_1, f_2] = E[f_1 \circ \gamma, f_2 \circ \gamma]$, for all $\gamma \in \Gamma$.   (3)

2. **Effect of Random Warpings**: Suppose we have found the optimal warping function $\gamma^* \in \Gamma$, defined by

$$\gamma^* = \operatorname*{argmin}_{\gamma \in \Gamma} E[f_1, f_2 \circ \gamma] \ .$$

Once this γ^* is found, the resulting matched height values are $f_1(t)$ and $f_2(\gamma^*(t))$, for all t. Now, let us suppose that we warp f_1 and f_2 by random functions, say γ_1 and γ_2. What is the optimal correspondence between these new functions? It will be given by

$$\tilde{\gamma}^* = \operatorname*{argmin}_{\gamma \in \Gamma} E[f_1 \circ \gamma_1, (f_2 \circ \gamma_2) \circ \gamma] = \operatorname*{argmin}_{\gamma \in \Gamma} E[f_1, f_2 \circ (\gamma_2 \circ \gamma \circ \gamma_1^{-1})],$$

where the last equality follows from the above *invariance property*. The last two equations immediately imply that $\gamma^* = \gamma_2 \circ \tilde{\gamma}^* \circ \gamma_1^{-1}$, or $\tilde{\gamma}^* = \gamma_2^{-1} \circ \gamma^* \circ \gamma_1$. More interestingly, the optimal registration of functions and the minimum value of E remains unchanged despite the presence of random γ_1 and γ_2, i.e.,

$$\min_{\gamma \in \Gamma} E[f_1, f_2 \circ \gamma] = \min_{\gamma \in \Gamma} E[f_1 \circ \gamma_1, (f_2 \circ \gamma_2) \circ \gamma] \ .$$

This important equality intimately depends on the invariance property [Eq. (3)] and the group structure of Γ. Without the invariance property one can expect the results of registration to be highly dependent on γ_1 and γ_2. For instance, this undesirable situation will occur if we use the $\mathbb{L}^2$ metric between the functions to define E.

3. **Inverse Symmetry**: We know that registration is a symmetric property. That is, if f_1 is registered to f_2, then f_2 is also registered to f_1. Similarly, if f_1 is optimally registered to $f_2 \circ \gamma$, then f_2 is optimally registered to $f_1 \circ \gamma^{-1}$. Therefore, the choice of E should be such that this symmetry is preserved. That is,

$$\gamma^* = \operatorname*{argmin}_{\gamma \in \Gamma} E[f_1, f_2 \circ \gamma] \Rightarrow \gamma^{*-1} = \operatorname*{argmin}_{\gamma \in \Gamma} E[f_1 \circ \gamma, f_2] \ .$$

This symmetry property has also been termed as *inverse consistency*. If the invariance property holds, then this inverse symmetry follows immediately from the group structure of Γ.

In addition to registering any two functions, it is often important to compare them and to quantify their differences. For this, we need a proper distance function, to be able to compare f_1 and $f_2 \circ \gamma^*$, where γ^* is the optimal warping of f_2 that registers it with f_1. One can always choose an unrelated distance function on the space, e.g., the $\mathbb{L}^2$ norm, that measures differences in the registered functions f_1 and $f_2 \circ \gamma^*$. However, this makes the process of registration an unrelated pre-processing step for the eventual comparison. Ideally, we would like to jointly solve these two problems, under a unified metric. Therefore, it will be useful if the quantity $\inf_{\gamma \in \Gamma} E[f_1, f_2 \circ \gamma]$ is also a proper distance in some sense. That is, in addition to symmetry, it also satisfies non-negativity and the triangle inequality. The sense in which we want it to be a distance is that the result does not change if we randomly warp the individual functions in arbitrary ways.

3 Extended Fisher-Rao Approach

In order to reach a cost function E that avoids the pinching effect, satisfies the *invariance property* and, as a consequence, allows inverse symmetry and invariance of E to random warping, we introduce a new mathematical representation for functions.

3.1 Mathematical Background

Our general framework for alignment of chromatograms is adapted from ideas in shape analysis of curves [5, 12] and is described more comprehensively in [8, 13, 15]. For a broader introduction to this theory, including asymptotic results and identifiability results, we refer the reader to these papers.

Starting fresh, this time we are going to restrict to those f that are absolutely continuous on $[0, 1]$; let $\mathscr{F}$ denote the set of all such functions. The new representation of functions is based on the following transformation. Define a mapping: $Q : \mathbb{R} \to \mathbb{R}$ according to:

$$Q(x) \equiv \begin{cases} \text{sign}(x)\sqrt{|x|}, & x \neq 0 \\ 0, & x = 0 \end{cases}. \tag{4}$$

Note that Q is a continuous map. For the purpose of studying the function f, we will represent it using the SRSF defined as follows:

Definition 1 (SRSF Representation of Functions). Define the SRSF of f to be the function $q : [0, 1] \to \mathbb{R}$, where

$$q(t) \equiv Q(\dot{f}(t)) = \text{sign}(\dot{f}(t))\sqrt{|\dot{f}(t)|}.$$

This representation includes those functions whose parameterization can become singular in the analysis. In other words, if $\dot{f}(t) = 0$ at some point, it does not cause any problem in the definition of $q(t)$. It can be shown that if the function f is absolutely continuous, then the resulting SRSF is square integrable [11]. Thus, we will define $\mathbb{L}^2([0, 1], \mathbb{R})$ (or simply $\mathbb{L}^2$) to be the set of all SRSFs. For every $q \in \mathbb{L}^2$ there exists a function f (unique up to a constant) such that the given q is the SRSF of that f. In fact, this function can be obtained precisely using the equation: $f(t) = f(0) + \int_0^t q(s)|q(s)|ds$. Thus, the representation $f \leftrightarrow (f(0), q)$ is invertible.

The next question is: If a function is warped, then how does its SRSF change? For an $f \in \mathscr{F}$ and $\gamma \in \Gamma$, let q be the SRSF of f. Then, what is the SRSF of $f \circ \gamma$? This can simply be derived as:

$$\tilde{q}(t) = Q\left(\frac{d}{dt}(f \circ \gamma)(t)\right) = \text{sign}\left(\frac{d}{dt}(f \circ \gamma)(t)\right)\sqrt{\left|\frac{d}{dt}(f \circ \gamma)(t)\right|} = q(\gamma(t))\sqrt{\dot{\gamma}(t)}.$$

In more mathematical terms, this denotes an action of group Γ on $\mathbb{L}^2$ from the right side: $\mathbb{L}^2 \times \Gamma \to \mathbb{L}^2$, given by $(q, \gamma) = (q \circ \gamma)\sqrt{\dot{\gamma}}$. One can show that this action of Γ on $\mathbb{L}^2$ is *compatible* with its action on $\mathcal{F}$ given earlier, in the following sense:

$$
\begin{array}{ccc}
 & f & \xrightarrow{\text{SRSF}} & q & \\
\text{action on } \mathcal{F} & \downarrow & & \downarrow & \text{action on } \mathbb{L}^2 \\
 & f \circ \gamma & \xrightarrow{\text{SRSF}} & (q \circ \gamma)\sqrt{\dot{\gamma}} &
\end{array}
$$

We can apply the group action and compute SRSF in any order, and the result remains the same. The most important advantage of using SRSFs in functional data analysis comes from the following result. Recall that the $\mathbb{L}^2$ inner-product is given by: $\langle v_1, v_2 \rangle = \int_0^1 v_1(t)v_2(t)\,dt$.

Lemma 1. *The mapping $\mathbb{L}^2 \times \Gamma \to \mathbb{L}^2$ given by $(q, \gamma) = (q \circ \gamma)\sqrt{\dot{\gamma}}$ forms an action of Γ on $\mathbb{L}^2$ by isometries.*

Proof. That the mapping is a group action has been mentioned earlier. The proof of isometry is an easy application of integration by substitution. For any $v_1, v_2 \in \mathbb{L}^2$,

$$\langle(v_1, \gamma), (v_2, \gamma)\rangle = \int_0^1 v_1(\gamma(t))\sqrt{\dot{\gamma}(t)}v_2(\gamma(t))\sqrt{\dot{\gamma}(t)}dt$$

$$= \int_0^1 v_1(\gamma(t))v_2(\gamma(t))\dot{\gamma}(t)dt = \int_0^1 v_1(s)v_2(s)ds = \langle v_1, v_2 \rangle.$$

$\square$

This lemma ensures that any framework based on this SRSF and $\mathbb{L}^2$ norm will satisfy the *invariance property* listed in the previous section.

It is well known that the geodesics in $\mathbb{L}^2$, under the $\mathbb{L}^2$ Riemannian metric, are straight lines and the geodesic distance between any two elements $q_1, q_2 \in \mathbb{L}^2$ is given by $\|q_1 - q_2\|$. Since the action of Γ on $\mathbb{L}^2$ is by isometries, the following result is automatic. Still, for the sake of completeness, we provide a short proof.

Lemma 2. *For any two SRSFs $q_1, q_2 \in \mathbb{L}^2$ and $\gamma \in \Gamma$, we have that $\|(q_1, \gamma) - (q_2, \gamma)\| = \|q_1 - q_2\|$.*

Proof. For an arbitrary element $\gamma \in \Gamma$, and $q_1, q_2 \in \mathbb{L}^2$, we have

$$\|(q_1, \gamma) - (q_2, \gamma)\|^2 = \int_0^1 (q_1(\gamma(t))\sqrt{\dot{\gamma}(t)} - q_2(\gamma(t))\sqrt{\dot{\gamma}(t)})^2 dt$$

$$= \int_0^1 (q_1(\gamma(t)) - q_2(\gamma(t)))^2\dot{\gamma}(t)dt = \|q_1 - q_2\|^2. \quad \square$$

An interesting corollary of this lemma is the following.

Corollary 1. *For any $q \in \mathbb{L}^2$ and $\gamma \in \Gamma$, we have $\|q\| = \|(q, \gamma)\|$.*

This implies that the action of Γ on $\mathbb{L}^2$ is actually a norm-preserving transformation. Conceptually, it can be equated with the rotation of vectors in Euclidean spaces. Due to the norm-preserving nature of warping in this representation, the pinching effect is completely avoided.

3.2 Pairwise Alignment Procedure

With this mathematical foundation, the pairwise alignment of chromatograms can be accomplished as follows. Let f_1, f_2 by functional forms of the two given spectra and let q_1, q_2 be the corresponding SRSFs. Then, the optimization problem is given by:

$$\inf_{\gamma \in \Gamma} \|q_1 - (q_2, \gamma)\|^2 = \inf_{\gamma \in \Gamma} \|q_2 - (q_1, \gamma)\|^2 . \tag{5}$$

This minimization is performed in practice using a numerical approach called *the dynamic programming algorithm* [1]. We have already mentioned (in Lemma 2) that the use of SRSFs and $\mathbb{L}^2$ norm satisfies the invariance property from Sect. 2.2. We now show that the remaining two properties—effect of random warpings and inverse symmetry—are also satisfied. Using Lemma 2 and the group structure of Γ one can show that if γ^* is a minimizer on the left side of Eq. (5), then $(\gamma^*)^{-1}$ is a minimizer on the right side! Furthermore, using the same tools one can show that:

$$\inf_{\gamma \in \Gamma} \|q_1 - (q_2, \gamma)\| = \inf_{\gamma \in \Gamma} \|(q_1, \gamma_1) - ((q_2, \gamma_2), \gamma)\| ,$$

for any $\gamma_1, \gamma_2 \in \Gamma$.

Figure 7 shows some results from using this alignment framework using some simple LC-MS chromatograms. In each of the four examples shown there, the left panel shows the original chromatograms f_1, f_2, the middle panel shows the aligned chromatograms $f_1, f_2 \circ \gamma^*$, and the last panel shows the optimal warping function γ^*. In the first row, the level of misalignment is relatively small, and a small warping is able to align the peaks. However, as we go down this figure, the level of misalignment increases and it takes an increasing amount of warping to align the peaks. This increasing warping is visible in the corresponding warping functions in terms of their deviations from γ_{id}, the 45° line. Noticeably, the algorithm is quite successful in alignment of peaks for all these datasets. Furthermore, it is fully automatic and requires no parameter tuning or any kind of manual intervention.

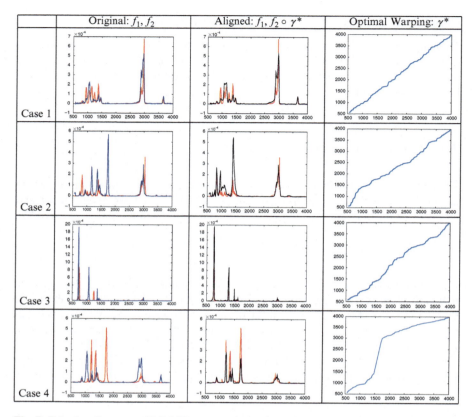

Fig. 7 Pairwise alignment of LC-MS spectra using optimization given in Eq. (5). In each *row* we display the original and the aligned chromatograms, along with the optimal warping functions

3.3 Multiple Alignment Algorithm

With the ability to align chromatograms in a pairwise fashion, we now extend this idea to simultaneous alignment of multiple chromatograms. We will use a template-based approach, where we iteratively define a template chromatogram and align the given chromatograms to this template in a pairwise manner [using Eq. (5)]. The template is created by taking an average of the aligned chromatograms in the SRSF space, at each iteration. The full alignment algorithm is as follows:

Algorithm 1 (Alignment of Multiple Chromatograms). *Given a set of chromatograms in functional form $f_1, f_2, \ldots f_n$ on $[0, 1]$, let $q_1, q_2, \ldots, q_n$ denote their SRSFs, respectively.*

1. Initialize $\tilde{q}_i = q_i$ for $i = 1, 2, \ldots, n$.
2. Compute their mean according to $\mu = \frac{1}{n} \sum_{i=1}^{n} \tilde{q}_i$.
3. For $i = 1, 2, \ldots, n$, find γ_i^* by solving:

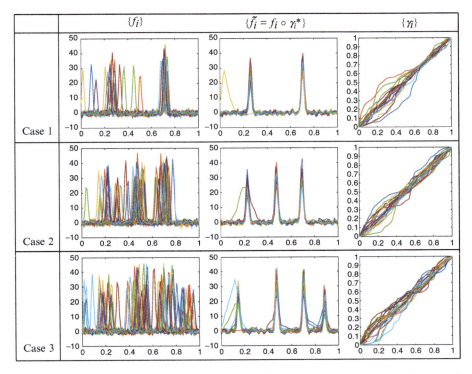

Fig. 8 Alignment of simulated chromatograms using Algorithm 1. In each *row* we see the original functions, the aligned functions, and the corresponding warping functions

$$\gamma_i^* = \operatorname*{arginf}_{\gamma \in \Gamma} \| \mu - (q_i, \gamma) \|^2 .$$

This minimization is approximated using the dynamic programming algorithm.
4. *Compute the aligned SRSFs* $\tilde{q}_i = (q_i, \gamma_i^*)$ *and aligned functions* $\tilde{f}_i = f_i \circ \gamma_i^*$.
5. *Check for convergence. If not converged, go to Step 2.*
6. *Return the template* μ, *the warping functions* $\{\gamma_i^*\}$, *and the aligned functions* $\{\tilde{f}_i\}$.

We show some illustrations of this example using simulated data in Fig. 8. In this experiment, we simulate chromatograms as superpositions of Gaussian probability density functions with random shifts in heights and locations. Each chromatogram is made up of either two (top row), three (middle row), or four (bottom row) Gaussian probability density functions, and we use 20 such chromatograms in each case. Additionally, we corrupt these chromatograms by adding white Gaussian noise at each time t. As can be seen in the middle panels, the algorithm is quite successful in finding and aligning the corresponding peaks across chromatograms. In the process, the algorithm discovers non-trivial, nonlinear warping functions that are required for alignments. We reemphasize that this algorithm is fully automated and does not require any manual input, nor does it involve any parameter tuning for superior performance. Interestingly, it not only does a good job in aligning major (taller) peaks but it also aligns smaller peaks that can be attributed mainly to noise.

4 Experimental Results on Real Data

In this section we present some alignment results on several LC-MS datasets taken from various sources.

1. As the first result, we utilize the proteomics dataset introduced by Koch et al. [7]. As described there, protein profiling can be used to study changes in protein expression in reference to therapeutic treatments for diseases, and this data involves protein profiles of patients with Acute Myeloid Leukemia. The original data with markers corresponding to the key peaks in the data is presented in Fig. 9 (left panel). An interesting part about this dataset is that it comes with an expert-labeling that provides a unique number to each of the major peaks. This numbering, from 1 to 14 for each spectrum, is used only to study the alignment but are not used in the alignment process itself. As can be seen in the left panel, the peaks in the data are not well aligned as the corresponding numbers demonstrate. The results of applying our alignment method are presented in Fig. 9 (right panel). The aligned functions exhibit good registration with almost all of the peaks lining up. There are a few exceptions involving peaks numbered 1 and 2. Since they have a very low amplitude, their registration is relatively difficult.

   We can also quantify the alignment performance using the decrease in the cumulative cross-sectional variance of the aligned functions. For any functional dataset $\{g_i(t), i = 1, 2, \ldots, n, t \in [0, 1]\}$, let

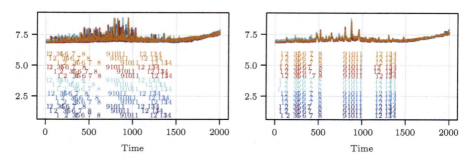

Fig. 9 Alignment of proteomics data using the square-root slope framework with original data in *left panel* and aligned functions in *right panel*. The aligned functions exhibit good registration of marked peaks. Picture taken from [16]

$$\mathrm{Var}(\{g_i\}) = \frac{1}{n-1} \int_0^1 \sum_{i=1}^n \left(g_i(t) - \frac{1}{n}\sum_{i=1}^n g_i(t) \right)^2 dt \, ,$$

denote the cumulative cross-sectional variance in the given data. For the proteomics data, we found

$$\text{Original Variance} = \mathrm{Var}(\{f_i\}) = 4.05, \quad \text{Aligned Variance} = \mathrm{Var}(\{\tilde{f}_i\}) = 1.13$$

$$\text{Warping Variance} = \mathrm{Var}(\{\mu_f \circ \gamma_i^*\}) = 3.04 \, .$$

where $\{f_i\}$ is the set of original functions, $\{\tilde{f}_i\}$ is the set of aligned functions, μ_f is the mean of the aligned functions, and $\{\mu_f \circ \gamma_i^*\}$ is the result of applying the warping functions $\{\gamma_i^*\}$ to μ_f. From the decrease in the aligned variance and increase in the warping variance we can quantify the level of alignment.

Figure 10 presents a zoom-in on a region of the data on the time interval [615, 911]. The top panel is the original data where we see very poor alignment of the peaks. The bottom panel is the corresponding aligned data using the extended Fisher-Rao framework, where very tight alignment of the peaks and valleys have occurred.

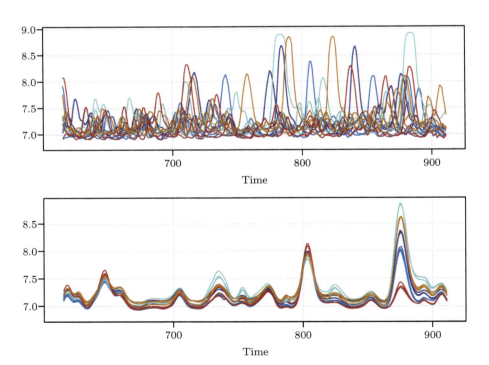

Fig. 10 Zoom-in of the region (600–910) to demonstrate accurate alignment of smaller as well larger peaks. The *top picture* is before and *bottom* is after alignment. Picture taken from [16]

2. Another experiment involves a set of eight chromatograms shown in the top panel in Fig. 11. In this data, some of the major peaks are not aligned, especially in the domain range [15, 25]. The outcome of Algorithm 1 on this data is shown in the second row where the peaks appear to be sharply aligned throughout the spectrum. To emphasize the quality of alignment we look at a couple of smaller intervals in the spectrum more carefully. The zoom-ins of these smaller regions are shown in the last two row of the figure. In each of the last two rows, we show the before and after alignment spectra for these two domains: 0–20 and 28–50. It can be seen in these zoom-ins that the algorithm aligns both the major and minor peaks remarkably well. Once again, this procedure does not require any prior peak detection or matching to reach this alignment.
3. In the third and final example, we study a set of 14 chromatograms associated with urine samples collected at NIST. The top row of Fig. 12 shows the full chromatograms before and after alignment. Since the misalignments in this examples are relatively small, compared with the full range of retention times, it is difficult to evaluate the quality of alignment in this full view. In the bottom row, we look at magnified view of a smaller region—5 to 15—and find the peaks are very closely aligned after the algorithm has been applied.

5 Conclusions

The problem of alignment of mass spectrometry data is both important and challenging task. We have utilized a recent comprehensive approach that treats chromatograms as real-valued functions, uses extended Fisher-Rao metric to perform alignment of peaks and valleys in these functions. The key idea is to form SRSFs of the given chromatograms and then to use the standard $\mathbb{L}^2$ norm between these functions to perform both pairwise and groupwise alignment. We demonstrate this framework using a number of examples involving real and simulated database taken from different spectrometry applications. The success of this alignment procedure is clearly visible in all experiments where the peaks are nicely aligned across observation. This procedure is fully automated and does not require any user input. Furthermore, it aligns full chromatograms (functions) rather than simply matching a few dominant peaks.

Acknowledgements The author is very thankful to the people who provided data for experiments presented in this paper—Prof. I. Koch of Adelaide, South Australia and Dr. Yamil Simon of National Institute of Standards and Technology (NIST), Gaithersburg, Maryland. This research was supported in part by the grants NSF DMS-1208959 and NSF CCF 1319658, and support from the Statistics Division at NIST.

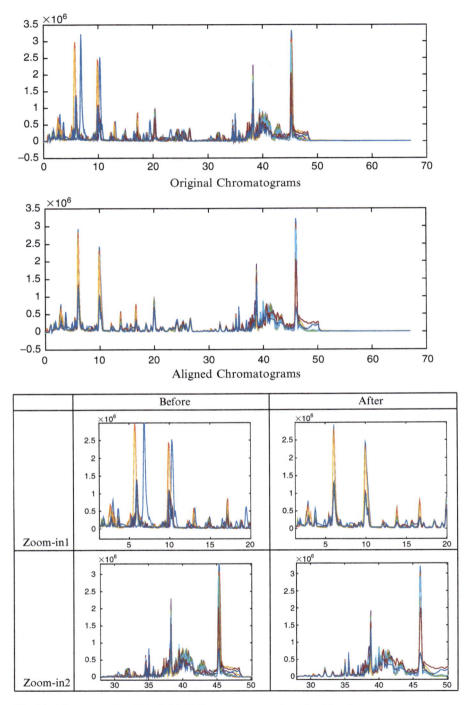

Fig. 11 Alignment of multiple LC-MS chromatograms using Algorithm 1

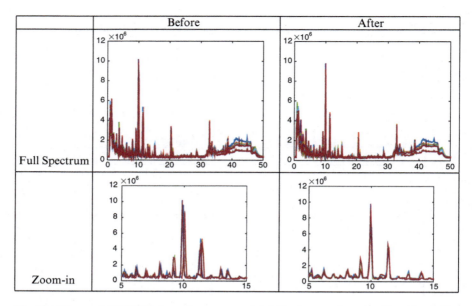

Fig. 12 Alignment of LC-MS chromatograms associated 14 urine samples using Algorithm 1. The *top row* shows the full chromatograms while the *bottom row* shows a magnified region to facilitate a closer look at the alignment performance

References

1. Bertsekas, D. P. (1995). *Dynamic programming and optimal control*. Boston: Athena Scientific.
2. Bloemberg, T. G., Gerretzen, J., Lunshof, A., Wehrens, R., & Buydens, L. M. (2013). Warping methods for spectroscopic and chromatographic signal alignment: A tutorial. *Analytica Chimica Acta, 781*, 14–32.
3. Browne, W. J., Dryden, I. L., Handley, K., Mian, S., & Schadendorf, D. (2010). Mixed effect modelling of proteomic mass spectrometry data by using Gaussian mixtures. *Journal of the Royal Statistical Society. Series C (Applied Statistics), 59*(4), 617–633.
4. James, G. (2007). Curve alignments by moments. *Annals of Applied Statistics, 1*(2), 480–501.
5. Joshi, S. H., Klassen, E., Srivastava, A., & Jermyn, I. H. (2007). A novel representation for Riemannian analysis of elastic curves in $\mathbb{R}^n$. In *Proceedings of IEEE CVPR* (pp. 1–7).
6. Kneip, A., & Ramsay, J. O. (2008). Combining registration and fitting for functional models. *Journal of the American Statistical Association, 103*(483), 1155–1165.
7. Koch, I., Hoffmann, P., & Marron, J. S. (2013). Proteomics profiles from mass spectrometry. *Electronic Journal of Statistics, 8*(2), 1703–1713.
8. Kurtek, S., Srivastava, A., & Wu, W. (2011). Signal estimation under random time-warpings and nonlinear signal alignment. In *Proceedings of Advances in Neural Information Processing Systems (NIPS), Grenada, Spain* (pp. 676–683).
9. Liu, X., & Muller, H. G. (2004). Functional convex averaging and synchronization for time-warped random curves. *Journal of the American Statistical Association, 99*, 687–699.
10. Marron, J. S., Ramsay, J. O., Sangalli, L. M., & Srivastava, A. (2015). Functional Data analysis of amplitude and phase variation. *Statistical Science, 30*(4), 468–484.
11. Robinson, D. (2012, August). *Functional Analysis and Partial Matching in the Square Root Velocity Framework*. PhD thesis, Florida State University.

12. Srivastava, A., Klassen, E., Joshi, S. H., & Jermyn, I. H. (2011, July). Shape analysis of elastic curves in Euclidean spaces. *IEEE Transactions on Pattern Analysis and Machine Intelligence, 33*(7), 1415–1428.
13. Srivastava, A., Wu, W., Kurtek, S., Klassen, E., & Marron, J. S. (2011). Registration of functional data using Fisher-Rao metric. arXiv:1103.3817v2 [math.ST].
14. Tang, R., & Muller, H. G. (2008). Pairwise curve synchronization for functional data. *Biometrika, 95*(4), 875–889.
15. Tucker, J. D., Wu, W., & Srivastava, A. (2013). Generative models for functional data using phase and amplitude separation. *Computational Statistics and Data Analysis, 61*, 50–66.
16. Tucker, J. D., Wu, W., & Srivastava, A. (2014). Analysis of proteomics data: Phase amplitude separation using an extended Fisher-Rao metric. *Electronic Journal of Statistics, 8*(2), 1724–1733.
17. Wallace, W. E., Srivastava, A., Telu, K. H., & Simon-Manso, Y. (2014). Pairwise alignment of chromatograms using an extended Fisher-Rao metric. *Analytica Chimica Acta, 841*, 10–16.
18. Wong, J. W., Cagney, G., & Cartwright, H. M. (2005). SpecAlign – processing and alignment of mass spectra datasets. *Bioinformatics, 21*(9), 2088–2090.

The Analysis of Peptide-Centric Mass-Spectrometry Data Utilizing Information About the Expected Isotope Distribution

Tomasz Burzykowski, Jürgen Claesen, and Dirk Valkenborg

1 Introduction

In shotgun proteomics, much attention and instrument time is dedicated to the generation of tandem mass spectra. These spectra contain information about the fragments of, ideally, one peptide and are used to infer the amino acid sequence of the scrutinized peptide. This type of spectrum acquisition is called a product ion scan, tandem MS, or MS2 spectrum. Another type of spectrum is the, often overlooked, precursor ion scan or MS1 spectrum that catalogs all ionized analytes present in a mass spectrometer. While MS2 spectra are important to identify the peptides and proteins in the sample, MS1 spectra provide valuable information about the quantity of the analyte. In this chapter, we describe some properties of MS1 spectra, such as the isotope distribution, and how these properties can be employed for low-level signal processing to reduce data complexity and as a tool for quality assurance. Furthermore, we describe some cases in which advanced modeling of the isotope distribution can be used in quantitative proteomics analysis.

T. Burzykowski (✉)
I-BioStat, Hasselt University, Diepenbeek, Belgium

IDDI, Louvain-la-Neuve, Belgium
e-mail: tomasz.burzykowski@uhasselt.be

J. Claesen
I-BioStat, Hasselt University, Diepenbeek, Belgium

D. Valkenborg
I-BioStat, Hasselt University, Diepenbeek, Belgium

VITO, Mol, Belgium

Center for Proteomics, University of Antwerp, Antwerp, Belgium

© Springer International Publishing Switzerland 2017
S. Datta, B.J.A. Mertens (eds.), *Statistical Analysis of Proteomics, Metabolomics, and Lipidomics Data Using Mass Spectrometry*, Frontiers in Probability and the Statistical Sciences, DOI 10.1007/978-3-319-45809-0_3

Mass-spectrometry (MS) data contain a lot of noise and redundant information. A high-resolution MS1 mass spectrum from, e.g., an orbitrap mass spectrometer typically contains thousands of data points. Depending on the density of the biological sample, a number of, say, 80 peptides can be found in such a mass spectrum. Each peptide can be represented by an abundance measure and a monoisotopic mass. In other words, the numerous data points can contain information about just 80 peptides. This means that there is a lot of irrelevant information in MS data.

In high-resolution MS, a peptide appears in a mass spectrum as a series of locally correlated peaks, which exhibit a specific characteristic profile related to the *isotope distribution* of the peptide. The use of the (expected) isotope peak-patterns can increase effectiveness of selecting the relevant information from mass spectra and subsequent statistical analysis of the data. In this chapter, we describe the necessary methodological background and present several examples of the use of the information about the isotope distribution in the analysis of peptide-centric MS data.

The remainder of the chapter is structured as follows. In Sect. 2, we discuss the isotope distribution in more detail. In Sect. 3, we present various applications, in which the analysis of the MS data is enhanced by the use of the expected isotope distribution. Concluding remarks are presented in Sect. 4.

2   Isotope Distribution

Under natural conditions, particular chemical elements appear as different stable or radioactive variants. These variants have different numbers of neutrons in their atomic nucleus and are known as isotopes. For example, there are two stable isotopes of carbon (C) that appear in nature: ^{12}C and ^{13}C. In the notation, the upper-left index indicates the mass number or nucleon number and represents the total number of protons and neutrons, that is, nucleons, in the atomic nucleus. Due to the difference in the total number of nucleons, isotopes differ in mass. They also differ in their abundance, i.e., the frequency of occurrence. Table 1 presents the stable isotopes, their masses, and abundance for carbon (C), hydrogen (H), nitrogen (N), oxygen (O), and sulfur (S), as defined by the IUPAC 1997 standard [20]. These are the five chemical elements which are predominantly present in peptides and proteins.

Terrestrial molecules incorporate elemental isotopes according to their natural abundances. Thus, a molecule can have different isotope variants with different masses that depend on the number of different isotopes in the atomic composition of the molecule. The probability of occurrence of these isotope variants can be calculated given the atomic composition and the known elemental isotope abundances. The result of this calculation is known as the isotope distribution.

For example, consider a molecule of methane (C_1H_4). Its *monoisotopic variant* is composed solely of the atoms of the most abundant isotopes of C and H, i.e., ^{12}C and ^1H, respectively. It is the lightest variant with the molecular mass (see Table 1) of $1 \times 12 + 4 \times 1.008 = 16.032$. The probability of occurrence of the monoisotopic variant is equal to $0.9893 \times (0.9999)^4 = 0.9889$. However, another possible

Table 1 Isotope variants of C, H, N, O, and S as defined by the IUPAC 1997 standard

Chemical element	Isotope	Atomic mass	Abundance
Carbon	^{12}C	12.0000000000	98.93
	^{13}C	13.0033548378	1.07
Hydrogen	^{1}H	1.0078250321	99.9885
	^{2}H	2.0141017780	0.0115
Nitrogen	^{14}N	14.0030740052	99.632
	^{15}N	15.0001088984	0.368
Oxygen	^{16}O	15.9949146000	99.757
	^{17}O	16.9991312000	0.038
	^{18}O	17.9991603000	0.205
Phosphor	^{31}P	30.9737620000	100
Sulfur	^{32}S	31.9720707000	94.93
	^{33}S	32.9714584300	0.76
	^{34}S	33.9678666500	4.29
	^{36}S	35.9670806200	0.02

Table 2 The isotope distribution of methane (C_1H_4)

^{12}C	^{13}C	^{1}H	^{2}H	Mass	Probability	Nucleons
1	0	4	0	16.032	0.988904	16
0	1	4	0	17.035	0.010696	17
1	0	3	1	17.038	0.000099	17
0	1	3	1	18.041	0.000001	18
1	0	2	2	18.044	$<10^{-8}$	18
0	1	2	2	19.047	$<10^{-9}$	19
1	0	1	3	19.050	$<10^{-12}$	19
0	1	1	3	20.053	$<10^{-13}$	20
1	0	0	4	20.056	$<10^{-16}$	20
0	1	0	4	21.059	$<10^{-17}$	21

variant is composed of one ^{13}C-atom and four ^{1}H-atoms. This variant has the molecular mass of $1 \times 13.0032 + 4 \times 1.008 = 17.035$ and occurs with probability $(1 - 0.9893) \times (0.9999)^4 = 0.0107$. In total, there are ten different isotope variants of a molecule of methane. Table 2 presents the isotope distribution of methane, i.e., the molecular masses of all isotope variants and their probabilities of occurrence (computed using the data from Table 1). Additionally, the total number of nucleons for each variant has been provided. Note that, for methane, which is a simple molecule with a small atomic mass, the monoisotopic variant is also the most abundant one. However, for more complex and heavier molecules, this is not necessarily the case.

All the isotope variants of methane, presented in Table 2, differ in mass. However, it can be seen that some variants have a very similar mass, close to integer values of 16, 17, 18, 19, 20, and 21, corresponding to the total number of nucleons. If we "aggregate" the variants with the same nucleon-content, we obtain the *aggregated isotope distribution* of methane, with only six *aggregated isotope variants*. The probabilities of occurrence of the first three variants, with masses of (about) 16, 17, and 18, are equal to 0.988904, 0.010795 (=0.010696 + 0.000099), and 0.000001, respectively.

From the aforementioned considerations it follows that peptide molecules can have different isotope variants. In a high-resolution mass spectrum, a peptide produces a series of peaks that are separated by roughly 1 Thomson and that correspond to the aggregated isotope distribution (small mass differences between the isotope variants corresponding to a particular aggregated variant are not visible in a spectrum). A convenient algorithm to compute the aggregated isotope distribution on computers is the BRAIN algorithm developed by Claesen et al. [2]. A complete overview of methods for the in silico generation of isotope distributions can be found in the review by Valkenborg et al. [28].

For instance, the left-hand side panel of Fig. 1 presents a full scan of a tryptic-digest of bovine cytochrome C. The right-hand side panel shows a close-up of the region near the mass of 743 m/z. In the close-up we can clearly observe a series of locally correlated peaks, approximately separated by 0.5 m/z (as the peptide is doubly charged) that correspond to the aggregated isotope distribution of a peptide. It is this type of peak-clusters that would be of interest when analyzing a spectrum to reduce complexity and asses spectral quality.

To decide whether a peak-cluster may be due to a peptide, one could check whether the observed pattern (of the form presented in the right-hand side panel of Fig. 1) corresponds to an aggregated isotope distribution of a peptide. This raises an important practical issue, though: to compute the distribution, one has to know the atomic composition of a molecule. Thus, to decide whether a particular peak-cluster may be due to a peptide, one would have to compare the observed pattern to the aggregated isotope distributions of all known peptides with a similar mass. Obviously, this is not feasible, not only for computational reasons, but also because, e.g., the observed pattern may correspond to a modified (e.g., phosphorylated) version of a lighter peptide.

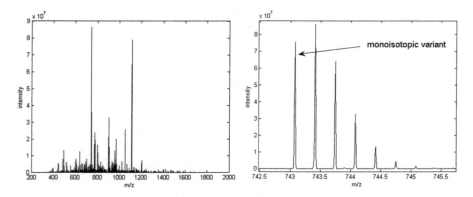

Fig. 1 A full scan of a tryptic-digest of bovine cytochrome C (*left-hand side panel*), with a close-up of the region near 743 m/z (*right-hand side panel*)

An alternative, more practical solution is to compare the observed peak-cluster pattern to the *expected aggregated isotope distribution* for peptides with the monoisotopic mass corresponding to the mass of the first peak in the observed cluster. The expected distribution can be effectively computed using a model developed by Valkenborg et al. [26]. In particular, Valkenborg et al. [26] represented an aggregated isotope distribution by the set of ratios (termed *isotopic ratios*) of the probability of occurrence of the $(m + 1)$-th aggregated isotope variant to the probability of occurrence of the m-th variant (with $m = 1$ denoting the monoisotopic variant). Subsequently, each ratio is modeled by a fourth-order polynomial-regression model using the monoisotopic mass as the explanatory variable. To estimate the coefficients of the model for the first six ratios, Valkenborg et al. [26] used the monoisotopic mass of the theoretical "averagine" peptide proposed by Senko et al. [22]. Alternately, Ghavidel et al. [6] estimated the coefficients by using an in silico digest of the actual proteins included in the Human HUPO database. By using the estimated coefficients, the expected values of the isotopic ratios for a particular mass can be computed and compared to the ratios obtained for the observed peak-cluster. The similarity between the expected and observed ratios can be measured by using Pearson's chi-squared statistic [6, 26]. If the value of the statistic is smaller than a selected threshold, the observed peak-cluster can be regarded as resulting from a peptide.

3   Applications

In this section, we present several applications, in which the analysis of the MS data is enhanced by the use of the expected isotope distribution.

3.1   *Enzymatic Labeling*

MS measurements are influenced by different sources of variability, which can obstruct the detection of differentially expressed proteins or peptides. To reduce the effect of the variability on the data, a labeling approach can be considered. In isotope labeling, peptides from one sample are coded with stable isotope tags and mixed together with another, unlabeled sample. The stable isotope tag will result in an increase of the peptide's mass. Due to this increased mass, a peptide from the labeled sample is discernable from its unlabeled counterpart in a precursor scan. Quantification of the relative abundance is based on the observed intensities.

A relatively low-cost and open-source technique for stable isotope labeling is the enzymatic ^{18}O-labeling, where the two ^{16}O atoms in the carboxyl-terminus of a peptide are replaced with oxygen isotopes from heavy-oxygen water. Enzymatic labeling is performed in two steps. In the first step, protein digestion is done in normal water. In the second step, the labeling is done in heavy-oxygen water.

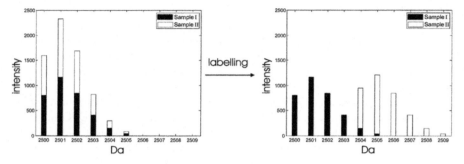

Fig. 2 Chemical reaction scheme for the two-step enzymatic ^{18}O-labeling procedure

Fig. 3 Effect of enzymatic ^{18}O-labeling in a mass spectrum in "stick" representation on a pooled sample. *Left panel*: "sticks" can be seen as a representation of the distribution of the isotope variants of the peptide. *Right panel*: labeling causes accumulation of different isotope variants in a spectrum

The oxygen replacements in the carboxyl-terminus are a continuous process and are enzymatically catalyzed by a proteolytic reagent. This labeling reaction is schematically depicted in Fig. 2. We assume that both oxygen-atoms on the carboxyl-terminus will react equally favorable with the ^{18}O-atoms. Hence, in ideal circumstances, the labeling should lead to an increase of the mass of the peptide molecule by 4 daltons (Da).

For example, the labeled peptides from Sample II can now be pooled together with the unlabeled peptides from Sample I and processed simultaneously by MS. Without the enzymatic ^{18}O-labeling, the isotope peaks, corresponding to the isotope distribution of a peptide present in both samples, would appear at the same location in the resulting pooled MS1 or *joint mass spectrum*. This is graphically illustrated in the left panel of Fig. 3. In this situation, no distinction could be made between the contributions of the different biological samples to the peptide peaks observed in the joint spectrum. However, with the enzymatic ^{18}O-labeling, the isotope peaks which correspond to the labeled peptide will shift 4 Da (in a singly charged spectrum) to the right in the mass spectrum, as shown in the right-hand-side panel of Fig. 3. This allows making a distinction between the peaks related to peptides from different samples. Consequently, a direct comparison of the peptide abundance in the two samples is possible because the abundance measurements are affected by the same amount of machine noise. A "naïve" approach to compute the relative abundance of the peptide in the two samples would be to take the ratio of the heights of the first and fifth peak observed for the peptide in the mass spectrum (see the right-hand-side panel of Fig. 3), as these peaks would correspond to the monoisotopic variants of the peptide in the unlabeled and labeled sample, respectively. However, as it can be observed from Fig. 3, some isotope peaks of the unlabeled peptide will still overlap with the monoisotopic peak of the labeled peptide. Thus, even in this ideal setting, where a mass shift of 4 Da is acquired, the ratio would yield a biased estimate of the relative abundance, because it does not take into account the overlap of the isotope peaks.

In practice, however, there are more problems related to the use of the enzymatic ^{18}O-labeling strategy. First, the heavy-oxygen water does not contain 100 % pure ^{18}O-water. It can also contain ^{16}O- and ^{17}O-atoms. We term these *water impurities*. Note that, if the two carboxyl-terminus oxygen atoms are replaced by, e.g., ^{17}O-atoms, the peptide molecule becomes heavier by only 2, and not 4 Da, as it ideally would be the case in 100 % pure ^{18}O-water. Second, the speed of the enzymatic reaction, i.e., the oxygen incorporation rate, depends on multiple unobserved factors and therefore can differ for different peptides. As a result, at the end of the enzymatic reaction, not all peptide molecules from Sample II may have been actually labeled. The isotope peaks for these molecules will overlap with the peaks from Sample I, which results in a biased estimate of the relative abundance.

These problems imply that the peaks, observed for a peptide in a spectrum, will correspond to a complex mixture of shifted and overlapping isotope peaks that are related to the isotope distributions of the peptide molecules in the unlabeled and labeled samples. In order to obtain an unbiased estimate of the relative abundance of the peptide in the two samples, the overlap of the isotope peaks has to be taken into account [33].

Several methods that address the issue at the data analysis stage have been developed [3, 12, 13, 17, 27]. In particular, Valkenborg and Burzykowski [27] developed a model-based approach that combines the regression framework considered by Mirgorodskaya et al. [13] with a probabilistic model, which describes the kinetics of the enzymatic ^{18}O-labeling reaction. An important advantage of the method is that it allows estimating the peptide's isotope distribution directly from the observed data, which in turn can be used to validate whether the observed peaks are indeed originating from a bonafide peptide. This implies that no additional MS steps are required for quantification, while the information is unbiasedly extracted from the observed spectra. The method is able to accommodate additional joint mass spectra for a given peptide, which can arise from, e.g., technical replicates. Moreover, the method can account for the possible presence of ^{16}O- and ^{17}O-atoms in the heavy-oxygen water.

Figure 4 illustrates the general concept underlying the method. The figure presents a part of a joint mass spectrum corresponding to a certain peptide. The fifth peak of the observed part of the joint spectrum (plot at the right-hand side) is composed out of the unobserved isotope variants of the peptide in Sample I and Sample II before the labeling (plots at the left-hand side). Note that the first five isotope variants of the peptide in Sample II (labeled sample) contribute to the fifth peak of the observed part of the joint spectrum via the mass-shift probabilities

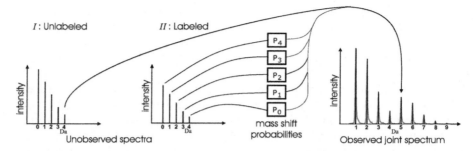

Fig. 4 The height of the fifth peak of the observed part of the joint spectrum can be defined in terms of the unobserved peptide peak intensities before labeling and the mass-shift probabilities. Due to the imprecise labeling of a peptide, five potential mass shifts can occur, with probabilities P_0, P_1, P_2, P_3, and P_4. In this way, the set of isotope peaks from the labeled peptide can contribute to the fifth peak in the joint spectrum via the mass-shift probabilities

induced by the labeling. Hence, to estimate the relative abundance, Q say, of the peptide in Sample II as compared to Sample I, we need information about the theoretical isotope distribution of the peptides before the labeling.

Toward this aim, Valkenborg and Burzykowski [27] proposed a model, which assumes that the expected values of the observed peak intensities in a peptide-specific part of the joint spectrum are expressed as a linear function of the abundance of the (aggregated) isotope variants of the peptide in Sample I and the relative abundance Q. In particular, assume that there are M aggregated isotope variants of the peptide. Let R_m $(m = 1, 2, \ldots, M)$ denote the ratio of the probability of occurrence of the m-th variant relative to the monoisotopic one $(m = 1)$; put $R_1 = 1$. Let H denote the abundance of the peptide in Sample I. The model is defined by assuming that the intensity y_k $(k = 1, 2, \ldots, M + 4)$ of the k-th peak observed in the peptide-specific part of the joint spectrum can be expressed as follows:

$$y_k = HR_kI \, (k \le M) + HQ \sum_{j=\max(k-M,0)}^{\min(k-1,4)} P_jR_{k-j} + \varepsilon_k,$$

where the residual errors ε_k follow a normal distribution with mean 0 and variance σ^2, and $I(A)$ is the indicator function equal to 1 when condition A is fulfilled and 0 otherwise. The coefficients of the linear function are the mass-shift probabilities P_j $(j = 0, 1, \ldots, 4)$, where P_j denotes the probability that the isotope distribution of a labeled peptide will be shifted by j Da due to the incorporation of a combination of ^{16}O-, ^{17}O-, and ^{18}O-atoms from the heavy-oxygen water. The mass-shift probabilities P_j themselves are computed by using a Markov-chain model with a transition-probability matrix that depends on the (assumed known) presence of the ^{16}O- and ^{17}O-atoms in the heavy-oxygen water and the peptide-specific incorporation rate λ (assumed constant in time), which gives the number of the oxygen-exchange reactions per time unit.

Valkenborg and Burzykowski [27] applied the proposed method to a set of six joint mass spectra obtained from the tryptic peptides of bovine cytochrome C from LC Packings. The sample containing all peptides resulting from the tryptic-digest was divided into two parts. One part was enzymatically labeled (overnight) with a stable ^{18}O-isotope, with trypsin as a catalyst, while the other part remained unlabeled. Next, the labeled and unlabeled peptides were mixed according to a ratio $Q = 1/3$. Six samples from the resulting mixture were automatically spotted on one stainless steel plate by a robot. The plate was processed by a 4800 MALDI-TOF/TOF analyzer (Applied Biosystems) mass spectrometer, resulting in six joint spectra.

The analysis was restricted to the parts of the six joint spectra corresponding to three bovine cytochrome C peptides. The amino acid compositions of these peptides were as follows: peptide CC1 (mass 1167.61 Da)—TGPNLHGLFGR; peptide CC2 (mass 1455.66 Da)—TGQAPGFSYTDANK; and peptide CC3 (mass 1583.75 Da)—KTGQAPGFSYTDANK. Panel (a) of Fig. 5 presents the heights of the consecutive peaks observed in the parts of the joint spectra corresponding to peptide CC1 in the stick representation.

The estimated values of Q were equal to 0.5518 (standard error, SE = 0.0318) for peptide CC1, 0.3340 (SE = 0.0129) for peptide CC2, and 0.3318 (SE = 0.0068) for peptide CC3. The estimates for peptides CC2 and CC3 were in a very good agreement with the targeted value of 1/3. Also, the estimated isotope distributions were virtually identical to the theoretical ones. Thus, the model adequately described the data for the two peptides. However, for peptide CC1, the estimated relative abundance was markedly different from 1/3 and the estimated isotope distribution was statistically significantly different from the theoretical one. This suggested a possible failure of the experimental procedure for the peptide.

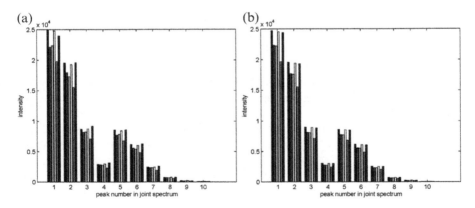

Fig. 5 Observed (panel (**a**)) and estimated (panel (**b**)) peak heights in the parts of the six joint spectra corresponding to the tryptic bovine cytochrome C peptide CC1 with mass 1167.61 Da and $Q = 1/3$. Note that the peaks are grouped per isotope peak

The "naïve" estimates, obtained by taking the mean value of the ratio of the heights of the first and fifth peak observed in the peptide-specific part of the joint mass spectrum, were equal to 0.5336, 0.3473, and 0.3556, respectively. Thus, in this case, there was not much difference between the model-based and "naïve" estimates. Valkenborg and Burzykowski [27] argued that this was likely due to the efficient ^{18}O-labeling (overnight) and the use of highly purified heavy-oxygen water, which caused a clear separation between the labeled and unlabeled spectra. In a realistic setting, however, labeling might be inefficient and it would be cost-efficient to use less purified oxygen labels, which would then favor the use of the proposed model.

The model developed by Valkenborg and Burzykowski [27] assumed, among others, that the observed peak intensities in the joint spectrum were normally distributed with a constant residual variance. The homoscedasticity assumption was relaxed in the model developed by Zhu, Valkenborg, and Burzykowski [32]. For the latter model, Zhu and Burzykowski [31] proposed a Bayesian formulation.

3.2 Hydrogen/Deuterium Exchange

The isotope distribution can also be used to analyze data from hydrogen/deuterium exchange mass spectrometry (HDXMS). HDXMS is a method to study the conformation and dynamics of proteins and peptides. Hydrogen atoms from NH-groups of peptide bonds can be replaced by deuterium. Unprotected hydrogens located at the surface of the protein will exchange faster than hydrogens buried in the hydrophobic core of the protein under study. Therefore, monitoring the exchange of labile hydrogen atoms over time makes it possible to draw conclusions about the conformation and dynamics of proteins. After exchange of one hydrogen, the molecular mass of the protein under study will increase by approximately 1 Da. The intensities of the isotope distribution will also change (Fig. 6).

Several methods have been proposed to determine the deuteration-level of a protein. The simplest approach [29] calculates the number of exchanged H-atoms by subtracting the average mass of the undeuterated ion from the average mass of the (partially) deuterated protein. There are two potential issues with this approach. First, the resolution and mass accuracy of the mass spectra can complicate the calculation of the average mass of the (partially) deuterated protein. Second, the amount of information provided by the differences in mass can be insufficient to draw valid conclusions regarding the structure and/or dynamics of the protein under study.

Other methods use the isotope distribution to estimate the deuteration-level. They are based on two ideas:

- The first idea assumes that the observed isotope distribution is a combination of the natural, non-deuterated isotope distribution, and a deuteration distribution. Palmblad et al. [15] proposed to estimate the exchange probability (P_{exch}), linked to this deuteration distribution, with a deconvolution-based approach.

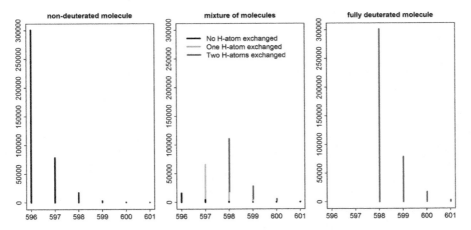

Fig. 6 HDX mass spectra for a molecule with two exchangeable hydrogens. The bars in the *middle panel* show the cumulative intensity resulting from an overlap of isotope distributions of molecules with a different number of exchanged hydrogens

The amount of deuteration, expressed as the probability of exchange, is determined by a modified version of the fast Fourier transform approach of Rockwood and Van Orden [19]:

$$DI_{calc} = F^{-1} \{ F(I_C)^{nc} * F(I_H)^{nH} * F(I_O)^{no} * F(I_S)^{ns} * F(I_{sol})^{nsol} * F(I_D) \} \quad (1)$$

where DI_{calc} are the intensities of the expected isotope distribution of a (partially) deuterated protein, $F(I_x)$ denotes the Fourier-transformed intensity distribution of chemical element X, I_{sol} are the intensities of too-fast-exchanging hydrogens, and I_D are the intensities of the deuteration distribution which is assumed to be binomial:

$$I_D = \text{Bin}\,(P_{exch}, n_{exch}), \quad (2)$$

with n_{exch} denoting the number of hydrogens available for exchange. This approach, implemented in AUTOHD [15], assumes that each exchangeable hydrogen has the same exchange probability. The probability is estimated by minimizing the differences between the observed and the expected isotope distribution of the deuterated protein/peptide.

• The second idea is actually very similar to the approach used in the analysis of the enzymatically ^{18}O-labeled mass spectra. It considers the (partially) deuterated isotope distribution as a linear combination of $n + 1$ mass-shifted natural isotope distributions (Fig. 7):

$$DI_i = \sum_{j=0}^{i} \omega_j * I_{i-j} \quad (3)$$

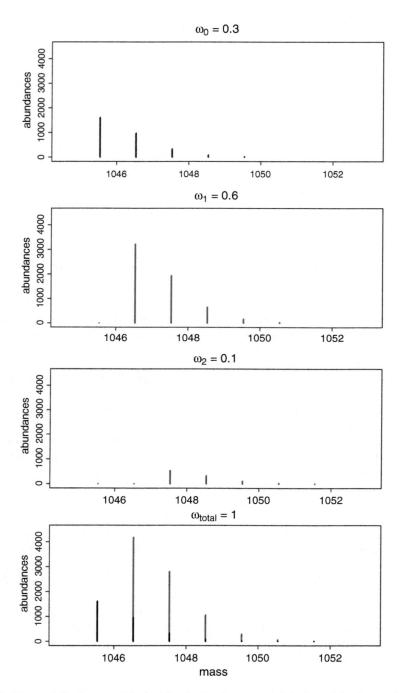

Fig. 7 The partially deuterated isotope distribution (*bottom panel*) is the weighted sum of three isotope distributions with different levels of deuteration, i.e., no deuterium (*top panel*), one deuterium atom (*second panel from the top*), and two deuterium atoms (*third panel from the top*)

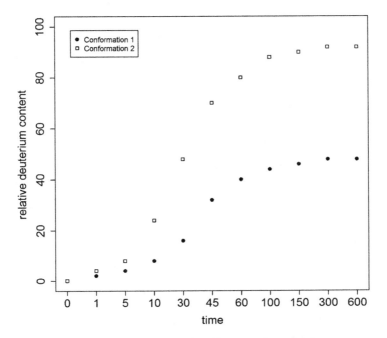

Fig. 8 Deuteration/incorporation plot of a protein with two conformational states

where DI_i denotes the normalized intensity of the deuterated ith isotope variant, I_{i-j} is the intensity of the undeuterated $(i-j)$th isotope variant, and ω_j is the abundance of j exchanged deuteriums. In contrast to the method of Zhang and Smith [29] and Palmblad et al. [15], estimating ω_j returns, for each time point, a distribution of relative deuterium-content. Zhang et al. [30] discuss three different methods to determine ω_j, i.e., directly solving Eq. (3), ordinary least squares minimization, or maximum-entropy method.

Plotting the determined deuteration content against the labeling time can provide clues about the 3D structure and/or dynamics of the protein. These deuteration incorporation plots (Fig. 8) give an intuitive idea about the kinetic exchange rate of a protein.

For each exchangeable hydrogen a kinetic exchange rate can be estimated:

$$D_t = N - \sum_{i=1}^{N} e^{-k_i t}, \tag{4}$$

where N is the total number of exchangeable hydrogen atoms, and k_i the exchange rate of hydrogen i.

A commonly applied method to estimate these exchange rates is the model proposed by Zhang and Smith [29] and Smith et al. [24]. The method heavily relies on the determined deuterium-content without taking the uncertainty linked to these estimates into account (Eq. 4). Another approach is to directly estimate the exchange rates from the observed isotope distribution [4, 18]. For each time point t, the states of all exchangeable hydrogens, N, are represented by 2^N labile binary vectors, containing 1's and 0's, where "1" corresponds to exchanged, and "0" to non-exchanged hydrogen. The vectors represent all possible combinations of exchanging hydrogens in the analyzed protein or fragment, ranging from no exchange to a full exchange of all H-atoms. The overall probability of each vector $p_{tot;t}$ is calculated as the product of individual exchange probabilities:

$$p(D_i)_t = 1 - e^{-k_i t} = 1 - p(H_i)_t,$$

where $p(D_i)_t$ denotes the probability of deuterium exchange of the ith H-atom by time t.

Binary vectors with the same number of exchanged hydrogens are combined, and their probabilities are summed. These summed overall probabilities are compared with the normalized intensities of the protein or fragment under study using the overall mean square deviance or the Kullback–Leibler distance as the overall divergence measure. The exchange rates of each exchangeable H-atom are then estimated by minimizing the overall divergence measure.

3.3 Other Application Domains

The previous sections present the cases for peptide-centric quantification and structural elucidation of proteins by controlled stable isotope labeling of the scrutinized molecules [5]. One of the key features of the observed spectra is that they can be interpreted as resulting from overlapping isotope distributions.

However, there are many more experimental protocols that can result in overlapping isotope distributions. The data analysis strategies explained earlier can be generalized to investigate any type of MS data with intentional or artefactual overlapping isotope distributions. This section provides a non-exhausting overview of alternative labeling strategies and presents cases where, for example, degradative chemical reactions, such as deamidation, can cause the overlap of isotope distributions.

It is paramount to note that any observed spectrum that contains overlapping or non-overlapping isotope profiles can be expressed as a linear combination of the unobserved theoretical isotope distribution of the compounds that are composing the mass spectrum (see Fig. 9). Hence, a linear model can be defined to disentangle and quantify the overlapping and non-overlapping isotope patterns present in a spectrum. Equation (5) presents an example of a linear model that could be used for a part of an observed spectrum. Assume that the observed isotope peaks Y_1 to Y_{20} are the result of a labeling protocol. The labeling causes overlapping isotope patterns in peaks Y_1 to Y_5 due to the three different labels (L_1 to L_3). The fourth label (L_4) does not introduce overlap in the spectrum, as it is well separated in the mass domain from the cluster of overlapping peptides. Nevertheless, the observed

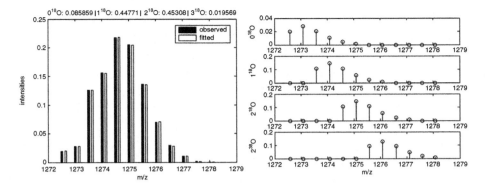

Fig. 9 An observed spectrum (*left-hand-side plot, black bars*) that can be expressed as a linear combination (*left-hand side, white bars*) of the shifted theoretical isotope distributions (*right-hand-side plots*) of the compounds that are composing the mass spectrum

peaks from the fourth label can be taken into account by the model. A simple fitting scheme will yield the quantitative values corresponding to the labels (L_1 to L_4). It is worth noting that in this general modeling framework, the number of compounds in the spectrum (columns of the matrix), the overlap caused by the labels (leading zeros in the columns), and the molecular composition of the compounds (isotope distributions $I_{11}–I_{43}$) should be known.

$$
\begin{bmatrix} Y_1 \\ Y_2 \\ Y_3 \\ Y_4 \\ Y_5 \\ \vdots \\ Y_{18} \\ Y_{19} \\ Y_{20} \end{bmatrix}
=
\begin{bmatrix}
I_{11} & 0 & 0 & 0 \\
I_{12} & I_{21} & 0 & 0 \\
I_{13} & I_{22} & I_{31} & 0 \\
0 & I_{23} & I_{32} & 0 \\
0 & 0 & I_{33} & 0 \\
\vdots & \vdots & \vdots & \vdots \\
0 & 0 & 0 & I_{41} \\
0 & 0 & 0 & I_{42} \\
0 & 0 & 0 & I_{43}
\end{bmatrix}
\times
\begin{bmatrix} L_1 \\ L_2 \\ L_3 \\ L_4 \end{bmatrix}
\qquad (5)
$$

An example of how model (5) can be employed is given in Fig. 9. Assume that the objective of the experiment is to quantify the extent of protein degradation for different wetlab protocols. The plot at the top of the right-hand-side panel of the figure presents the theoretical isotope profile of a biomolecule in a doubly charged mass spectrum. Due to protein degradation in the ^{18}O-water, the molecule starts incorporating ^{18}O-atoms, with shifts of 1 m/z (corresponding to 2 Da) as a result. The shifted profiles, resulting from the incorporation of 1, 2, or 3 ^{18}O-atoms, are presented in the right-hand-side panel of Fig. 9. The black bars in the left-hand-side panel of Fig. 9 depict the profile observed in the overall mass spectrum, resulting from the overlap of the profiles shown in the right-hand-side panel. Based on this overall profile and using a model similar in spirit to the one given in Eq. (5), one can estimate the proportion of molecules incorporating 1, 2, or 3 ^{18}O-atoms. The resulting "fitted" profile, corresponding to a particular set of proportions (shown at the top of the plot in the left-hand-side panel of Fig. 9), is shown as white bars in the plot in the left-hand-side panel of Fig. 9.

The enzymatic ^{18}O-labeling approach can be categorized in the subclass of chemical labeling. Chemical labeling is the intentional modification of a peptide or amino acid with a particular and well-known mass tag. To date, several strategies are available for chemical labeling, for example, isotope-coded protein labels (ICPL; [21]), Isotope-Coded Affinity Tags (ICAT; [7]), or synthetic AQUA peptides [25]. Although the latter labels are meticulously engineered to minimize the overlap of the isotope distributions, it can happen that due to isotope impurities or incomplete labeling a certain amount of overlap occurs that requires a modeling approach for the accurate quantification of the label's molecules.

Special cases of chemical labeling are tandem mass tags (TMT, Thermo Scientific) and isobaric tag for relative and absolute quantification (iTRAQ, SCIEX). These labels are constructed in such a way that no mass difference is present between, e.g., the six labels in a TMT 6-plex; hence the term isobaric labeling. However, the labels are developed so that, upon fragmentation, a charged reporter ion and neutral balancer will be released from the label causing 1 Da separated reporter ions that can be used as a proxy for their relative abundance in the pooled sample. Despite the very low mass of these reporter ions, they do exhibit a pronounced isotope distribution due to isotope impurities present during the synthesis of these labels. These isotope impurities are known and communicated by the vendor in the data sheets. The model in Eq. (5) can be adapted to correct for the systematic biases introduced by these impurities that cause the isotope distributions of the reporter ions to overlap. A similar approach has already been proposed by Shadforth et al. [23].

Metabolomic labeling, e.g., SILAC [14], is a popular stable isotope labeling strategy that labels an entire organism by growing the cell culture in a medium that contains isotopically modified essential amino acids. Typically, arginine or lysine is modified with stable isotopes of carbon and nitrogen, i.e., ^{13}C and ^{15}N. For arginine, mass shifts of 4 Da (^{15}N$_4$), 6 Da (^{13}C$_6$), and 10 Da (^{15}N$_4$ + ^{13}C$_6$), compared to the unlabeled or light variants of the organism, are common. The isotopically modified arginines can be combined with lysine, which introduces mass shifts of 2 Da (^{15}N$_2$) and 6 Da (^{13}C$_6$), or the combination thereof that leads to an 8 Da shift (^{15}N$_2$ + ^{13}C$_6$), compared to the light variant. When using a single modified amino acid, four cell cultures can be grown with a different label incorporation. After inducing an experimental condition to the cell culture, the four experiments can be pooled together and processed simultaneously during sample preparation and LC-MS measurement. The advantage of this procedure, as compared to chemical labeling, is that in a very early stage the samples can be aggregated such that variability introduced at the level of protein extraction, reduction, trypsinization, etc., is the same for the pooled samples. A disadvantage is that only peptides containing, for example, an arginine are labeled. In case of missed cleavages or when using proteases that do not cleave at arginine it may happen that peptides become multiply labeled by the modified amino acid. Another attention point is that some organisms convert arginine to proline. This arginine-to-proline conversion causes additional modified amino acids that could be present in the spectrum [16]. In contrast to chemical labeling, metabolomics labeling can induce different variants of high/low mass n-tuples in the mass spectral data. Therefore, often fragment information or equivalently, peptide identification, is required to group the high/low mass n-tuples together. Disregarding the complexity of the data analysis, metabolomics labeling is popular because it allows multiplexing of four samples [8]. Furthermore, the minimum mass shift between the overlapping distributions is 4 Da that relaxes the requirement of complex isotope modeling.

A recent and interesting observation is that not only peptide or reporter ions can overlap in a mass spectrum [9]. More specifically, the modern fragmentation mechanism called Electron-Transfer Dissociation (ETD) is of such a complexity that multiple reaction products may occur before a protein or peptide fragmentation occurs. These alternative reaction mechanisms are currently being investigated [10]. From the observed isotope pattern of the fragment ion one can hypothesize that three additional reaction pathways exist that provoke the isotope patterns to shift by steps of 1 Da. These reaction pathways are called Electron-Transfer but no Dissociation (ETnoD), proton transfer reaction (PTR), and hydrogen transfer reaction (HTR). Former reactions lead to loss of hydrogen, charge, or a combination of both, causing the reaction products to overlap with the desired fragmentation results of the ETD reaction. To gain insight in the reaction mechanism and to deal with the nuisance from the unwanted reactions, a model similar in spirit to Eq. (5) can be developed to disentangle the information and remove the interference [11].

Another interesting and prevalent modification with clinical relevance is deamidation of proteins and peptides. Protein deamination is a biologically relevant phenomenon, but can also be induced unintentionally by, e.g., sample handling or protein degradation. The modification appears in mass spectrometry by shifting the analytes isotope profile by 1 Da causing an overlap with the unmodified counterpart. Software algorithms to deconvolute the spectral information often disregard potential deamidation and therefore the monoisotopic mass is defectively inferred from the observed isotope cluster. Atlas and Datta [1] presented an algorithm based on the linear model presented schematically in Eq. (5) to correctly determine the monoisotopic mass from an overlapping cluster of deamidated peptides.

Artefactual modifications also occur in the fields of MS-based lipidomics. Lipids with unsaturated bonds are 1 Da lighter than their saturated counterparts and multiple bonds can be saturated in a lipid (see Fig. 10). Apart from the bonds, the composition of the overlapping lipids remains unchanged, making this molecule very appropriate for the presented modeling approach as the mass shift, number of mass shifts, and composition is often known or not needed since an approach similar to the one used for ^{18}O-labeling can be applied with isotope information inferred from the observed isotope peaks.

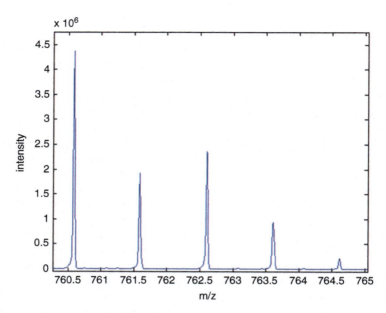

Fig. 10 Lipids bonds can cause overlap of isotope profiles in the observed mass spectrum

4   Concluding Remarks

In peptide-centric MS, the isotope distribution offers important information about the quantification of the biologically relevant signal in MS1 spectra. By using this information, the processing and analysis of MS data can be made more efficient.

Information about the isotope distribution can also be applied at the stage of the interpretation of the results of an MS experiment. For instance, it can be used to improve the quality and confidence of peptide identifications from database-search strategies [6].

In Sect. 3.3 we discussed the analysis of data from MS-experiments applying labeling techniques. In such experiments, information about the form of the overlapping isotope distributions, their number, and their mass shifts is known. This allows the use of linear models to analyze the data. Overlapping isotope profiles can also occur in spectra due to co-eluting peptide-homologues. However, in this case the linear-model-based strategy is more difficult to apply, because the information about, e.g., the number of overlapping isotope profiles or their mass shifts is not known. Hence, disentangling arbitrary patterns of overlapping peptides is a tedious task, though some progress can be made by using, e.g., the Bayesian approach [34].

References

1. Atlas, M., & Datta, S. (2009). A statistical technique for monoisotopic peak detection in a mass spectrum. *Journal of Proteomics and Bioinformatics, 2*, 202–216.
2. Claesen, J., Dittwald, P., Burzykowski, T., & Valkenborg, D. (2012). An efficient method to calculate the aggregated isotopic distribution and exact center-masses. *Journal fo the American Society for Mass Spectrometry, 23*, 753–763.
3. Eckel-Passow, J. E., Oberg, A. L., Therneau, T. M., Mason, C. J., Mahoney, D. W., Johnson, K. L., et al. (2006). Regression analysis for comparing protein samples with ^{16}O/^{18}O stable-isotope labeled mass-spectrometry. *Bioinformatics, 2*, 305–318.
4. Geller, O., & Lifshitz, C. (2004). Applying a new algorithm to H/D exchange of multiply protonated cytochrome c. *International Journal of Mass Spectrometry, 223*, 125–129.
5. Gevaert, K., Impens, F., Ghesquière, B., Van Damme, P., Lambrechts, A., & Vandekerckhove, J. (2008). Stable isotopic labeling in proteomics. *Proteomics, 8*, 4873–4885.
6. Ghavidel, F. Z., Mertens, I., Baggerman, G., Laukens, K., Burzykowski, T., & Valkenborg, D. (2014). The use of the isotopic distribution as a complementary quality metric to assess tandem mass spectra results. *Journal of Proteomics, 98*, 150–158.
7. Gygi, S. P., Rist, B., Gerber, S. A., Turecek, F., Gelb, M. H., & Aebersold, R. (1999). Quantitative analysis of complex protein mixtures using isotope-coded affinity tags. *Nature Biotechnology, 17*, 994–999.
8. Impens, F., Colaert, N., Helsens, K., Ghesquière, B., Timmerman, E., De Bock, P. J., et al. (2010). A quantitative proteomics design for systematic identification of protease cleavage events. *Molecular and Cellular Proteomics, 9*, 2327–2333.
9. Lermyte, F., Konijnenberg, A., Williams, J. P., Brown, J. M., Valkenborg, D., & Sobott, F. (2014). ETD allows for native surface mapping of a 150 kDa noncovalent complex on a commercial Q-TWIMS-TOF instrument. *Journal of the American Society for Mass Spectrometry, 25*, 343–350.
10. Lermyte, F., Verschueren, T., Brown, J. M., Williams, J. P., Valkenborg, D., & Sobott, F. (2015). Characterization of top-down ETD in a travelling-wave ion guide. *Methods, 89*, 22–29.
11. Lermyte, F., Łąckic, M. K., Valkenborg, D., Baggerman, G., Gambin, A., & Sobott, F. (2015). Understanding reaction pathways in top-down ETD by dissecting isotope distributions: A mammoth task. *International Journal of Mass Spectrometry, 390*, 146–154.
12. Lopez-Ferrer, D., Ramos-Fernandez, A., Martinez-Bartolome, S., Garca-Ruiz, P., & Vazquez, J. (2006). Quantitative proteomics using ^{16}O/^{18}O labeling and linear ion trap mass spectrometry. *Proteomics, 6*, S4–S11.
13. Mirgorodskaya, O. A., Kozmin, Y. P., Titov, M. I., Korner, R., Sonksen, C. P., & Roepstorff, P. (2000). Quantitation of peptides and proteins by matrix-assisted laser desorption/ionization mass spectrometry using ^{18}O-labeled internal standards. *Rapid Communications in Mass Spectrometry, 14*, 1226–1232.
14. Ong, S. E., Blagoev, B., Kratchmarova, I., Kristensen, D. B., Steen, H., Pandey, A., et al. (2002). Stable isotope labeling by amino acids in cell culture, SILAC, as a simple and accurate approach to expression proteomics. *Molecular and Cellular Proteomics, 1*, 376–386.
15. Palmblad, M., Buijs, J., & Hakansson, P. (2001). Automatic analysis of Hydrogen/Deuterium exchange mass spectra of peptides and proteins using calculations of isotopic distributions. *Journal of the American Society for Mass Spectrometry, 12*, 1153–1162.
16. Park, S. K., Jin, L. L., Kim, Y., & Yates, J. R., III. (2009). A computational approach to correct arginine-to-proline conversion in quantitative proteomics. *Nature Methods, 6*, 184–185.
17. Rao, K. C., Carruth, R. T., & Miyagi, M. (2005). Proteolytic ^{18}O-labeling by peptidyl-Lys metalloendopeptidase for comparative proteomics. *Journal of Proteome Research, 4*, 507–514.
18. Reuben, B., Ritov, Y., Geller, O., McFarland, M., Marshall, A., & Lifshitz, C. (1993). Applying a new algorithm for obtaining site specific rate constants for H/D exchange of the gas phase proton-bound arginine dimer. *Chemical Physics Letters, 380*, 88–94.

19. Rockwood, A. L., & Van Orden, S. L. (1996). Ultrahigh-speed calculation of isotope distributions. *Analytical Chemistry, 68,* 2027–2030.
20. Rosman, K. J. R., & Taylor, P. D. P. (1998). Isotopic compositions of the elements 1997. *Pure and Applied Chemistry, 70,* 217–235.
21. Schmidt, A., Kellermann, J., & Lottspeich, F. (2005). A novel strategy for quantitative proteomics using isotope-coded protein labels. *Proteomics, 5,* 4–15.
22. Senko, M. W., Beu, S. C., & McLafferty, F. W. (1995). Determination of monoisotopic masses and ion populations for large biomolecules from resolved isotopic distribution. *Journal of the American Society for Mass Spectrometry, 6,* 229–233.
23. Shadforth, I. P., Dunkley, T. P. J., Lilley, K. S., & Bessant, C. (2005). i-Tracker: For quantitative proteomics using iTRAQ™. *BMC Genomics, 6,* 145.
24. Smith, D., Deng, Y., & Zhang, Z. (1997). Probing the non-covalent structure of proteins by amide hydrogen exchange and mass spectrometry. *Journal of Mass Spectrometry, 32,* 135–146.
25. Stemmann, O., Zou, H., Gerber, S. A., Gygi, S. P., & Kirschner, M. W. (2001). Dual inhibition of sister chromatid separation at metaphase. *Cell, 107,* 715–726.
26. Valkenborg, D., Jansen, I., & Burzykowski, T. (2008). A model-based method for the prediction of the isotopic distribution of peptides. *Journal of the American Society for Mass Spectrometry, 19,* 703–712.
27. Valkenborg, D., & Burzykowski, T. (2011). A Markov-chain model for the analysis of high-resolution enzymatically ^{18}O-labeled mass spectra. *Statistical Applications in Genetics and Molecular Biology,* 10, article 1.
28. Valkenborg, D., Mertens, I., Lemière, F., Witters, E., & Burzykowski, T. (2012). The isotopic distribution conundrum. *Mass Spectrometry Reviews, 31,* 96–109.
29. Zhang, Z., & Smith, D. L. (1993). Determination of amide Hydrogen exchange by mass spectrometry: A new tool for protein structure elucidation. *Protein Science, 2,* 522–531.
30. Zhang, Z., Guan, S., & Marshall, A. (1997). Enhancement of the effective resolution of mass spectra of high-mass biomolecules by maximum-entropy based deconvolution to eliminate the isotopic natural abundance distribution. *Journal of the American Society for Mass Spectrometry, 8,* 659–670.
31. Zhu, Q., & Burzykowski, T. (2011). A Bayesian Markov-chain-based heteroscedastic regression model for the analysis of ^{18}O-labelled mass spectra. *Journal of the American Society for Mass Spectrometry, 22,* 499–507.
32. Zhu, Q., Valkenborg, D., & Burzykowski, T. (2010). A Markov-chain-based heteroscedastic regression model for the analysis of high-resolution enzymatically ^{18}O-labeled mass spectra. *Journal of Proteome Research, 9,* 2669–2677.
33. Ye, X., Luke, B., Andresson, T., & Blonder, J. (2009). [18]O stable isotope labeling in MS-based proteomics. *Briefings in Functional Genomics and Proteomics, 8,* 136–144.
34. Zhu, Q., Kasim, A., Valkenborg, D., & Burzykowski, T. (2011). A Bayesian model-averaging approach to the quantification of overlapping peptides in a MALDI-TOF mass spectrum. *International Journal of Proteomics,* article ID 928391.

Probabilistic and Likelihood-Based Methods for Protein Identification from MS/MS Data

Ryan Gill and Susmita Datta

1 Introduction

The main goal of proteomic studies is to detect biomarker proteins for the early detection of cancer. Tandem mass spectrometry (MS/MS) plays a significant role in the discovery of these biomarker proteins. The first step of an experiment using tandem mass spectrometry (MS/MS) is the digestion of a mixture of proteins by an enzyme, often trypsin. Each of the proteins is separated into peptides which are subsequently ionized. Then selected peptide ions are fragmented in the gas phase, and the mass-to-charge ratios and abundances or intensities of the small fragmented ions are recorded in an MS/MS spectra.

Technological improvements have led to a greater abundance of tandem mass spectrometry data as well as an increase in the size of generated data sets [1]. Consequently, it is not feasible to manually attempt to identify the peptides present in the sample, and hence software tools are needed to perform this task. SEQUEST [2] is a popular software tool which uses the precursor ion mass for each observed spectrum to find candidate peptides from a database of protein sequences which are sufficiently close in mass to the spectrum. Each observed spectrum is preprocessed by finding the highest intensity peaks in each of a set of pre-specified bins and normalizing those values to obtain an observed n-dimensional vector u. The theoretical spectrum, denoted by v, is also computed for each of the candidate peptides, and each theoretical spectrum is then preprocessed the same way that the

R. Gill (✉)
Department of Mathematics, University of Louisville, Louisville, KY 40292, USA
e-mail: ryan.gill@louisville.edu

S. Datta
Professor, Department of Biostatistics, University of Florida, Gainesville, FL, USA
e-mail: susmita.datta@ufl.edu

© Springer International Publishing Switzerland 2017
S. Datta, B.J.A. Mertens (eds.), *Statistical Analysis of Proteomics, Metabolomics, and Lipidomics Data Using Mass Spectrometry*, Frontiers in Probability and the Statistical Sciences, DOI 10.1007/978-3-319-45809-0_4

observed spectrum was preprocessed. Then, each observed spectrum is compared
with each theoretical spectrum in its candidate list by a preliminary score S_p based
on the number of predicted fragment ions that match ions in the spectrum and their
abundances as well as the number of predicted sequence ions. Finally, a further score

$$Xcorr = R_0 - \frac{1}{151} \sum_{\tau=-75}^{75} R_\tau$$

is computed for each of the top 500 candidate spectra where $R_\tau = \sum u_i v_{i+\tau}$ is
the discrete cross-correlation with lag τ. Here R_0 is the scalar dot product between
the observed and theoretical spectra. As described in [2] and [3], *Xcorr* gives a
measure of spatial similarity to assess the coherence of the observed and each
theoretical spectra by not only computing R_0 but also by including a correction
factor to account for background correlation between the observed and theoretical
spectra by using offset values. The highest *Xcorr* score is reported by SEQUEST
as a peptide-spectrum match (PSM). SEQUEST is a commercial program, but there
are alternate implementations of the original SEQUEST algorithm such as Crux [3]
and Tide [4] which are freely available and which also reportedly lead to drastic
increase in speed compared with the original SEQUEST algorithm. Software for
other database search algorithms such as X!TANDEM [5], Mascot [6], MS-Tag [7],
and MS-GF [8–10] are also available.

In spite of the developments of the above-mentioned search algorithms, there
still remain uncertainties associated with the peptide and protein identifications.
Experimental errors and lack of adequate search algorithms can, sometimes,
lead to highly erroneous peptide and protein identifications from a tandem mass
spectrometry experiment; in fact, without proper filtering, it is possible that 80–
90 % of identified proteins may not be correct [11, 12]. The situation becomes
more complicated in the presence of "degenerate" peptides. A peptide is referred
to as "degenerate" if it is generated by multiple proteins. Degenerate peptides
create additional challenges for protein identification because even if the peptide
identification were known to be correct with no uncertainty, the identity of the
protein that generated it is not clearly determined. A typical situation of degeneracy
is explained through Fig. 1 (adapted from figures in [13–15]).

Figure 1 summarizes the steps in an MS/MS experiment for three proteins P_1,
P_2, and P_3 in red, green, and blue, respectively; each of these proteins generates
two peptides: P_1 generates p_1 and p_2, P_2 generates p_3 and p_4, and P_3 generates p_2
and p_5. Since p_2 is generated by two distinct proteins P_1 and P_3, it is referred to as a
degenerate peptide. Note that only some peptide ions are selected for fragmentation;
some might be selected multiple times like peptide p_1 in Fig. 1, while others might
not be selected at all, like peptides p_3 and p_5. Of course, errors can occur during
peptide identification; in Fig. 1, peptide p_4 is misidentified as p_x. This also leads to
the incorrect conclusion that protein P_x is present in the sample, and the incorrect
decision that P_2 is not present because of the misidentification of peptide p_4 and
the fact that p_3 is not sampled. Degeneracy of peptides can also be an issue at the
database search stage as the example illustrates with protein P_y; if an algorithm

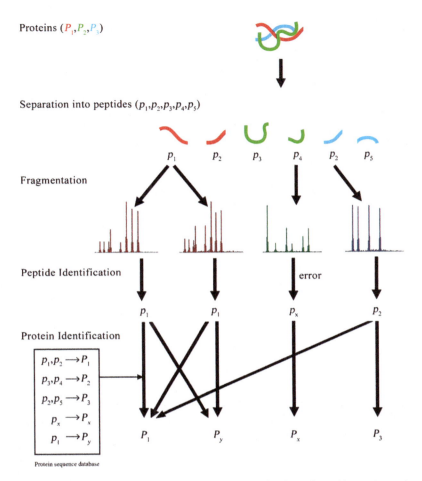

Proteins (P_1, P_2, P_3)

Separation into peptides (p_1, p_2, p_3, p_4, p_5)

Fragmentation

Peptide Identification   error

Protein Identification

$$p_1, p_2 \longrightarrow P_1$$
$$p_3, p_4 \longrightarrow P_2$$
$$p_2, p_5 \longrightarrow P_3$$
$$p_x \longrightarrow P_x$$
$$p_1 \longrightarrow P_y$$

Protein sequence database

Fig. 1 The steps of an MS/MS experiment for the identification of peptides and proteins by database search methods. This figure is adapted from several sources [13–15]

includes proteins for which its peptides are present, then protein P_y is incorrectly identified as present since its peptide p_1 is correctly identified. The proteins P_1 and P_3 are identified, though there may be concerns about these conclusions since other proteins generate each of the peptides used to identify the proteins.

In recent years, several efforts have been made for providing confidences to the peptide and protein identification. This paper first describes PeptideProphet and ProteinProphet, among the first and still among the most popular probabilistic two-step methods for peptide and protein identification. Then we discuss a couple of likelihood-based one-step methods (hierarchical statistical model (HSM) and nested mixture model (NMM)) which attempt to improve on some of the weaknesses of two-step procedures. Finally, we compare the methods reviewed in this paper in the Discussion section.

2   Two-Step Process

In this section, we describe a highly regarded two-step process. In this process, peptides are identified first using database search scores such as those provided by SEQUEST through an algorithm called PeptideProphet. Then the resulting identified peptides along with their confidences can be used to attempt to determine the proteins which are present in the sample using an algorithm called ProteinProphet. A brief description of each of these two algorithms is given below.

2.1   PeptideProphet

Suppose that $x_1, \ldots, x_s$ are database search scores for a peptide. In [16], four SEQUEST search scores are used:

1. $Xcorr' = \begin{cases} \frac{\ln(Xcorr)}{\ln(N_L)} & \text{if } L < L_c \\ \frac{\ln(Xcorr)}{\ln(N_C)} & \text{if } L \geq L_c \end{cases}$, where L is the number of amino acids in the peptide, N_L is the expected number of fragment ions for a peptide with L amino acids, L_c is a threshold beyond which $Xcorr$ does not depend on the length, and N_C is the expected number of fragment ions for a peptide with more than L_C amino acids.
2. ΔC_n, the relative difference between the two highest $Xcorr$ scores.
3. $\ln(SpRank)$, the natural logarithm of the rank of the preliminary score S_p.
4. d_M, the absolute difference of the masses of the precursor ions for the spectrum and the assigned peptide.

Linear discriminant analysis can be used to combine these search scores into a single score

$$F(x_1, \ldots, x_s) = c_0 + \sum_{i=1}^{s} c_i x_i$$

where the weights $c_0, c_1, \ldots, c_s$ are selected to optimize the probability of a correct classification of the peptide assignments as being correct or incorrect based on training data where it is known which peptide assignments are correct. See Chapter 4 of [17] for details on linear discriminant analysis as well as other linear and nonlinear methods for classifying categorical data.

This score is used to help compute the probability that a peptide score is correct. Let $F|+$ and $F|-$ be the distributions of the discriminant scores for correct and incorrect peptide assignments, respectively. The number of tryptic termini (NTT) also provides useful information regarding the probability that the peptide score is correct as discussed in [16], so let $G|+$ and $G|-$ denote the NTT distributions for correct and incorrect peptide assignments, respectively. Then the probability that the peptide assignment is correct given the discriminant scores can be computed using Bayes' formula

$$
\begin{aligned}
p\left(+\big|F,G\right) &= \frac{p\left(F,G\big|+\right)p\left(+\right)}{p\left(F,G\big|+\right)p\left(+\right) + p\left(F,G\big|-\right)p\left(-\right)} \\
&= \frac{p\left(F\big|+\right)p\left(G\big|+\right)p\left(+\right)}{p\left(F\big|+\right)p\left(G\big|+\right)p\left(+\right) + p\left(F\big|-\right)p\left(G\big|-\right)p\left(-\right)}
\end{aligned}
\tag{1}
$$

where $p\left(+\right)$ and $p\left(-\right)$ are prior probabilities of correct and incorrect peptide assignments, respectively. The prior probabilities used in [16] are the observed proportions in the training data, and the conditional distributions of the discriminant scores for the correct and incorrect assignments are modeled by a Gaussian distribution with estimated mean μ and variance σ^2 and by shifted Gamma distribution with estimated shape, scale, and location parameters, respectively. In (1), it is assumed that the discriminant scores and NTT distributions are independent when conditioned on the peptide assignment status; empirical evidence is provided in [16] to support this assumption.

An alternative to directly using the observed proportions from the training data as the prior probabilities in (1) is to use a mixture model which simultaneously estimates the prior and conditional probabilities based on a two-step iterative process using the expectation-maximization (EM) algorithm [18]. Let N be the number of spectra in the data set. Starting with initial estimates of $p\left(+\right)$, $p\left(-\right)$, $p\left(F\big|+\right)$, $p\left(F\big|-\right)$, $p\left(G\big|+\right)$, and $p\left(G\big|-\right)$, the first step of each iteration of the EM algorithm computes estimates of $p\left(+\big|F,\ G\right)$ based on (1). The second step of each iteration updates the estimates of $p\left(+\right)$, $p\left(-\right)$, $p\left(F\big|+\right)$, $p\left(F\big|-\right)$, $p\left(G\big|+\right)$, and $p\left(G\big|-\right)$ under the assumption that the contribution of each of the N spectra to the distribution of correct/incorrect peptide assignments is proportional to the current computed probability that it is correctly/incorrectly assigned. Specifically, the prior probabilities are updated by the formulas

$$
p\left(+\right) = \sum_{i=1}^{N} p\left(+\big|F_i,\ G_i\right)
$$

and $p\left(-\right) = 1 - p\left(+\right)$ where F_i and G_i refer to the respective values for the ith spectrum. The parameters of the conditional distributions are computed using estimates reweighted using weights proportional to the probability of a correct/incorrect peptide assignment conditioned on the spectrum; for the Gaussian distribution which models the distribution of the discriminant scores for positive peptide assignments, the estimated parameters are $\mu = \sum_{i=1}^{N} p\left(+\big|F_i,\ G_i\right) F_i / (Np(+))$ and

$$\sigma^2 = \sum_{i=1}^{N} p\left(+\big|F_i,\ G_i\right)(F_i - \mu)^2 / (Np(+)).$$

2.2 ProteinProphet

Once estimates of the probabilities of the peptide assignments are obtained, then the goal is to estimate the probability that a protein is present in the sample. The probabilities of peptide assignments need not be made using PeptideProphet, or even a database search method, but it is only reasonable to expect that the estimates of the probability of the presence of the protein might be good if the estimates of the probabilities for the peptide assignments are good. Temporarily ignore the possibility of degenerate peptides for which there are multiple corresponding proteins. Let D_i^j includes peptide information for the jth assignment to the ith peptide—such as discriminant scores and number of tryptic termini as used in (1). If the independence of events for peptide assignments is assumed, then the formula

$$\widetilde{P} = 1 - \prod_i \prod_j \left(1 - p\left(+\big|D_i^j\right)\right)$$

gives the probability that at least one peptide assignment corresponding to the protein is correct.

Instead, ProteinProphet [13] makes the conservative estimate that, for each peptide, the probability of all assignments being incorrect is equal to the minimum (one minus the maximum) of the probabilities of incorrect assignments among all assignments. Hence, the inside product in the formula for $\widetilde{P}$ is replaced by $1 - \max_j p\left(+\big|D_i^j\right)$ and, ProteinProphet [13] estimates the probability that a protein is present in the sample using the formula

$$P = 1 - \prod_i \left\{1 - \max_j p\left(+\big|D_i^j\right)\right\}. \tag{2}$$

ProteinProphet's use of only the maximum assignment score for each peptide when estimating the protein probabilities may be overoptimistic since high scores for incorrect peptide identifications may occur by chance particularly when the peptides are assigned more than once.

To adjust for multihit proteins (proteins which correspond to multiple correctly assigned peptides), the estimated probabilities in (1) can be modified by conditioning on another random variable, the estimated number of sibling peptides for each given peptide; it is seen in [13] that correct peptide assignments tend to correspond to multihit proteins, while incorrect peptide assignments are more likely to occur with proteins for which there are no correct peptide assignments. Some further refinements are also proposed as part of the iProphet multi-level models [19] which update the probabilities computed by PeptideProphet conditioning on number of sibling searches, number of replicate searches, number of sibling experiments, number of sibling ions, and number of sibling modifications via Bayes' theorem. Figure 1 in [19] provides a nice figure illustrating the multi-level approach.

The description of ProteinProphet presented in [13] also provides a method for attempting to handle degenerate peptides using the EM algorithm. Let N_s be the number of proteins that the ith peptide is assigned to, and let P_s be the probability that the sth protein is present. Then Eq. (2) is modified so that

$$P_n = 1 - \prod_i \left\{ 1 - w_i^n \max_j p\left(+ \middle| D_i^j \right) \right\}$$

with weights

$$w_i^n = \frac{P_n}{\sum_{s=1}^{N_s} P_s}$$

which give the probability that the ith peptide corresponds to the nth protein. The algorithm begins by using uniform weights and then proceeds by iteratively updating and recomputing the above equations until the values converge.

Since the intensity measurements in the spectra are subject to noise, incorrect peptide identifications will likely lead to incorrect protein identifications. Moreover, using knowledge of probabilities of the presence of proteins in a sample can affect the probabilities that the peptide identification are correct, and it is clear that appropriate inclusion of feedback in modeling peptides and proteins is critical in making good inferences about each. In the following section, two one-step processes are presented which attempt to simultaneously determine the proteins which are present and the peptides which are correctly identified.

3   One-Step Processes

For the two one-step likelihood-based methods (HSM and NMM) reviewed in this section, it is important to note that HSM handles the possibility of degenerate peptides by assuming that a peptide will be in the sample if at least one of the proteins that generate it is present in the sample. On the other hand, NMM does not account for degeneracy, which can cause problems, particularly when estimating the probabilities that proteins are present in complex high-level organisms.

3.1   Hierarchical Statistical Model

The hierarchical statistical model (HSM) proposed in [14] assumes a parametric multilayer joint distribution of five random vectors Y, V, Z, W, and S representing N proteins with at least one peptide hit and M peptides assigned to at least one spectrum.

In this model, $Y = (Y_1, \ldots, Y_N)$ is a vector of indicators for the presence/absence of the proteins in the sample where $Y_i = 1$ indicates that the ith protein is present in the sample. Letting ρ be the probability that a protein is present in the sample, HSM assumes that $Y_1, \ldots, Y_N$ are independent Bernoulli random variables with probability mass function

$$f(y_i) = \rho^{y_i}(1 - \rho)^{1-y_i}$$

for $i = 1, \ldots, N$.

The HSM also considers a vector of independent Bernoulli variables $V = (V_1, \ldots, V_N)$ for each protein indicating whether the number of peptide hits for the protein exceeds a specified threshold h; in particular, $V_i = 1$ indicates that the ith protein has more than h peptide hits. Then the probability mass functions for the Bernoulli random variables can be expressed as

$$f\left(v_i \middle| y_i\right) = \gamma_1^{y_i v_i}(1 - \gamma_1)^{y_i(1-v_i)}\gamma_0^{(1-y_i)v_i}(1 - \gamma_0)^{(1-y_i)(1-v_i)}$$

where γ_1 and γ_0 are parameters for the Bernoulli distributions in the cases where the protein is present or absent, respectively.

Next, $Z = (Z_1, \ldots, Z_M)$ is a vector of indicators for the presence/absence of the peptides in the sample, and each Z_i is modeled conditionally on Y with specific parameters based on the type and number of cleavages. It is assumed that $Z_j \middle| Y$ follows a Bernoulli distribution with parameters based on the type and number of cleavages contained in a five-dimensional vector of probabilities $\alpha = (\alpha_n, \alpha_s, \alpha_{nn}, \alpha_{ns}, \alpha_{ss})$ where an n in the index of a component of α indicates a non-specific cleavage and an s indicates a specific cleavage (so, for example, if a protein with a constituent peptide that is generated with one non-specific and one specific cleavage, then α_{ns} is the probability $P\left(Z_j = 1 \middle| Y_i = 1\right)$ that the peptide will be present in the sample given that the protein is present in the sample). Letting C_j be the set of proteins that might generate peptide j, the conditional probability mass function for the presence of the jth peptide is

$$f\left(z_j \middle| y\right) = \left(\prod_{i \in C_j}\left(1 - P\left(Z_j = 1 \middle| Y_i = 1\right)\right)^{y_i}\right)^{1-z_j} \left(1 - \prod_{i \in C_j}\left(1 - P\left(Z_j = 1 \middle| Y_i = 1\right)\right)^{y_i}\right)^{z_j}.$$

In the next layer of the HSM model, $W = (W_{11}, \ldots, W_{1T_1}, \cdots, W_{M1}, \ldots, W_{MT_M})$ is a double-indexed vector of indicators of correct assignments of present peptides to a spectrum where $W_{jk} = 1$ indicates that the kth assignment of the jth peptide to a spectrum is correct and T_j is the number of assignments of the jth peptide to a spectrum. Then the conditional probabilities that particular assignments are correct given that the jth peptide is present is assumed to be Bernoulli with probability τ so that the conditional probability mass function of W_{jk} given Z_j is

$$f\left(w_{jk}\big|z_j\right) = z_j \tau^{w_{jk}}(1-\tau)^{1-w_{jk}}.$$

Finally, $S = (S_{11}, \ldots, S_{1T_1}, \cdots, S_{M1}, \ldots, S_{MT_M})$ is a double-indexed vector of matching scores for each peptide and potential assignment. The HSM also allows the density to be based on an additional factor Q_{jk} and assumes that there are different density functions depending on whether the assignment of the kth assignment of the jth peptide to a spectrum is correct so that

$$f\left(s_{jk}\big|w_{jk}=w, q_{jk}=q\right) = f_{q,w}\left(s_{jk}; \beta_{qw}\right)$$

for $w = 0, 1$. Combining all of these components of the HSM, the joint density of Y, V, Z, W, and S based on the model is assumed to have the form

$$f(y, z, w, s, v) = \prod_{i=1}^{N} f(y_i) \prod_{i=1}^{N} f\left(v_i\big|y_i\right) \prod_{i=1}^{M} f\left(z_i\big|y\right) \prod_{j=1}^{M}\prod_{k=1}^{T_j} f\left(w_{jk}\big|z_j\right) f\left(s_{jk}\big|w_{jk}\right).$$

The EM algorithm is used to iteratively update the parameters of the marginal and conditional distributions and model the latent variables Y, Z, and W to attempt to maximize the joint distribution. Finally, the joint distribution is used to obtain the desired outputs: the conditional probabilities $P\left(Z_j = 1\big|S, V; \hat{\theta}\right)$ that the jth peptide is present for $j = 1, \ldots, M$, and the conditional probabilities $P\left(Y_i = 1\big|S, V; \hat{\theta}\right)$ that the ith protein is present for $i = 1, \ldots, N$ using the estimated values of the model parameters $\hat{\theta}$.

3.2 Nested Mixture Model

The nested mixture model (NMM) proposed in [21] assumes a mixture model for the joint density of the random variables Y, P, n, and X. Here $Y = (Y_1, \ldots, Y_N)$ is a vector of indicators for the presence/absence of the proteins in the sample where $Y_k = 1$ indicates that the kth protein is present in the sample, $P = (P_{1,1}, \ldots, P_{1,n_1}, \cdots, P_{N,1}, \ldots, P_{N,n_N})$ is a double-indexed vector of indicators of correct assignments of present peptides to a spectrum where $P_{k,i} = 1$ indicates that the ith peptide of the kth protein is correctly identified, n_k is the number of peptide identifications for the kth protein, $n = (n_1, \ldots, n_N)$, and $X = (x_{1,1}, \ldots, x_{1,n_1}, \cdots, x_{N,1}, \ldots, x_{N,n_N})$ is a double-indexed vector of scores for each peptide assignment. Letting π_1^* denote the probability of a protein being present in the model, NMM assumes that $Y_1, \ldots, Y_N$ are independent Bernoulli random variables with probability mass function

$$f(y_k) = \left(\pi_1^*\right)^{y_k} \left(1 - \pi_1^*\right)^{1-y_k}$$

for $k = 1, \ldots, N$. Letting π_1 be the probability that the ith peptide is correctly identified given that the kth protein is present, it is also assumed that the conditional distribution of $P_{k,i}$ given Y_k has probability mass function

$$f\left(p_{k,i}\big|y_k\right) = \{1 - (1 - y_k)p_{k,i}\}\,\pi_1^{(1-p_{k,i})y_k}(1 - \pi_1)^{p_{k,i}y_k}.$$

Then the conditional distribution of the scores for the kth protein given Y_k is modeled by the mixture distribution

$$g_t(x_{k,1}, \ldots, x_{k,n_k}) = \prod_{i=1}^{n_k}\sum_{p=0}^{1} f\left(p\big|y\right) f_p(x_{k,i})$$

where f_0 is the probability density function for a Normal random variable with mean μ and variance σ^2 and f_1 is the probability density function for a shifted gamma random variable with shape parameter α, scale parameter β, and shift parameter γ. Finally, [21] assumes that the conditional distribution of n_k given Y_k follows a truncated Poisson distribution with probability mass function

$$h_y(n_k) = \frac{e^{-c_jl_k}\left(c_jl_k\right)^{n_k}}{n_k!\left(1 - e^{-c_jl_k}\right)}$$

for $n_k = 1, 2, \ldots$, where c_j represents the average number of incorrect/correct peptide identification per unit protein length for $j = 0, 1$, respectively. Then combining these components of the NMM, the joint density of Y, P, n and X is assumed to have the form

$$f(y, z, w, s, v) = \prod_{i=1}^{N}f(y_i)\prod_{i=1}^{N}f\left(v_i\big|y_i\right)\prod_{i=1}^{M}f\left(z_i\big|y\right)\prod_{j=1}^{M}\prod_{k=1}^{T_j}f\left(w_{jk}\big|z_j\right)f\left(s_{jk}\big|w_{jk}\right).$$

Let ψ denote the vector of all model parameters. Then the EM algorithm is used to estimate ψ, and these estimates are used to obtain $P\left(Y_k = 1\big|x_{k,1}, \ldots, x_{k,n_k}, n_k\right)$, the probability that the kth protein is present given the scores and number of peptide hits for that protein, and to obtain

$P\left(P_{k,i} = 1 \middle| x_{k,1}, \ldots, x_{k,n_k}, n_k\right)$, the probability that the ith peptide for the kth protein is present given the scores and number of peptide hits for that protein.

4 Discussion

Proper inference from data produced from tandem mass spectrometry experiments regarding proteins present in tissues and fluids can assist in providing important biological information. Several popular probabilistic and likelihood-based methods for protein identification from MS/MS data have been reviewed: the benchmark two-step process of PeptideProphet followed by ProteinProphet and two likelihood-based one-step processes HSM and NMM. It is important to note that there are many other approaches available in the literature. See [22–24] for review of some other two-step methods for peptide and protein identification. There are also several other one-step protein identification procedures proposed in the literature and a few will be discussed here briefly. ProteinFirst [25] is a two-dimensional target decoy method which simultaneously controls the false discovery rates of proteins and peptide-to-spectrum match levels by modifying PSM scores based on the confidence in the protein identification score. A couple of other methods also consider feedback from proteins when determining the peptides that are present. An iterative procedure to compute peptide and protein probabilities simultaneously is considered by [26] which uses the PeptideProphet results as input for confidence concerning the peptides. Alternately, the method in [27] uses a different mechanism for feedback, starting with peptide identification results from a database search; these results are used to obtain a list of proteins which are further used to obtain a peptide adjacency matrix. Then peptide identification probabilities are estimated based on a logistic regression model and subsequently used to update the protein list and adjacency matrix. Another approach proposed in [28] uses a tripartite graph with three layers corresponding to the spectrum, peptide, and protein levels and uses machine learning techniques in a single optimization procedure for protein identification via a Barista model. The number of true proteins identified by this method exceeds that of ProteinProphet for six different data sets in [28] over a wide range of false discovery rate levels. A promising recent full Bayesian approach (BHM) is proposed in [20] that incorporates the fact that proteins which share the same biological pathway may not be independent. Instead, BHM groups the proteins that are functionally related and uses this fact as prior information for protein identification. Moreover, BHM fully handles the degeneracy issue and considers full posterior inference via a Gibbs sampling scheme. Methods of integrating additional information outside the MS/MS experiment have also been considered and are briefly reviewed in [15].

Various criteria have been used to evaluate the performance of peptide and protein identification procedures, and the performance of the methods has been analyzed and compared using several data sets in the literature. In [16], a training dataset from [11] with ESI-MS/MS spectra generated from a control sample with 18 purified proteins was used, and the results of PeptideProphet based on SEQUEST database search scores are thoroughly analyzed. In this application, peptide assignments with known validity were generated in a training dataset using SEQUEST with a database including the sequences of the 18 control proteins and a *Drosophilia* peptide database. Test data was generated using the control peptides and a human peptide database. It is shown in Figure 3 of [16] that the estimated distribution for the discriminant score is very close to the true distribution for the test data. Also, the accuracy of the probability estimates of the peptide assignments for the test data is illustrated in Figure 4 of [16] by comparing the true probability with the computed probability. Finally, a pair of graphs in Figure 5 of [16] illustrated the tradeoff between the fraction of identified peptide assignments which are actually correct (sensitivity) and the fraction of identified peptide assignments which are actually incorrect for various thresholds used to classify the peptide assignments and the relationship between these fractions and the threshold.

Some similar analyses were also performed in [13] using the data from [11] to evaluate the ability of ProteinProphet to make protein identifications. Figures 5 and 6 of [13] compare the true probability with the computed probability for the presence of proteins and the relationships between the sensitivity, error rates, and threshold for declaring a protein to be identified. One important additional consideration in these plots was the comparison of results with or without using the number of sibling peptides; all figures clearly showed that the results were better when this information was included.

Comparisons have been made between PeptideProphet/ProteinProphet, HSM, and NMM by comparing the empirical FDR (false discovery rate) versus the estimated FDR, the sensitivity versus the specificity, and the number of true positive proteins versus the number of false positive proteins. In [14], MS/MS spectra data generated based on standard protein mixture [29] were studied, and peptide and protein identification was performed using HSM, PeptideProphet, and ProteinProphet. The results were evaluated using decoy data from *Shewanella oneidensis*, and the empirical FDR was compared with the estimated FDR for each of the methods in Figure 4 of [14]. It was found that PeptideProphet significantly underestimates the FDR. HSM and ProteinProphet both were slightly optimistic at low values of the empirical FDR, and conservative at high FDR. Receiver operating characteristic curves were also presented in Figure 5 of [14] for these methods to compare the sensitivity with the specificity (fraction of false peptides not identified), and the sensitivity of HSM was best for sufficiently high levels of specificity shown, followed by ProteinProphet and PeptideProphet. Additionally, HSM and ProteinProphet were also compared in [14] for processed MS/MS spectra data from [14] generated from a yeast (*Saccharomyces cerevisiae*) dataset with peptide fragments obtained from a QSTAR mass spectrometer [30]. Database search scores were again obtained using SEQUEST for the true data and decoy data from *Caenorhabditis elegans*. It was found that ProteinProphet selects more proteins at each threshold, and that ProteinProphet was optimistic in estimating the FDR for low values of the empirical FDR, while HSM was always conservative in estimating the FDR. MS/MS spectra data generated based on standard protein mixture [29] were also analyzed in [21], and peptide identification was performed using PeptideProphet, ProteinProphet, HSM, and NMM with SEQUEST database search scores. It was found that the NMM was much more conservative than the other methods (see Figure 6b of [21]). Furthermore, it was seen that the sensitivity of NMM far exceeded that of PeptideProphet and HSM when the specificity was large (see Figure 6a of [21]).

Additionally in [21], NMM and PeptideProphet/ProteinProphet were compared on another yeast dataset from [31] and scores for the peptide assignments were computed using SEQUEST. The decoy data was created by permuting the target sequences; [21] was unable to run HSM for this dataset because of the large protein group sizes. Probability thresholds were selected based on the number of decoy peptides, and it is seen in Figure 7a of [21] that the number of true peptides identified by NMM clearly exceeded the number identified by PeptideProphet. However, when the threshold was selected to control the number of decoy proteins, ProteinProphet slightly outperformed NMM in terms of the number of true positive proteins for most thresholds.

NMM and HSM were also compared in [21] for the processed MS/MS spectra data from [14] generated from the yeast dataset. Thresholds were selected based on the number of false positive proteins, and it is seen in Figure 8 of [21] that the number of true detected proteins of NMM exceeded that of HSM for each threshold (fixed number of false positives). The same data was also shown in [15], and the number of true positives for BHM exceeded NMM and HSM for most thresholds. The number of true positives for BHM also exceeded that of NMM and ProteinProphet in an example in [20] and [15] on *Haemophilus influenzae* data from [13] (see Figure 6 of [15]).

References

1. Yates, J. R., Ruse, C. I., & Nakorchevsky, M. (2009). Proteomics by mass spectrometry: Approaches, advances, and applications. *Annual Review of Biomedical Engineering, 11*(1), 49–79.
2. Eng, J. K., McCormack, A. L., & Yates, J. R., III. (1994). An approach to correlate tandem mass spectral data of peptides with amino acid sequences in a protein database. *Journal of the American Society for Mass Spectrometry, 5*(11), 976–989.
3. Eng, J. K., Fischer, B., Grossmann, J., & Maccoss, M. J. (2008). A fast SEQUEST cross correlation algorithm. *Journal of Proteome Research, 7*(10), 4598–4602.
4. Diament, B. J., & Noble, W. S. (2011). Faster SEQUEST searching for peptide identification from tandem mass spectra. *Journal of Proteome Research, 10*(9), 3871–3879.
5. Craig, R., & Beavis, R. C. (2004). TANDEM: Matching proteins with tandem mass spectra. *Bioinformatics, 20*(9), 1466–1467.
6. Perkins, D. N., Pappin, D. J., Creasy, D. M., & Cottrell, J. S. (1999). Probability-based protein identification by searching sequence databases using mass spectrometry. *Electrophoresis, 20*(18), 3551–3567.

7. Clauser, K. R., Baker, P., & Burlingame, A. L. (1999). Role of accurate mass measurement (+/− 10 ppm) in protein identification strategies employing MS or MS/MS and database searching. *Analytical Chemistry, 71*(14), 2871–2882.
8. Kim, S., Gupta, N., & Pevzner, P. A. (2008). Spectral probabilities and generating functions of tandem mass spectra: A strike against decoy databases. *Journal of Proteome Research, 7*(8), 3354–3363.
9. Swaney, D. L., Wenger, C. D., & Coon, J. J. (2010). Value of using multiple proteases for large-scale mass spectrometry-based proteomics. *Journal of Proteome Research, 9*(3), 1323–1329.
10. Granholm, V., Kim, S., Navarro, J. C. F., Sjolund, E., Smith, R. D., & Kall, L. (2014). Fast and accurate database searches with MSGF+ Percolator. *Journal of Proteome Research, 13*(2), 890–897.
11. Keller, A., Purvine, S., Nesvizhskii, A. I., Stolyar, S., Goodlett, D. R., & Kolker, E. (2002). Experimental protein mixture for validating tandem mass spectral analysis. *Omics, 6*(2), 207–212.
12. Nesvizhskii, A. I., & Aebersold, R. (2004). Analysis, statistical validation and dissemination of large-scale proteomics data sets generated by tandem MS. *Drug Discovery Today, 9*(4), 173–181.
13. Nesvizhskii, A. I., Keller, A., Kolker, E., & Aebersold, R. (2003). A statistical model for identifying proteins by tandem mass spectrometry. *Analytical Chemistry, 75*(17), 4646–4658.
14. Shen, C., Wang, Z., Shankar, G., Zhang, X., & Li, L. (2008). A hierarchical statistical model to assess the confidence of peptides and proteins inferred from tandem mass spectrometry. *Bioinformatics, 24*(2), 202–208.
15. Sikdar, S., Gill, R., & Datta, S. (2015). Improving protein identification from tandem mass spectrometry data by one-step methods and integrating data from other platforms. *Briefings in Bioinformatics, 17*(2), 262–269.
16. Keller, A., Nesvizhskii, A. I., Kolker, E., & Aebersold, R. (2002). Empirical statistical model to estimate the accuracy of peptide identifications made by MS/MS and database search. *Analytical Chemistry, 74*(20), 5383–5592.
17. Hastie, T., Tibshirani, R., & Friedman, J. (2009). *The elements of statistical learning: Data mining, inference, and prediction.* New York: Springer.
18. Dempster, A. P., Laird, N. M., & Rubin, D. B. (1977). Maximum likelihood from incomplete data via the EM algorithm. *Journal of the Royal Statistical Society, Series B, 39*(1), 1–38.
19. Shteynberg, D., Deutsch, E. W., Lam, H., Eng, J. K., Sun, Z., Tasman, N., et al. (2011). iProphet: Multi-level integrative analysis of shotgun proteomic data improves peptide and protein identification rates and error estimates. *Molecular & Cellular Proteomics, 10*(12), 1–15.
20. Mitra, R., Gill, R., Sikdar, S., & Datta, S. (2015). Bayesian hierarchical model for protein identifications. *Under review.*
21. Li, Q., MacCoss, M., & Stephens, M. (2010). A nested mixture model for protein identification using mass spectrometry. *The Annals of Applied Statistics, 4*(2), 962–987.
22. Huang, T., Wang, J., Yu, W., & He, Z. (2012). Protein inference: A review. *Briefings in Bioinformatics, 13*(5), 586–614.
23. Nesvizhskii, A. I., Vitek, O., & Aebersold, R. (2007). Analysis and validation of proteomic data generated by tandem mass spectrometry. *Nature Methods, 4*(10), 787–797.
24. Serang, O., & Noble, W. (2012). A review of statistical methods for protein identification using tandem mass spectrometry. *Stat Interface, 5*(1), 3–20.
25. Bern, M. W., & Kil, Y. J. (2011). Two-dimensional target decoy strategy for shotgun proteomics. *Journal of Proteome Research, 10*(12), 5296–5301.
26. Shi, J., & Wu, F.-X. (2012). A feedback framework for protein inference with peptides identified from tandem mass spectra. *Proteome Science, 10*, 68.
27. Shi, J., Chen, B., & Wu, F.-X. (2013). Unifying protein inference and peptide identification with feedback to update consistency between peptides. *Proteomics, 13*(2), 239–247.
28. Spivak, M., Weston, J., Tomazela, D., Maccoss, M. J., & Noble, W. S. (2012). Direct maximization of protein identifications from tandem mass spectra. *Molecular & Cellular Proteomics, 11*(2), M111.012161.

29. Purvine, S., Picone, A. F., & Kolker, E. (2004). Standard mixtures for proteome studies. *OMICS, 8*(1), 79–92.
30. Elias, J. E., Haas, W., Faherty, B. K., & Gygi, S. P. (2005). Comparative evaluation of mass spectrometry platforms used in large-scale proteomics investigations. *Nature Methods, 2*(9), 667–675.
31. Kall, L., Canterbury, J., Weston, J., Noble, M. J., & MacCoss, W. S. (2007). A semi-supervised machine learning technique for peptide identification from shotgun proteomics datasets. *Nature Methods, 4*, 923–925.

An MCMC-MRF Algorithm for Incorporating Spatial Information in IMS Proteomic Data Processing

Lu Xiong and Don Hong

1 Introduction to IMS Data Analysis

Imaging mass spectrometry (IMS) is a technique mapping biological molecules like protein, lipids, and metabolites. It can be used to visualize the distribution of molecules in biological tissue [12]. Figure 1 is an illustration of IMS technique and IMS data. To generate IMS data, a tissue slice is divided into multiple pixels depending on the resolution needed. Then each tissue pixel is analyzed by mass spectrometry (MS) technique and MS data is obtained for this tissue pixel. Once we arrange these MS data by their pixel positions, we get a set of IMS data of a hyper-spectral type data cube. The IMS data at a specific mass-over-charge ratio (m/z value) makes up an intensity image which maps the distribution of corresponding molecules. At a specific pixel position, the IMS data gives the mass spectrum of this tissue pixel.

The advantage of IMS compared with MS is that IMS data provide not only MS information but also spatial information that visualizes the spatial distribution of molecules. Currently, the IMS technique is one of the very few techniques that can establish the spatial biochemical composition of the sample in the full molecular range [1]. Unlike MS data processing [4, 5], to fully utilize IMS data, it is

L. Xiong
Computational Science Program, College of Basic and Applied Sciences, Middle Tennessee State University, Murfreesboro, TN 37132, USA
e-mail: lu@sigmaactuary.com

D. Hong (✉)
Computational Science Program, College of Basic and Applied Sciences, Middle Tennessee State University, Murfreesboro, TN 37132, USA

North China University of Technology, Beijing, China
e-mail: Don.Hong@mtsu.edu

© Springer International Publishing Switzerland 2017
S. Datta, B.J.A. Mertens (eds.), *Statistical Analysis of Proteomics, Metabolomics, and Lipidomics Data Using Mass Spectrometry*, Frontiers in Probability and the Statistical Sciences, DOI 10.1007/978-3-319-45809-0_5

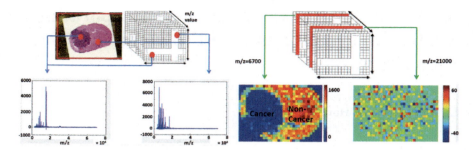

Fig. 1 A snap of IMS data. (*Left*) For a fixed IMS pixel, there is a corresponding mass spectrum (MS). (*Right*) If an m/z value (mass-to-charge ratio) is fixed, the corresponding MS intensities for all pixels that make up an image shows the spatial intensity of that protein

important to not only analyze mass spectra, but also incorporate spatial information in IMS proteomic data processing. Just recently, image fusion techniques were applied to IMS studies. "The IMS-microscopy fusion image is a predictive modality that delivers both the chemical specificity of IMS and the spatial resolution of microscopy in one integrated whole. Each source image measures a different aspect of the content of a tissue sample. The fused image predicts the tissue content as if all aspects were observed concurrently." [14].

2 Motivation of This Study

IMS data analysis methods such as principle component analysis (PCA) [6], support vector machine (SVM) [8], and multi-resolution analysis (MRA) method [16] did not include the spatial interaction relationships of spectra in consideration. Hong and Zhang proposed the elastic net algorithm for IMS data analysis (EN4IMS) [17] and weighted elastic net (WEN) model for IMS data processing [9], and showed classification accuracy improvement. A software package is formulated in [10]. The EN4IMS and WEN models have certain limitations in the robustness since the weights depend on data variance. The growth of a tumor is a continuous process starting from one spot and then spanning to its neighboring area. If a cell is spatially surrounded by cancer cells, then this cell should have a high probability to be cancerous. That is to say, the class of a cell (cancer or non-cancer) is highly determined by the class configuration of its neighboring cells. Such spatial property can be described as locality or Markovianity in the data space. Markov random field (MRF) [13] is a mathematical tool that describes such Markovianity in the space. Therefore, MRF is an ideal tool to incorporate spatial information in IMS data. We would like to use MRF as the prior distribution of pixel classes and use Markov chain Monte Carlo (MCMC) framework to estimate the true class label of each pixel based on the initial classification result from the MRA based methods [16].

The classification accuracy can be expected to be improved in this way since we fully utilize spatial information of IMS data with MRF to describe pixels classes' spatial relations.

The remainder of the paper is organized as follows. In Sect. 3, we give a basic description of the modeling of the study and a brief review on MRA methods for MS proteomic data analysis. Then, a brief introduction to MRF and Ising model, which is a specific type of MRF is given in Sect. 4. In Sects. 5 and 6, we develop the MCMC-MRF computation framework for IMS data classification and the parameter estimation methods. In Sect. 7, we will implement a data experiment using one data set to train parameters and another different data set to test the model, as well as discuss how the way the neighborhood system is defined matters to the classification result. Final conclusion and remarks are given in Sect. 8.

3   Modeling

3.1   Data Description and MCMC-MRF Modeling

This study is based on two IMS data sets extracted from two different mouse brains of the same species and implanted with the same type of tumor. One IMS data set has 24×34 pixels resolution; the other has 44×64 pixels resolution. We use one data set to train model parameters and test the model performance on another data set. To reduce the mistakes made by the boundary pixels, we select only the central part of cancer and non-cancer area as training and test data so that the selected pixel class is easy to see to judge performance of tested model. The red round areas shown in Fig. 2 are selected training and test data.

Each IMS pixel can be considered as a lattice point in MRF. Our goal is to classify each IMS pixel to either cancer or non-cancer. As we mentioned before, since the status (cancer or non-cancer) of a pixel is highly related to the status of its surrounding pixels, such Markovianity can be described well by MRF. We use MRA method [16] to do initial classification. Then each IMS pixel will get a binary status label, -1 for cancer, 1 for non-cancer. This can be considered as a MRF. Then we use MCMC algorithm on this MRF to do the second-round classification to improve classification accuracy of IMS data. Details on algorithm design are discussed in Sects. 5 and 6. MCMC-MRF implementation on IMS data analysis is given in Sect. 7. Figure 3 represents a result of this data process.

3.2   About MRA Method

MRA is a multiscale approximation method using wavelet transforms which represents a signal in a sequence of nested subspaces $\{V_j\}$ and corresponding orthogonal

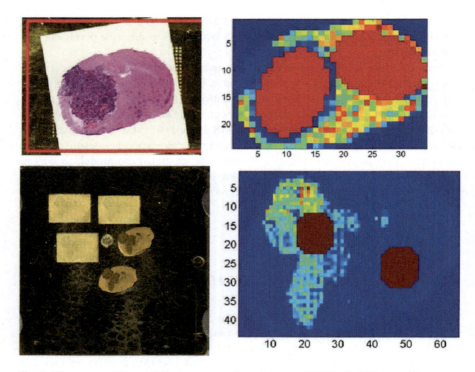

Fig. 2 We use two different IMS data set to train and test model. (*Top left*) Picture of the mouse brain tissue slice where training IMS data was generated from. The *dark area* is cancer area. (*Top right*) A snapshot of training IMS data set. The *round red areas* in *left side* and *right side* are cancer training data and non-cancer training data, respectively. Resolution for this IMS data set is 24 × 34 pixels. (*Down left*) Picture of the mouse brain tissue slice where test IMS data was generated from. The *dark areas* on brain slice are cancer areas. (*Down right*) A snapshot of test IMS data set. The *round red areas* in *left side* and *right side* are cancer test data and non-cancer test data, respectively. Resolution for this data set is 44 × 64 pixels

complement W_j of V_j in V_{j-1}. Space V_0 represents the one that contains the finest resolution data. The projection of the data on $V_j, j = 1, 2, \cdots$ has increasingly coarser resolution. The beautiful computation structure of wavelet coefficients under a selected wavelet basis representation provides efficient and effective algorithms for data denoising and reconstruction, as well as feature extraction. For applications of the MRA methods in MS and IMS data pre-processing, for example, one can check [4, 5, 10] and the references therein. In IMS data processing using MRA techniques, as discussed in [16], it first transforms each IMS pixel's mass spectrum to wavelet space. Then on wavelet space, the algorithm compares the statistical difference between wavelet coefficients from cancer and non-cancer groups to select the wavelet coefficients with large difference between two groups. These wavelet coefficients can be used as "biomarkers" to tell cancer and non-cancer pixels apart. The advantage of this method is that, compared with other IMS analysis methods

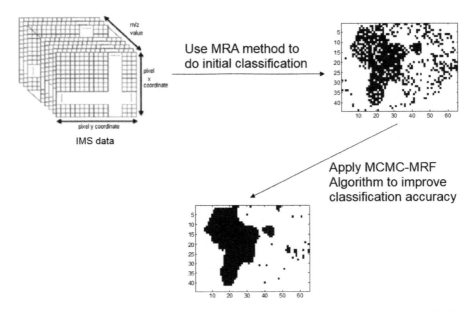

Fig. 3 The input of MCMC-MRF algorithm is the classification result of MRA method. MCMC-MRF algorithm improves the classification accuracy by incorporating spatial information

such as EN4IMS, SAM, EN, and PCA, MRA method contains a shorter list while still achieving high classification accuracy and can capture major biomarkers. The disadvantage of this method is that MRA doesn't use any spatial information of IMS data, since each IMS pixel's mass spectrum is considered independent to each other in the MRA method. However, in reality, IMS pixels are spatially related to each other, and their mass spectrum is related as well. MRF can be used to incorporate such spatial relations to enhance the classification accuracy for algorithms such as MRA, SVM, and PCA, which didn't consider spatial dependence of IMS data. In next three sections, we will first introduce the basic knowledge of MRF briefly, then describe the MCMC computing framework for the IMS data classification and parameters estimation of MRF prior and likelihood.

4   Introduction to Markov Random Field

4.1   Definition of Markov Random Field

The MRF is an n-dimensional random process defined on a discrete lattice. Usually, the lattice is a regular 2-dimensional grid in the plane, finite or infinite. In the $2D$ setting, assume that $S = \{1, 2, \ldots, N\} \times \{1, 2, \ldots, M\}$ is the set of $N \times M$ points

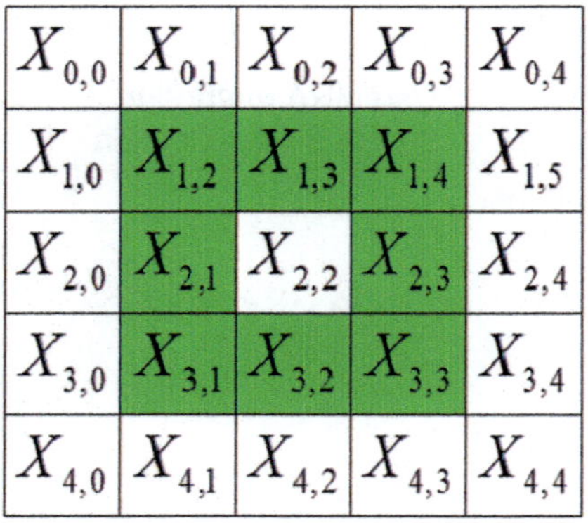

$X_{0,0}$	$X_{0,1}$	$X_{0,2}$	$X_{0,3}$	$X_{0,4}$
$X_{1,0}$	$X_{1,2}$	$X_{1,3}$	$X_{1,4}$	$X_{1,5}$
$X_{2,0}$	$X_{2,1}$	$X_{2,2}$	$X_{2,3}$	$X_{2,4}$
$X_{3,0}$	$X_{3,1}$	$X_{3,2}$	$X_{3,3}$	$X_{3,4}$
$X_{4,0}$	$X_{4,1}$	$X_{4,2}$	$X_{4,3}$	$X_{4,4}$

Fig. 4 In MRF, the value of, for example, $X_{2,2}$ is only determined by the values of its neighborhood (*green pixels*). Such Markov property fits the reality of cancer tissue classification problem. In a tissue with cancer and non-cancer area, if a position is surrounded by cancer areas, then this position has a high probability to be cancer. The same rule applies for non-cancer area. Therefore, MRF can be an ideal tool to deal with cancer tissue classification problem

[3], called sites. For a fixed site s, define a neighborhood ∂s. For example, the neighborhood of the site (i,j) could be $\partial(i,j) = \{(i-1,j),(i+1,j),(i,j-1),(i,j+1)\}$. Markov property of $X(S)$ is defined via local conditions

$$P(X_s|X_r, r \neq s) = P(X_s|X_t, t \in \partial r), \tag{1}$$

where s is a lattice point in S, X_s is the value of X at s, and ∂s consists of the neighboring points of s. The random field X, which has Markov property defined in formula (1), is called MRF. Formula (1) shows that in MRF, the probability of the value at any point is only determined by the values configuration of its neighboring points. For example, in the MRF shown in Fig. 4, the value of $X_{2,2}$ is only determined by values of its neighborhood (green pixels). Incorporation of prior information into the image processing has become a widely accepted model over recent years. There have been many improvements along this direction for image reconstruction including MCMC/EM approach and robust homogeneous prior methods. In general, the procedures require substantial computing effort. A model using inhomogeneous Gaussian random field as a general prior was discussed in [2]. The model allows rapid calculation and flexible determination of spatially varying prior parameters.

4.2 The Simplest MRF: Ising Model

Ising model is the simplest type of MRF where there are only two possible values at any site: $+1$ and -1. It was proposed by a German physicist named Ernst Ising from his magnetic substance research. Originally in magnetic substance research

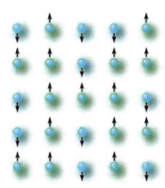

Fig. 5 Ising model is originally proposed from the research of magnetic substance to describe the macroscopic change of particle spin direction configuration with the microcosmic interaction of individual particle existing. $+1$ represents spinning up and -1 represents spinning down. We can use this model to describe the interaction between cancer and non-cancer pixels, with -1 representing cancer and $+1$ representing non-cancer

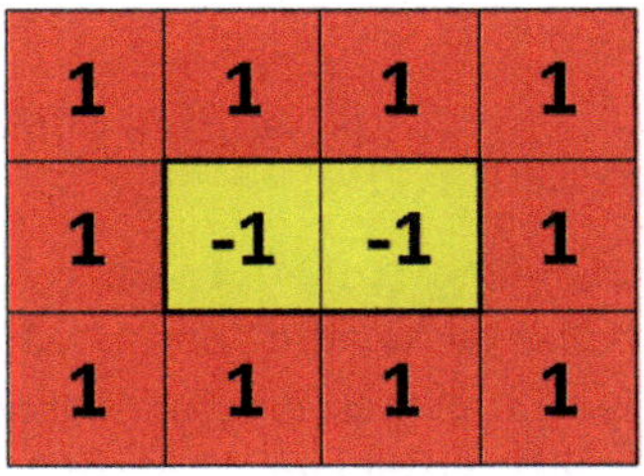

Fig. 6 Illustration of disagree edges number. The edge between two pixels with different values is disagree edge, which is marked as *bold black line* here. In this figure, number of disagree edges is 6

1	1	1	1
1	-1	-1	1
1	1	1	1

context, $+1$ represents the north polarity when a particle is up and -1 represents the north polarity when a particle is down, while the polarity direction of particles are interacting with each other [11]. Such interaction is similar to the interaction of cancer and non-cancer cells depending on the spatial position. Here in IMS data analysis, we can represent a cancer pixel by a -1 site and a non-cancer pixel by a $+1$ site. Figure 5 is an example of an Ising model.

The density function of Ising model is

$$P(X = x) \propto \frac{1}{Z}\exp\left(\beta \sum_{i \sim j} x_i x_j\right) \propto \frac{1}{Z}\exp(-2Jd_x), \qquad (2)$$

where J is a constant parameter that needs to be estimated using training data, d_x is the number of disagree edges (see Fig. 6), $i \sim j$ means pixel X_i and X_j are neighboring to each other, Z is scale constant which will be canceled in later Ising priors ratio.

5 MCMC Computation Framework for IMS Data Classification

We denote θ as true classification, which is our goal to approximate, and y as observed classification, which we already obtained from a previous existing classification algorithm [16] without incorporating spatial information. Our task is to estimate true classification θ while the observed classification y is given. Here we accomplish this task in a probabilistic way. We first obtain the distribution of true classification θ in the condition that the observed classification y is given, and then we estimate θ by its posterior distribution probability in each class. Therefore, we can not only get a classification result, but also a probability value being classified to each class. According to Metropolis–Hasting theorem, which is the key theory of MCMC sampling, if we use Metropolis–Hasting algorithm to simulate $\theta|y$, the simulated data will finally have a distribution converge to the true distribution of $\theta|y$, which is $f(\theta|y)$. In this way, we can estimate the true classification according to its probability distribution when only the observed classification is available. Here is the sampling rule of Metropolis–Hasting algorithm:

$$\alpha[(\theta'|y)|(\theta|y)] = \min\left(1, \frac{f(\theta'|y)}{f(\theta|y)}\right) = \min\left(1, \frac{L(y|\theta')P(\theta')}{L(y|\theta)P(\theta)}\right), \qquad (3)$$

where $\theta|y$ is the original value θ given observed classification y, $\theta'|y$ is the new proposed value θ' during MCMC sampling given observed classification y. The specific expressions for likelihoods' ratio $\frac{L(y|\theta')}{L(y|\theta)}$ and priors' ratio $\frac{P(\theta')}{P(\theta)}$ in this formula will be given later in formulas (4) and (5). Here are the steps of MCMC sampling algorithm:

Algorithm 1 ([15]). Start with the space of all configurations C in which each configuration $\theta\,|y$ is represented as a vector:

$$\theta\,|y = (\theta_1, \theta_2, \ldots, \theta_{n-1}, \theta_n, \theta_{n+1}, \ldots, \theta_{M \times N})$$

with the indexing $(i,j) \mapsto n = (i-1) \times N + j$. The MCMC sampling algorithm would have following steps:

Step 1   Start with $\theta\,|y \in C$. Usually, initially assign $\theta_0 = y$.

Step 2   Randomly select a pixel from $\theta\,|y$, for example, θ_n.

Step 3   Propose new value $\theta'|y$ as $\theta'|y = (\theta_1, \theta_2, \ldots, \theta_{n-1}, -\theta_n, \theta_{n+1}, \ldots, \theta_{M \times N})$ by changing the sign of the selected pixel's value.

Step 4   Generate an uniform random number $u \sim U(0,1)$. If $u < \alpha[(\theta'|y)|(\theta|y)]$, then accept $(\theta'|y)$ as new configuration. Otherwise, keep $(\theta|y)$ as current configuration.

Iterate above process until convergence.

Next, we will discuss the specific expressions for likelihoods' ratio and priors' ratio in formula (3). Since the conditional variables are independent to each other (this is because at each site, the observed class is completely determined by its true class and observation noise), then likelihoods' ratio can be written as:

$$\frac{L(y|\theta')}{L(y|\theta)} = \frac{\prod\limits_{m,n} L(y|\theta')}{\prod\limits_{m,n} L(y|\theta)}. \tag{4}$$

The prior is an Ising MRF. Then the priors' ratio can be obtained using the distribution formula of Ising model in formula (2):

$$\frac{P(\theta')}{P(\theta)} = \frac{\frac{1}{Z}\exp(-2Jd_{\theta'})}{\frac{1}{Z}\exp(-2Jd_{\theta})} = \exp[-2J(d_{\theta'} - d_{\theta})]. \tag{5}$$

The parameter J in above formula can be estimated by training data. We will discuss the details about how to estimate parameter J in next section. $d_{\theta'}$ and d_{θ} are the number of disagree edges in estimation of θ' and θ, respectively. The definition of d_{θ} was introduced in Sect. 4.2.

6   Parameters Estimation of MRF Prior and Likelihood

6.1   Ising MRF Prior Parameter Estimation Using MPL

In 1986, Geman and Graffigne [7] proved that the following pseudo likelihood method can be used to approximate the likelihood of Gibbs distribution, which is the general format of the distribution of MRF. This method converges to Gibbs distribution with probability 1. Here is how the pseudo likelihood is defined:

$$\text{PL}(X) \triangleq \prod_{i \in S} P(x_i|x_{N_i}) = \prod_{i \in S} \frac{P(x_i, x_{N_i})}{P(x_{N_i})} = \prod_{i \in S} \frac{P(x_i, x_{N_i})}{\sum\limits_{x_i \in L} P(x_i, x_{N_i})}, \tag{6}$$

where L is the labeling space (or class space). For instance, the label space for Ising model, a binary system, is $L = \{-1, 1\}$. i is the pixel index, S is the set of all pixels.

For the Ising model, plugging its distribution from formulas (2) into formula (6), we can obtain its pseudo likelihood:

$$\text{PL}(X) = \prod_{i \in S} \frac{\frac{1}{Z}\exp(-2Jd_{x_i})}{\sum\limits_{x_i \in L} \frac{1}{Z}\exp(-2Jd_{x_i})}. \tag{7}$$

To find its maximum, we take its natural log. Then,

$$\ln[\text{PL}(x)] = \sum_{i \in S} \left\{ -2Jd_{x_i} - \ln\left[\sum_{x_i \in L} \exp(-2Jd_{x_i})\right] \right\}. \tag{8}$$

Therefore, the Maximum Pseudo Likelihood (MPL) estimation of J is

$$\widehat{J} = \arg\max_{J}\{\ln[\text{PL}(X)]\}. \tag{9}$$

Using the one-dimensional optimization method, we can find the specific value of J that maximizes the pseudo likelihood in formula (8). This value is the estimation value for parameter J.

6.2 Likelihood Estimation

Using the training data whose observed classification and true classification are both known, we can estimate the likelihood. Here, we are doing binary classification. Therefore, there are only four types of likelihoods. They are

$$L(y = -1|\theta = -1), L(y = 1|\theta = -1), L(y = -1|\theta = 1), \text{ and } L(y = 1|\theta = 1).$$

For example, $L(y = 1|\theta = -1)$ denotes the probability that a cancer pixel is observed as a non-cancer pixel. We count the frequency of each instance in training data. Then we divide these frequencies with the total sample size to obtain the estimation of the above four likelihoods. For instance, if the case $y = 1|\theta = -1$ (which means the case that a cancer pixel is observed as a non-cancer pixel by initial algorithm without incorporating spatial information) happens 30 times, and the total number of computed pixel is 200, then the likelihood probability $L(y = 1|\theta = -1) = \frac{30}{200} = 0.15$.

7 Data Experiment

7.1 Model Parameter Estimations Using Training Data

First, we use the MPL method discussed in Sect. 6.1 to estimate the parameter J in the Ising model using formula (9). To compute this, we consider the class label configuration for all pixels and their neighboring relations. Figure 6 shows the true configuration of training data class labels. The blue pixels are cancer pixels ($X_i = -1$). Red pixels are non-cancer pixels ($X_i = +1$). Green pixels are margin blank space. There are 634 valid pixels for training data shown in Fig. 7. Plugging in values for training data to formula (8), (9) and using the one dimensional optimization method, we obtained the result that the PL(X) is maximized when $J = 1.0266$ (Fig. 8). Therefore, the MPL estimation value for J is 1.0266. Hence, for the experiment IMS data we use here, the priors' ratio in acceptance probability for MCMC sampling in formula (3) can be written as:

$$\frac{P(\theta')}{P(\theta)} = \exp[-2J(d_{\theta'} - d_{\theta})] = \exp[-2 \times 1.0266(d_{\theta'} - d_{\theta})]. \tag{10}$$

Second, we use the idea discussed in Sect. 6.2 to estimate the likelihood. We count the frequencies of four cases happening in selected training data computed by the initial algorithm: the case that a cancer pixel is classified as a cancer pixel [corresponding to $L(y = -1|\theta = -1)$]; the case that a cancer pixel is classified as a non-cancer pixel [corresponding to $L(y = 1|\theta = -1)$]; the case that a non-cancer pixel is classified as a cancer pixel [corresponding to $L(y = -1|\theta = 1)$]; and the case that a non-cancer pixel classified as a non-cancer pixel [corresponding to $L(y = 1|\theta = 1)$]. For example, we selected two round areas in training data, 323 IMS data pixels in total. According to the classification result by the initial algorithm [16] which did not incorporate spatial information, 122 cancer pixels are classified as non-cancer pixels, and the likelihood for this case is $L(y = 1|\theta = -1) = \frac{37}{323} =$

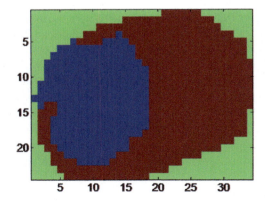

Fig. 7 True configuration of training data class labels: the *blue pixels* are cancer pixels. *Red pixels* are non-cancer pixels. *Green pixels* are margin blank space. This configuration can be used to estimate parameter J in the Ising MRF model using MPL method discussed in Sect. 6.1

Fig. 8 When $J = 1.0266$, formula (8) for training data takes maximum. Therefore, $J = 1.0266$ is the estimated value for corresponding parameter of Ising MRF priors' ratio in formula (5)

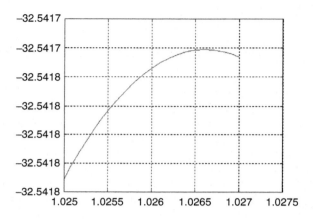

0.1146. Here are the likelihoods we computed from training data using the initial algorithm before optimization:

$$\begin{cases} L(y = -1|\theta = -1) = \frac{122}{323} = 0.3777, \\ L(y = 1|\theta = -1) = \frac{37}{323} = 0.1146, \\ L(y = -1|\theta = 1) = \frac{130}{323} = 0.4025, \\ L(y = 1|\theta = 1) = \frac{34}{323} = 0.1053. \end{cases} \tag{11}$$

7.2 Computation and Results on Test Data

We already estimated prior and likelihood for the acceptance probability in formula (3). Now we can start the MCMC simulation discussed in Sect. 5 for test data to estimate its true classification θ with its observed classification y, which is first computed by the initial algorithm that didn't consider spatial information for IMS data. The initial algorithm we use here is modified from the MRA method for IMS proposed by Xiong and Hong in 2015 [16], which didn't incorporate IMS data spatial information. Usually, in order for the MRA method to obtain high classification accuracy, there should be ten feature variables selected. Here, to leave some potential for optimization, we only select three feature variables so that the classification accuracy turns out to be 86%. We take this classification result as the initial classification rate. Together with the improved results, they are shown in Fig. 9. The acceptance probability in MCMC simulation for test data is

$$\alpha[(\theta'|y)|(\theta|y)] = \min\left(1, \frac{f(\theta'|y)}{f(\theta|y)}\right) = \min\left(1, \frac{L(y|\theta')P(\theta')}{L(y|\theta)P(\theta)}\right)$$

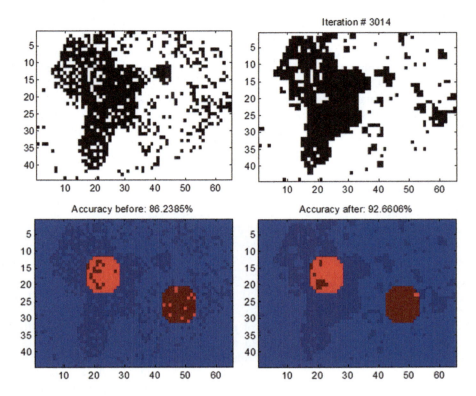

Fig. 9 (*Top left*) Classification result using an algorithm without incorporating spatial information. The *black pixels* (−1 in Ising MRF) are classified as cancer pixels and white pixels (+1 in Ising MRF) are classified as non-cancer pixels. (*Top right*) Classification result of optimized algorithm using the MCMC-MRF to incorporate spatial information. Lots of misclassifications have disappeared. (*Down left*) The classification accuracy for selected test area using an initial algorithm without incorporating spatial information is only 86.2385 %. (*Down right*) The optimized classification accuracy for the selected test area using the MCMC-MRF to incorporate spatial information is improved to 92.6606 %

$$= \min \left\{ 1, \frac{\prod\limits_{m,n} L(y|\theta')}{\prod\limits_{m,n} L(y|\theta)} \exp[-2 \times 1.0266(d_{\theta'} - d_{\theta})] \right\}. \quad (12)$$

The value of likelihood $L(y_{m,n}|\theta'_{m,n})$ is estimated in formula (11). Then we follow the MCMC simulation steps described in Algorithm 1. We start with initial estimation $\theta_0 = y$. Then propose a change of one pixel's class estimation; generate a uniform random number, and compare it with acceptance probability in formula (12) to determine whether accept this proposal or not. We iterate this process for a certain number of times until it converges. Afterwards, we gather statistics of the simulated data to compute the posterior probability $f(\theta_{m,n} = -1|y_{m,n})$ for each pixel. Then set

0.5 as the probability threshold. If $f(\theta_{m,n} = -1|y_{m,n}) > 0.5$, this means pixel $X_{m,n}$ has a higher chance to be cancer and we classify it as a cancer pixel. Otherwise, we classify $X_{m,n}$ as a non-cancer pixel. Figure 9 shows the result before and after applying Algorithm 1. We can see the classification accuracy is improved from 86.2385 % to 92.6606 %.

In the computation process above, the neighborhood system in the Ising model prior is defined as a 4-points neighborhood system. That is to say, for site (i, j) the neighborhood is

$$\partial(i,j) = \{(i-1,j), (i+1,j), (i,j-1), (i,j+1)\}. \tag{13}$$

Only the top, bottom, left, and right adjacent pixels are considered to be neighboring pixels in 4-point neighborhood system. However, it is more reasonable to also consider diagonally adjacent pixels as neighboring pixels because they also have impacts to the pixel at site (i, j). Therefore, we can define the neighborhood for site (i, j) in an 8-point neighborhood system (Fig. 10) as

$$\partial(i,j) = \{(i-1,j), (i+1,j), (i,j-1), (i,j+1), (i-1,j-1), (i-1,j+1),$$
$$(i+1,j-1), (i+1,j+1)\}. \tag{14}$$

After we modified the definition of neighborhood system in the Ising MRF prior from a 4-point neighborhood to an 8-point neighborhood, we apply the MCMC simulation described in Algorithm 1 to test the performance again. The result shows the performance under the 8-points neighborhood assumption is better than the one under the 4-point neighborhood. The classification accuracy is improved to 94.9541 %. This shows that a more realistic definition of neighborhood system, which describes the spatial impacts between cancel cells, leads to a better classification result (Fig. 11).

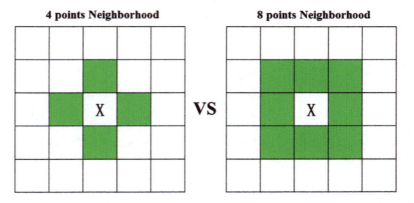

Fig. 10 4-points neighborhood system and 8-points neighborhood system in Ising MRF. As *left figure* shows, in 4-points neighborhood system, only up, down, left, right adjacent *green pixels* are considered to be neighboring to pixel X. But in 8-points neighborhood system as shown in the *right figure*, the diagonal adjacent pixels are also considered as neighboring to X

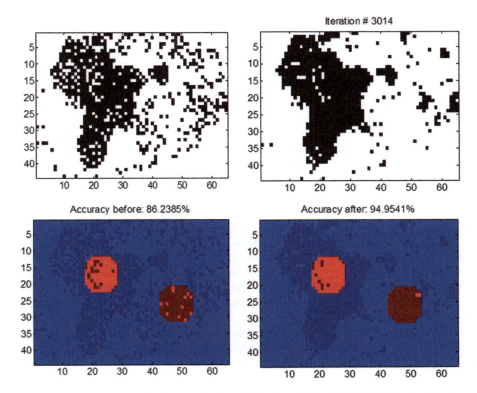

Fig. 11 Classification result using the 8-point neighborhood. (*Top left*) The initial classification result before optimization using an algorithm without incorporating spatial information. The *black pixels* (−1 in Ising MRF) are classified as cancer pixels and *white pixels* (+1 in Ising MRF) are classified as non-cancer pixels. (*Top right*) Classification result of optimized algorithm using the MRF-MCMC to incorporate spatial information with the 8-point neighborhood system. (*Down left*) The classification accuracy for selected test area before optimization is only 86.2385 %. (*Down right*) The optimized classification accuracy for selected test area using the 8-points neighborhood MRF is improved to 94.9541 %, better than the accuracy under the 4-point neighborhood

To take it one step further, we can define an even higher order neighborhood system for the Ising MRF prior. Because in reality, not only do the directly adjacent cells have impact on the surrounding central cell, but the cells close by (while are not directly adjacent) also impact on the central cell, even though the impacts could be weaker. Therefore, it is more reasonable to consider neighborhoods like Fig. 12 shows: the closest points are 1st order neighboring points; the second closest points are 2nd order neighboring points, etc. Then the Ising MRF prior probability for nth order neighborhood system is:

$$P(x) \propto \frac{1}{Z}\exp\left(\beta\sum_{i\sim j}x_ix_j\right) \propto \frac{1}{Z}\exp\left(c_1\sum_{i_1\sim j_1}x_{i_1}x_{j_1} + \ldots + c_n\sum_{i_n\sim j_n}x_{i_n}x_{j_n}\right)$$

$$\propto \frac{1}{Z}\exp[-2J(c_1d_{x,l=1} + \ldots + c_nd_{x,l=n})], \qquad (15)$$

where $d_{x,l=i}$ is the number of disagree edges between the ith order neighboring pixels and the central pixel, c_n is the impact coefficient from nth neighboring pixels. It is reasonable to assume that further pixels have less impact. Therefore, C_n is defined as inversely proportional to the Euclidean distance:

$$c_n \propto \frac{1}{D(x_{i_n}, x_{j_n})}. \qquad (16)$$

Here we define the 5th-order neighborhood system, as shown in Fig. 12. Then, we plug formula (15), formula (16) to the acceptance probability in formula 3 to update the corresponding computation in Algorithm 1 and retest the model on test data. It turns out that the result is even better than the 8-point neighborhood assumption. As shown in Fig. 13, the classification accuracy is improved to 95.4128 %, better than the ones using the 1st order 4-point and 8-point neighborhood systems. This once again shows that the more realistic the neighborhood system definition, which describes the spatial interactions between different areas in cancer tissue more precisely, the better classification result we can realize.

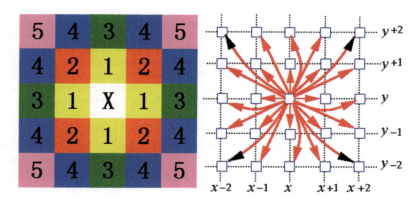

Fig. 12 (*Left*) The 5th-order neighborhood system. The closest pixels to X in the figure are the 1st order pixels. The furthest pixels of X in this figure are the 5th order pixels. (*Right*) The impact between neighboring pixels is defined in formula (15), inversely proportional to their Euclidean distance. In other words, the closer the pixel, the stronger the impact, and the further pixels have weaker impact on the central pixel

8 Conclusion and Remarks

The main idea of this work is to use MRF as prior knowledge in sampling to incorporate the spatial relationships between different parts of cancer tissue, and using the MCMC technique to estimate the true classification based on the observed classification (initial classification before optimization) by approximating the probability distribution of true classification. The MCMC-MRF model parameters are estimated using training data and the model is then tested using another data set. The data experiment shows that this method can improve the classification accuracy by more than 6 % compared with traditional IMS data algorithm like PCA, SVM, and MRA method that don't incorporate spatial information. Also, the test

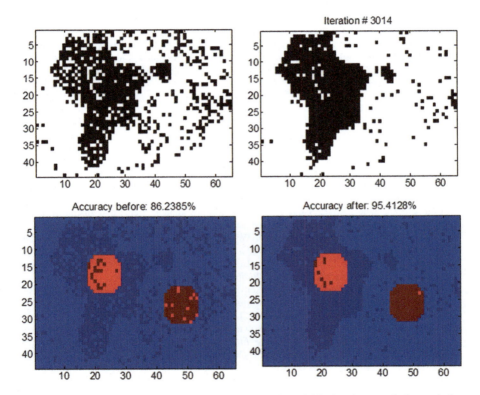

Fig. 13 Comparison of classification result of the 5th-order neighborhood system before and after the MRF-MCMC optimization. (*Top left*) The initial classification result before optimization. (*Top right*) Classification result after the MRF-MCMC optimization to incorporate spatial information with the 5th-order neighborhood system. (*Down left*) The classification accuracy for the selected test area before optimization is only 86.2385 %. (*Down right*) The optimized classification accuracy for the selected test area using the 5th-order neighborhood MRF has been improved to 95.4128 %, better than the accuracy rate under the 1st order neighborhood system of 4-point or 8-point neighborhood

result shows that the more realistic we define the neighborhood which precisely describes the interactions mechanism between different parts in cancer tissue, the better classification result we can obtain. Future work can be considered in three aspects. First, we need to consider faster computing methods, either coding-wise or mathematical algorithm-wise, since we experienced the time consuming of this MCMC-MRF simulation during the test, especially when the neighborhood system is defined as high order. Second, we can apply some statistical analysis to the simulation result to obtain variables such as confidence interval, standard deviation, so that we will have a better evaluation of the simulation. Instead of getting just one class label for each pixel, we can obtain more information using statistical analysis to the simulated data. Finally, we can consider defining a more complicated and more precise neighborhood system with impact coefficients estimated using more training data so that the model can describe the reality even better (Fig. 13).

Acknowledgements The authors would like to thank the anonymous referee for valuable suggestions on the paper. D. Hong was partially supported by the Beijing High-Caliber Overseas Talents Program and North China University of Technology, Beijing, China. We are also grateful to Vanderbilt Mass Spectrometry Research Center for providing us IMS data in the study.

References

1. Alexandrov, T., & Kobarg, J. H. (2011). Efficient spatial segmentation of large imaging mass spectrometry datasets with spatially aware clustering. *Bioinformatics, 27*(13), i230–i238
2. Aykroyd, R. G., & Zimeras, S. (1999). Inhomogeneous prior models for image reconstruction. *Journal of American Statistical Association (JASA), 94*(447), 934–946.
3. Bouman, C., Sauer, K., & Saquib, S. (1995). Markov random fields and stochastic image models. In *IEEE International Conference on Image Processing*.
4. Chen, S., Hong, D., & Shyr, Y. (2007). Wavelet-based procedures for proteomic MS data processing. *Computational Statistics and Data Analysis, 52*, 211–220.
5. Chen, S., Li, M., Hong, D., Billheimer, D., Li, H., Xu, B., et al. (2009). A novel comprehensive wave-form MS data processing method. *Bioinformatics, 25*(6), 808–814.
6. de Plas, R. V., De Moor, B., & Waelkens, E. (2007). *Imaging mass spectrometry based exploration of biochemical tissue composition using peak intensity weighted PCA*. Life Science Systems and Applications Workshop, 2007. LISA 2007. IEEE/NIH (pp. 209–212).
7. Geman, S., & Graffigne, C. (2011). Markov random field image models and their applications to computer vision. *Proceedings of the International Congress of Mathematicians, 4*(5), 1496–1517.
8. Gerhard, M., Deininger, S., & Schleif, F. (2007). Statistical classification and visualization of MALDI imaging data. *Proceedings of the 20th IEEE International Symposium on Computer-Based Medical Systems (CBMS 2007)*, pp. 403–405.
9. Hong, D., & Zhang, F. (2010) Weighted elastic net model for mass spectrometry imaging processing. *Mathematical Modelling of Natural Phenomena, 5(3)*, 115–133.
10. Liang, J., Hong, D., Zhang, F., & Zou, J. (2015). IMSmining: A tool for imaging mass spectrometry data biomarker selection and classification. In R. N. Mohapatra, D. R. Chowdhury, & D. Giri (Eds.), *Springer Proceedings in Mathematics & Statistics* (Vol. 139, pp.155–162). New York: Springer.
11. Lieb, E., Schultz, T., & Mattis, D. (1964). Two-dimensional Ising model as a soluble problem of many fermions. *Reviews of Modern Physics, 36*, 856–871.

12. Rohner, T., Staab, D., & Stoeckli, M. (2005). MALDI mass spectrometric imaging of biological tissue sections. *Mechanisms of Ageing and Development, 126*(1), 177–185.
13. Rozanov, Y. (1982). *Markov random fields*. New York: Springer.
14. Van de Plas, R., Yang, J., Spraggins, J., & Caprioli, R. M. (2015). Image fusion of mass spectrometry and microscopy: a multimodality paradigm for molecular tissue mapping. *Nature Methods, 12*, 366–372.
15. Wang, L., Liu, J., & Li, S. (2000). MRF parameter estimation by MCMC method. *Pattern Recognition, 33*(11), 1919–1925.
16. Xiong, L., & Hong, D. (2015). Multi-resolution analysis method for IMS data biomarker selection and classification. *British Journal of Mathematics and Computer Science, 5*(1), 64–80.
17. Zhang, F., & Hong, D. (2011). Elastic net-based framework for imaging mass spectrometry data biomarker selection and classification. *Statistics in Medicine, 30*, 753–768.

Mass Spectrometry Analysis Using MALDIquant

Sebastian Gibb and Korbinian Strimmer

1 Introduction

Mass Spectrometry (MS), a high-throughput technology commonly used in proteomics, enables the measurement of the abundance of proteins, metabolites, peptides, and amino acids in biological samples. The study of changes in protein expression across subgroups of samples and through time provides valuable insights into cellular mechanisms and offers a means to identify relevant biomarkers, e.g., to distinguish among tissue types, or for predicting health status. In practice, however, there still remain many analytic and computational challenges to be addressed, especially in clinical diagnostics [32]. Among these challenges the availability of open and easy-to-extend processing and analysis software is highly important [1, 34].

Here, we present MALDIquant [21], a complete open-source analysis pipeline for the R platform [48]. In the first half of this chapter we describe the methodology implemented and available in MALDIquant. In the second half we illustrate the versatility of MALDIquant by application to an experimental data set, showing, how raw intensity measurements are preprocessed and how peaks relevant for a specific outcome can be identified.

S. Gibb
Anesthesiology and Intensive Care Medicine, University Hospital Greifswald,
Ferdinand-Sauerbruch-Straße, 17475 Greifswald, Germany
e-mail: mail@sebastiangibb.de

K. Strimmer (✉)
Epidemiology and Biostatistics, School of Public Health, Imperial College London, Norfolk
Place, London W2 1PG, UK
e-mail: k.strimmer@imperial.ac.uk

© Springer International Publishing Switzerland 2017
S. Datta, B.J.A. Mertens (eds.), *Statistical Analysis of Proteomics, Metabolomics,
and Lipidomics Data Using Mass Spectrometry*, Frontiers in Probability and the
Statistical Sciences, DOI 10.1007/978-3-319-45809-0_6

Current documentation of specific version of MALDIquant can be found on its homepage at http://strimmerlab.org/software/maldiquant/ where we also provide instructions for installing the software. In addition, we provide a number of example R scripts on the MALDIquant homepage. Direct download of the MALDIquant software is also possible from the CRAN server at http://cran.r-project.org/package=MALDIquant.

2   Methodology Available in MALDIquant

2.1   General Workflow

The purpose of MALDIquant is to provide a complete workflow to facilitate the complex preprocessing tasks needed to convert raw two-dimensional MS data, as generated, for example, by MALDI or SELDI instruments, into a matrix of feature intensities required for high-level analysis. A typical workflow is depicted in Fig. 1.

Each analysis with MALDIquant consists of all or some of the following steps (see also [43, 45] for related analysis pipelines). First, the raw data is imported into the R environment. Subsequently, the data are smoothed to remove noise and also transformed for variance stabilization. Next, to remove chemical background noise a baseline correction is applied. This is followed by a calibration step to allow comparison of intensity values across different baseline-corrected spectra. As a next step a peak detection algorithm is employed to identify potential features and also to reduce the dimensionality of the data. After peaks have been identified a peak

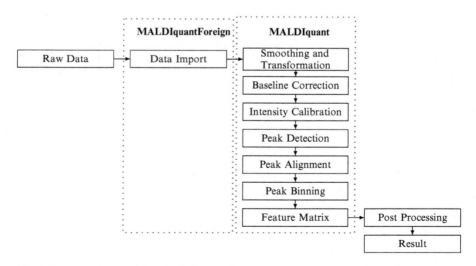

Fig. 1 Preprocessing workflow for MS data using MALDIquantForeign and MALDIquant

alignment procedure is applied as the *mass-to-charge ratios (m/z)* typically differ across different measurements and need to be adjusted accordingly. Finally, after feature binning an intensity matrix is produced that can be used as starting point for further statistical analysis, for example, for variable selection or classification.

In the following subsections we discuss each of these steps in more detail.

2.2   Import of Raw Data

A prerequisite of any analysis is to import the raw data into the *R* environment. Unfortunately, nearly every vendor of mass spectrometry machinery has its own native and often proprietary data format. This complicates the exchange of experimental data between laboratories, the use of analysis software, and the comparison of results. Fortunately, there is now much effort to create generic and open formats, such as *mzXML* [46] and its successor *mzML* [39] or *imzML* [54] for Mass Spectrometry Imaging (MSI) data. Nevertheless, the support of these formats is still limited and often conversion is needed to get the data into a suitable format for subsequent analysis [11].

Importing of raw data in `MALDIquant` is performed by its sister R package `MALDIquantForeign`. It offers import routines for numerous native and public data formats. In addition to the open XML formats (*mzXML, mzML*) it supports *Ciphergen XML, ASCII, CSV, NetCDF*, and *Bruker Daltonics *flex Series* files. It can also read MSI formats like *imzML* and *ANALYZE 7.5* [49].

A very useful feature of `MALDIquantForeign` is that it reads and traverses whole directory trees containing supported file formats so that simultaneous import of many spectra is straightforward. Furthermore, `MALDIquantForeign` allows to import data from remote resources so the spectral data can be read over an Internet connection from a website or database.

After importing the raw spectra an important step is quality control. This includes checking the mass range, the length of each spectra, and also visual exploration of spectra to find and remove potentially defective measurements. `MALDIquant` provides functions to facilitate this often neglected task.

2.3   Intensity Transformation and Smoothing

The raw data obtained from mass spectrometry experiments are counts of ionized molecules, with intensity values approximately following a *Poisson* distribution [17, 56]. Consequently, the variance depends on the mean, as mean and variance are identical for a *Poisson* distribution. However, by applying a square root transformation ($f(x) = \sqrt{x}$) we can convert the Poisson distributed data to approximately normal data, with constant variance independent of mean, which is an important requirement for many statistical tests [47]. In the preprocessing noise models other than the Poisson may be also assumed, which lead to different variance-stabilizing functions such as the logarithmic transformation [13, 61]. These can be easily applied in MALDIquant as well.

Subsequently, the transformed spectral data is smoothed to reduce small and high-frequent variations and noise. For this purpose MALDIquant offers the moving average smoother and the *Savitzky–Golay*-filter [53]. The latter is based on polynomial regressions in a moving window. In contrast to the moving average, the Savitzky–Golay filter preserves the shape of the local maxima.

Note that both algorithms require the specification of window size, which according to Bromba and Ziegler [9] should be chosen to be smaller than twice the Full Width at Half Maximum (FWHM) of the peaks.

2.4 Baseline Correction

The elevation of the intensity values in a typical Matrix-Assisted Laser Desorption/Ionization—Time-of-Flight Mass Spectrometer (MALDI-TOF) spectrum is called *baseline* and is caused by chemical noise such as matrix-effects and pollution. It is recommended to remove these background effects to reduce their influence in quantification of the peak intensities.

In the last few years many algorithms to adjust for the baseline have been developed, ranging from simple methods like the subtraction of the absolute minimum [20] or the moving minimum or median [38] to more elaborate methods such as fitting a LOWESS curve, a spline or an exponential function against the moving minima, respectively, median values [25, 26, 33, 37, 61, 69]. Other authors prefer morphological filters such as *TopHat* [52], iterative methods as the Statistics-sensitive Non-linear Iterative Peak-clipping algorithm (SNIP) [50], or the convex hull approach [36].

Unfortunately, there is no automatic way to select among the available procedures to find the baseline correction method that is most suitable for a given spectrum at hand. Instead, it is recommended to investigate multiple baseline estimations by visual inspection [69]. As shown in Fig. 2 the algorithms can indeed differ substantially.

MALDIquant provides three complex baseline correction algorithms that have been selected for inclusion in MALDIquant because of their favorable properties, such as respecting peak form and non-negativity of intensity values:

1. The *convex hull* algorithm [3] doesn't need a tuning parameter and is often very effective to find the baseline. Unfortunately, for concave matrix-effects as common in MALDI-TOF spectra this algorithm cannot be applied—see Fig. 2b ($m/z \approx 1500\,\mathrm{Da}$).
2. *TopHat* [23, 65] is a morphological filter combining a moving minimum (erosion filter) followed by a moving maximum (dilation filter). In contrast to the convex hull approach it has an additional tuning parameter, the window size of the

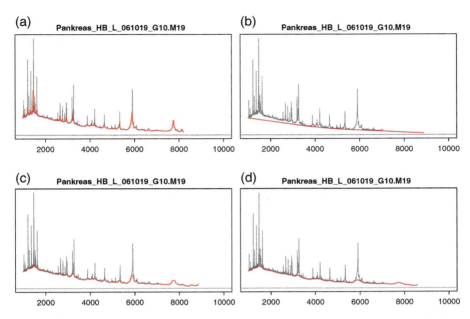

Fig. 2 Estimated baselines for a raw MALDI-TOF spectrum from Fiedler et al. [18]. The following algorithms were applied: (**a**) moving median, (**b**) convex hull, (**c**) *TopHat*, (**d**) SNIP

moving window, that controls the smoothness of the estimated baseline. The narrower the window the more of the baseline is removed but also of the peak heights. A wider window will preserve the peak intensities and produce a smoother baseline but will also cause some local background variation to remain (Fig. 2c).

3. The default baseline correction algorithm in MALDIquant is SNIP [50]. Essentially, this is a local window-based algorithm in which a baseline is reconstructed by replacing the intensities in a window by the mean of the surrounding points, if the mean is smaller than the local intensity, with window size decreasing iteratively starting from a specified upper limit [42].

In addition to the above MALDIquant also supports the moving-median algorithm (Fig. 2a) which is commonly used in the literature but may lead to negative intensity values after baseline subtraction.

2.5   Intensity Calibration

The intensity values in mass spectrometry data represent the relative amount of analytes, such as peptides. The measured intensity strongly depends on preanalytical and environmental factors like sample collection, sample storing, room temperature, air humidity, crystallization, etc. [4, 31, 32].

Further confounders are introduced by the so-called batch effects. These are systematical differences that hide the true biological effect and that are caused by different experimental conditions, for instance, a different preanalytical processing, measurements on different days by different operators in different laboratories on different devices [24, 27, 31].

The systematic errors can be stronger than the real biological effect, and are best minimized already at the stage of data acquisition by strictly adhering to a standardized preanalytical and experimental protocol [4].

Note that unlike in other omics data, such as gene expression data, batch effects and other systematic errors can be the source of shifts both on the x-axis (m/z values) and on the y-axis (intensity values). Hence, to ensure the validity of any subsequent statistical analysis, great care must be taken to address both of these shifts in preprocessing, by *intensity calibration* (often called normalization) and by *peak alignment/warping* (see also Sect. 2.7).

Methods to calibrate peak intensities can be divided into *local* and *global* approaches [41]. In a *local* calibration each single spectrum is calibrated on its own, by matching a specified characteristic such as the median, the mean, or the Total Ion Current (TIC) [7, 10, 41]. In contrast, *global* approaches use information across multiple spectra, e.g., employing linear regression normalization [10], quantile normalization [6], or Probabilistic Quotient Normalization (PQN) [15].

MALDIquant supports two *local* and one *global* method. Specifically, it implements the TIC and median calibration as well as PQN. In PQN all spectra are calibrated using the TIC calibration first. Subsequently, a median reference spectrum is created and the intensities in all spectra are standardized using the reference spectrum and a spectrum-specific median is calculated for each spectrum. Finally, each spectrum is rescaled by the median of the ratios of its intensity values and that of the reference spectrum [15].

It has been shown that applying intensity calibration is an essential step in preprocessing [41]. Despite its simplicity TIC is often the best choice, especially to account for effects between technical replicates [41, 55].

2.6 Peak Detection

Peak detection is a further step in processing mass spectrometry data, serving both to identify potential relevant features and to reduce the dimensionality of the data.

MALDIquant provides the most commonly used peak detection method based on finding local maxima [33, 44, 57, 61, 64, 70]. First a window is moved across the spectra and local maxima are detected. Subsequently these local maxima are compared against a noise baseline which is estimated by the Median Absolute Deviation (MAD) or alternatively Friedman's SuperSmoother [19]. If a local

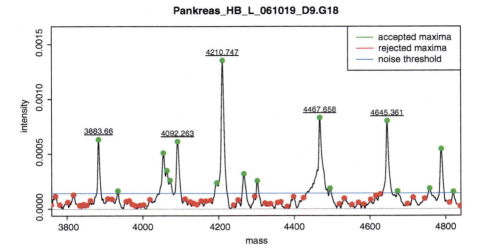

Fig. 3 Detail view of a MALDI-TOF spectrum from Fiedler et al. [18]. Local maxima are marked with points (*red*: rejected maxima, *green*: peaks). The *blue line* represents the estimated noise baseline as estimated by MAD

maximum is above a given Signal-to-Noise Ratio (SNR) it is considered a peak, whereas local maxima below the SNR threshold are discarded (Fig. 3).

Some authors advocate peak detection methods based on *wavelets* [16, 30]. These methods are implemented in the Bioconductor R packages `MassSpecWavelet` and `xcms` [57] and thus are readily available if needed.

2.7   Peak Alignment

As already noted above in Sect. 2.5 not only the intensities but also the m/z values differ across spectra, as result of the many possible sources of variation in the acquisition of mass spectrometry data. Methods to recalibrate the m/z values of the spectra are referred to as *peak alignment* or *warping*.

A simple approach is the Correlation Optimized Warping (COW) algorithm [43, 66, 67]. COW is based on pairwise comparisons of spectra and maximizes the correlation to find an optimal shift. The advantage of this approach is that correlation is fast to compute and with the use of a reference spectrum the method is also applicable to simultaneous alignment of multiple spectra. However, in actual data the location shifts are typically of a nonlinear nature [25], thus methods based on global linear shifts will often be ineffective in achieving an optimal alignment. A possible workaround is to divide the spectrum into several parts and perform local linear alignment instead.

An alternative, and much more flexible approach, is Dynamic Time Warping (DTW) which is based on dynamic programming [12, 29, 62, 63]. DTW is a pairwise alignment approach that is guaranteed to find the optimal alignment by comparing each point in the first spectrum to every other point in the second spectrum, and optimizing a distance score. Dynamic programming techniques are used to substantially shortcut computational time by means of an underlying decision tree, [e.g., 51]. Still, DTW is a computationally very expensive algorithm that also requires substantial computer memory, especially for multiple alignment.

As a compromise, recently the Parametric Time Warping (PTW) approach has been suggested [5, 25, 28, 35, 68] where polynomial functions are used to stretch or shrink a spectrum to increase the similarity between them. PTW is a very fast methods that is also able to correct for nonlinear shifts. As in all previously mentioned method a reference spectrum is necessary for multiple alignment. Note that the use of a reference spectrum requires prior calibration of the intensity values [58].

Finally, another simple strategy to align peaks is based on clustering, respectively, creation of bins of similar m/z values [61, 64, 71]. This is a fast and easy to implement approach, and in contrast to DTW, COW, and PTW it offers the possibility to align all spectra simultaneously. However, the clustering approach is valid only when there are relatively small shifts around the true peak position, hence this approach is only applicable if there are only mild distortions in the m/z values.

In MALDIquant we use nonlinear warping of peaks [25, 68]. First we align the m/z values of the peaks using PTW and subsequently we employ binning to identify common peak positions across spectra. Note that in contrast to the standard version of PTW we work on the peak level rather than on the whole spectrum.

Our peak alignment algorithm in MALDIquant starts by looking for stable peaks, which are defined as high peaks in defined, coarse m/z ranges that are present in most spectra. The m/z of the peaks is averaged and used as reference peak list (also known as anchor or landmark peaks—see Wang et al. [67]). Next, MALDIquant computes a LOcally WEighted Scatterplot Smoothing (LOWESS) curve or polynomial-based function to warp the peaks of each spectra against the reference peaks (Fig. 4). As not all reference peaks are found in each spectrum, the number of matched peaks out of all reference peaks is reported by MALDIquant for information.

Due to using the peaks instead of the whole spectral data the alignment approach implemented in MALDIquant is much faster than traditional PTW, still the results are comparable (Fig. 5). Another important advantage of our approach is that only m/z values are used for calibration, which implies that our approach does not require perfectly calibrated intensity values as is the case for full spectrum-based alignment methods.

After performing alignment, peak positions of identical features across spectra will become very similar but in general not numerically identical. Thus, as final step grouping of m/z values into bins is needed. For this purpose MALDIquant uses the following simple clustering algorithm: The m/z values are sorted in ascending order and split recursively at the largest gap until all m/z values in the resulting bins are

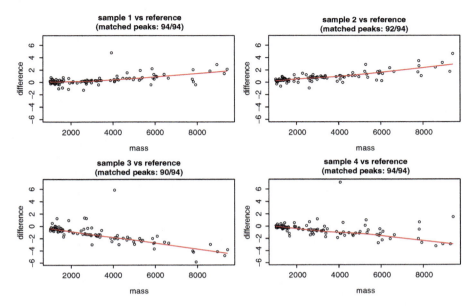

Fig. 4 Example warping function for four different peak lists. The *x*-axis represents the peak position and the *y*-axis the difference from the reference peak list. The *red line* shows the calculated warping function. The number of matched peaks out of all reference peaks is also shown for each spectrum

from different samples and their individual *m/z* values are in a small user-defined tolerance range around their mean. The latter becomes the new *m/z* value for all corresponding peaks in the associated bin.

2.8   Subsequent Statistical Analysis

With peak alignment the task of MALDIquant to transform raw mass spectrometry data into a matrix containing intensity measurements of potentially useful *m/z* values is complete.

Subsequently, the resulting feature intensity matrix can be used with any preferred univariate or multivariate analysis technique, e.g., to identify peaks that are useful for predicting a desired outcome, or simply to rank features with regard to group separation, e.g., [22].

In the following section we will describe in detail how such an analysis may be conducted. For more examples please see the MALDIquant homepage.

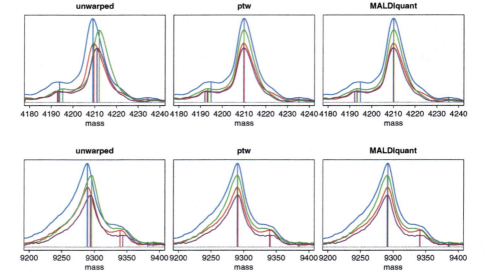

Fig. 5 Comparison of two peaks (*top row* and *bottom row*) present in four MALDI-TOF spectra from Fiedler et al. [18]. (**a, d**) unaligned; (**b, e**) warped using the PTW algorithm; (**c, f**) warped using MALDIquant's peak based PTW

3   Case Study

3.1   Dataset

For illustration how to use MALDIquant in practical data analysis we now show in detail how to use the software by application to the mass spectromety data published in Fiedler et al. [18]. The aim of this study was to determine proteomic biomarkers to discriminate patients with pancreas cancer from healthy persons. As part of their study the authors collected serum samples of 40 patients with diagnosed pancreas cancer as well as 40 healthy controls as training dataset. For each sample four technical replicates were obtained. These 320 samples were processed following a standardized protocol for serum peptidomics and subsequently analyzed in a linear MALDI-TOF mass spectrometer. For details on the experimental setup we refer to the original study.

Half of the patients and controls were recruited at the University Hospital Heidelberg and the University Hospital Leipzig. Due to the presence of strong batch effects we restrict ourselves to the samples from Heidelberg, leading to a raw data set containing 160 spectra for 40 probands, of which 20 were diagnosed with pancreatic cancer and 20 are healthy controls. Fiedler et al. [18] found marker peaks at *m/z* 3884 (double charged) and 7767 (single charged) and correspondingly suggested

Platelet Factor 4 (PF4) as potential marker, arguing that PF4 is down-regulated in blood serum of patients with pancreatic cancer.

3.2   Preparations

Prior to preprocessing the data we first need to set up our *R* environment by install the necessary packages, namely `MALDIquant` [21], `MALDIquantForeign`, `sda`, and `crossval`, and also download the data set:

```
install.packages(c("MALDIquant", "MALDIquantForeign",
                   "sda", "crossval"))
```

```
## load packages
library("MALDIquant")

Loading required package: methods
  This is MALDIquant version 1.13 Quantitative Analysis
of Mass Spectrometry Data See '?MALDIquant' for more
information about this package.

library("MALDIquantForeign")

## download the raw spectra data (approx. 90 MB)
githubUrl <-
paste0("https://raw.githubusercontent.com/sgibb/",
        "MALDIquantExamples/master/inst/",
        "extdata/fiedler2009/")
downloader::download(paste0(githubUrl,
     "spectra.tar.gz"), "fiedler2009spectra.tar.gz")
## download metadata
downloader::download(paste0(githubUrl,
     "spectra_info.csv"), "fiedler2009info.csv")
```

3.3   Import Raw Data and Quality Control

The first step in the analysis comprises importing the raw data into the *R* environment. As the raw data set contains both the samples from Heidelberg and Leipzig we filter out the samples from Leipzig, so that our final data set only contains the Heidelberg patients and controls:

```
## import the spectra
spectra <- import("fiedler2009spectra.tar.gz",
verbose=FALSE)

## import metadata
spectra.info <- read.csv("fiedler2009info.csv")

## keep data from Heidelberg
isHeidelberg <- spectra.info$location == "heidelberg"

spectra <- spectra[isHeidelberg]
spectra.info <- spectra.info[isHeidelberg,]
```

After importing the raw data it is recommend to perform basic checks for quality control. Below, we test whether all spectra contain the same number of data points, are not empty, and are regular, i.e., whether the differences between subsequent *m/z* values are constant:

```
table(lengths(spectra))


 42388
  160

any(sapply(spectra, isEmpty))

 [1] FALSE

all(sapply(spectra, isRegular))

 [1] TRUE
```

Next, we ensure that all spectra cover the same *m/z* range. The "trim" function automatically determines a suitable common *m/z* range if it is called without any additional arguments:

```
spectra <- trim(spectra)
```

Finally, it is advised to inspect the spectra visually to discover any obviously distorted measurements. Here, for reasons of space we only plot a single spectrum (Fig. 6):

```
plot(spectra[[47]], sub="")
```

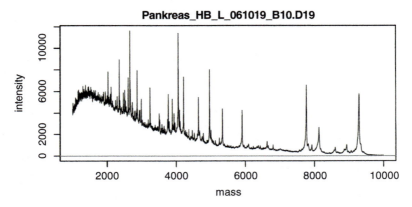

Fig. 6 Example of a raw, uncalibrated mass spectrum

3.4 Transformation and Smoothing

Next, we perform variance stabilization by applying the square root transformation to the raw data, and subsequently use a 41 point *Savitzky–Golay*-Filter [53] to smooth the spectra:

```
spectra <- transformIntensity(spectra, method="sqrt")

spectra <- smoothIntensity(spectra, method=
"SavitzkyGolay",halfWindowSize=20)
```

3.5 Baseline Correction

In the next step we address the problem of matrix-effects and chemical noise that result in an elevated baseline. In our analysis we use the *SNIP* algorithm [50] to estimate the baseline for each spectrum. Subsequently, the estimated baseline is subtracted to yield baseline-adjusted spectra (Figs. 7, 8, and 9):

```
baseline <- estimateBaseline(spectra[[1]], method="SNIP",
                             iterations=150)
plot(spectra[[1]], sub="")
lines(baseline, col="red", lwd=2)
```

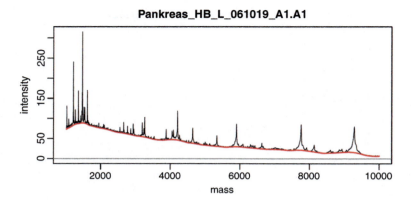

Fig. 7 Baseline estimated using the SNIP method (*red line*)

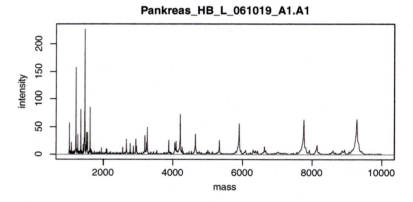

Fig. 8 Mass spectrum after baseline correction

```
spectra <- removeBaseline(spectra, method="SNIP",
                          iterations=150)
plot(spectra[[1]], sub="")
```

3.6 Intensity Calibration and Alignment

After baseline correction we calibrate each spectrum by equalizing the TIC across spectra. After normalizing the intensities we also need to adjust the mass values. This is done by the peak- based warping algorithm implemented in MALDIquant.

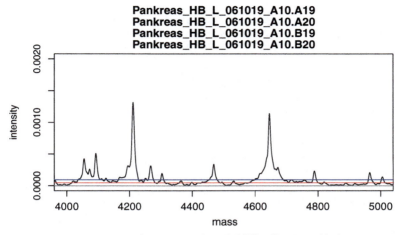

averaged spectrum composed of 4 MassSpectrum objects

Fig. 9 Thresholds based on signal to noise ratio (SNR): SNR=1 (*red line*) and SNR=2 (*blue line*)

In the example code the function `alignSpectra` acts as a simple wrapper around more complicated procedures. For a finer control of the underlying procedures the function `determineWarpingFunctions` may be used alternatively:

```
spectra <- calibrateIntensity(spectra, method="TIC")
spectra <- alignSpectra(spectra)
```

Next, we average the technical replicates before we search for peaks and update our meta information accordingly:

```
avgSpectra <-
  averageMassSpectra(spectra,
  labels=spectra.info$patientID)
avgSpectra.info <-
  spectra.info[!duplicated(spectra.info$patientID), ]
```

3.7 Peak Detection and Computation of Intensity Matrix

Peak detection is the crucial step to identify features and to reduce the dimensionality of the data. Before performing peak detection we first estimate the noise of selected spectra to investigate suitable settings for the *signal-to-noise ratio* (SNR):

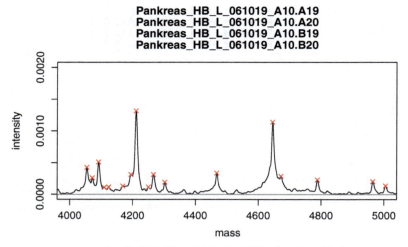

averaged spectrum composed of 4 MassSpectrum objects

Fig. 10 Peaks identified above SNR=2 threshold

```
noise <- estimateNoise(avgSpectra[[1]])
plot(avgSpectra[[1]], xlim=c(4000, 5000), ylim=c(0,
0.002))
lines(noise, col="red")                        # SNR == 1
lines(noise[, 1], 2*noise[, 2], col="blue") # SNR == 2
```

In this case we decide to set an *SNR* of 2 (blue line) and then run the peak detection algorithm (Fig. 10):

```
peaks <- detectPeaks(avgSpectra, SNR=2,
halfWindowSize=20)
```

```
plot(avgSpectra[[1]], xlim=c(4000, 5000), ylim=c(0,
0.002))
points(peaks[[1]], col="red", pch=4)
```

After the alignment the peak positions (mass) are very similar but not numerically identical. Consequently, binning is required to achieve identity:

```
peaks <- binPeaks(peaks)
```

In peak detection we choose a very low signal-to-noise ratio to keep as many features as possible. Using the information about class labels we can now filter out false positive peaks, by removing peaks that appear in less than 50 % of all spectra in each group:

```
peaks <- filterPeaks(peaks, minFrequency=c(0.5, 0.5),
                     labels=avgSpectra.info$health,
                     mergeWhitelists=TRUE)
```

As final step in MALDIquant we create the feature intensity matrix, and for convenience label the rows with the corresponding patient ID:

```
featureMatrix <- intensityMatrix(peaks, avgSpectra)
rownames(featureMatrix) <- avgSpectra.info$patientID
dim(featureMatrix)

[1]   40 166
```

This matrix is the final output of MALDIquant and contains the calibrated intensity values for identified features across all spectra. It forms the basis for higher-level statistical analysis.

3.8   Feature Ranking and Classification

The Fiedler et al. [18] data set contains class labels for each spectrum (healthy versus cancer), hence it is natural to perform a standard classification and feature ranking analysis. A commonly used approach is Fisher's linear discriminant analysis (LDA), see Mertens et al. [40] for details and applications to mass spectrometry analysis. Many other classification approaches may also be applied, such as based on Random Forests [8] or peak discretization [22].

Here, we use a variant of LDA implemented in the R package sda [2]. In particular, we use diagonal discriminant analysis (DDA), a special case of LDA with the assumption that the correlation among features (peaks) is negligible. Despite this simplification this approach to classification is very effective, especially in high dimensions [60]. In order to identify the most important class discriminating peaks we use standard t-scores, which are the natural variable importance measure in DDA.

As a first step in our analysis we therefore compute the ranking of features by t-scores, and list the 10 top-ranking features in Table 1:

Table 1 The ten top-ranking peaks as identified in the analysis

	idx	Score	t.cancer	t.control
8936.97	158.00	90.69	9.52	−9.52
4468.07	116.00	80.80	8.99	−8.99
8868.27	157.00	80.06	8.95	−8.95
4494.8	117.00	67.00	8.19	−8.19
8989.2	159.00	66.19	8.14	−8.14
5864.49	135.00	37.56	−6.13	6.13
5906.17	136.00	34.43	−5.87	5.87
2022.94	49.00	33.30	5.77	−5.77
5945.57	137.00	32.66	−5.71	5.71
1866.17	44.00	32.12	5.67	−5.67

```
library("sda")
colnames(featureMatrix) <-
    round(as.double(colnames(featureMatrix)),2)
Ytrain <- avgSpectra.info$health
ddar <- sda.ranking(Xtrain=featureMatrix, L=Ytrain,
fdr=FALSE,diagonal=TRUE)
```

To illustrate that feature selection based on the above feature ranking is indeed beneficial for subsequent analysis we apply hierarchical cluster analysis based on the Euclidean distance first to the data set containing all features (Figs. 11 and 12):

```
distanceMatrix <- dist(featureMatrix, method="euclidean")

hClust <- hclust(distanceMatrix, method="complete")

plot(hClust, hang=-1)
```

Next, we repeat the above clustering on the data set containing only the best two top-ranking peaks:

```
top <- ddar[1:2, "idx"]

distanceMatrixTop <- dist(featureMatrix[, top],
                          method="euclidean")

hClustTop <- hclust(distanceMatrixTop, method="complete")

plot(hClustTop, hang=-1)
```

As can be seen by comparison of the two trees, as a result of the feature selection we obtain a nearly perfect split between the Heidelberg pancreas cancer samples (labeled "HP") and the Heidelberg control group (labeled "HC").

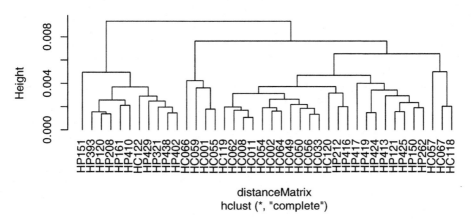

Fig. 11 Hierarchical clustering of patient samples using all features

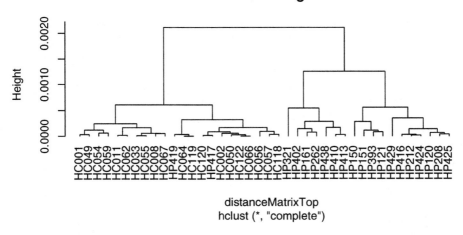

Fig. 12 Hierarchical clustering of patient samples using only the best two top-ranking peaks

The strong predictive capabilities of the first two discovered peaks can be further quantified by conducting a cross-validation analysis to estimate the prediction error. We use the `crossval` [59] package to perform a ten-fold cross validation using the predictor containing only the two selected peaks:

```r
library("crossval")
# create a prediction function for the cross validation
predfun.dda <- function(Xtrain, Ytrain, Xtest, Ytest,
                        negative) {
  dda.fit <- sda(Xtrain, Ytrain, diagonal=TRUE,
  verbose=FALSE)
  ynew <- predict(dda.fit, Xtest, verbose=FALSE)$class
  return(confusionMatrix(Ytest, ynew,
  negative
  =negative))
}

# set seed to get reproducible results
set.seed(1234)

cv.out <- crossval(predfun.dda,
                   X=featureMatrix[, top],
                   Y=avgSpectra.info$health,
                   K=10, B=20,
                   negative="control",
                   verbose=FALSE)
diagnosticErrors(cv.out$stat)

    acc      sens      spec      ppv       npv       lor
 0.9500000 0.9000000 1.0000000 1.0000000 0.9090909 Inf
```

As a result of the above analysis, we conclude that the identified peaks with mass *m/z* 8937 and 4467 allow for the construction of a very low-dimensional predictor function that is highly effective in separating cancer and control group with both high accuracy and high sensitivity.

4   Conclusion

The large-scale acquisition of mass spectrometry data is becoming routine in many experimental settings. In MALDIquant we have put together a robust *R* pipeline for preprocessing these data to allow subsequent high-level statistical analysis. All methods implemented in MALDIquant have been selected both for computational efficiency and for biological validity. In this chapter we have given an overview over the most commonly used procedures of MALDIquant as well as demonstrated their application in detail.

A topic that has not been covered here is Mass Spectrometry Imaging (MSI), which combines spectral measurements with spatial information [14]. MALDIquant also enables some simple MSI analysis, for practical examples in *R* we refer to the homepage of `MALDIquant` at http://strimmerlab.org/software/maldiquant/ as well as the associated web page https://github.com/sgibb/MALDIquantExamples/.

References

1. Aebersold, R., & Mann, M. (2003). Mass spectrometry-based proteomics. *Nature, 422*, 198–207.
2. Ahdesmäki, M., & Strimmer, K. (2010). Feature selection in omics prediction problems using cat scores and false nondiscovery rate control. *The Annals of Applied Statistics, 4*(1), 503–519.
3. Andrew, M. A. (1979). Another efficient algorithm for convex hulls in two dimensions. *Information Processing Letters, 9*, 216–219. Amsterdam: Elsevier.
4. Baggerly, K. A., Morris, J. S., & Coombes, K. R. (2004). Reproducibility of SELDI-TOF protein patterns in serum: Comparing datasets from different experiments. *Bioinformatics, 20*, 777–785.
5. Bloemberg, T. G., Gerretzen, J., Wouters, H. J. P., Gloerich, J., van Dael, M., Wessels, H. J. C. T., et al. (2010). Improved parametric time warping for proteomics. *Chemometrics and Intelligent Laboratory Systems, 104*, 65–74.
6. Bolstad, B. M., Irizarry, R. A., Astrand, M., & Speed, T. P. (2003). A comparison of normalization methods for high density oligonucleotide array data based on variance and bias. *Bioinformatics, 19*, 185–193.
7. Borgaonkar, S. P., Hocker, H., Shin, H., & Markey, M. K. (2010). Comparison of normalization methods for the identification of biomarkers using MALDI-TOF and SELDI-TOF mass spectra. *OMICS, 14*, 115–126.
8. Breiman, L. (2001). Random forests. *Machine Learning, 45*, 5–32.
9. Bromba, M. U. A., & Ziegler, H. (1981). Application hints for Savitzky–Golay digital smoothing filters. *Analytical Chemistry, 53*(11), 1583–1586.
10. Callister, S. J., Barry, R. C., Adkins, J. N., Johnson, E. T., Qian, W.-J., Webb-Robertson, B.-J. M., et al. (2006). Normalization approaches for removing systematic biases associated with mass spectrometry and label-free proteomics. *Journal of Proteome Research, 5*, 277–286.
11. Chambers, M. C., Maclean, B., Burke, R., Amodei, D., Ruderman, D. L., Neumann, S., et al. (2012). A cross-platform toolkit for mass spectrometry and proteomics. *Nature Biotechnology, 30*(10), 918–920.
12. Clifford, D., Montoliu, G. S. I., Rezzi, S., Martin, F.-P., Guy, P., Bruce, S., et al. (2009). Alignment using variable penalty dynamic time warping. *Analytical Chemistry, 81*, 1000–1007.
13. Coombes, K. R., Tsavachidis, S., Morris, J. S., Baggerly, K. A., Hung, M.-C., & Kuerer, H. M. (2005). Improved peak detection and quantification of mass spectrometry data acquired from surface-enhanced laser desorption and ionization by denoising spectra with the undecimated discrete wavelet transform. *Proteomics, 5*, 4107–4117.
14. Cornett, D. S., Reyzer, M. L., Chaurand, P., & Caprioli, R. M. (2007). MALDI imaging mass spectrometry: Molecular snapshots of biochemical systems. *Nature Methods, 4*, 828–833.
15. Dieterle, F., Ross, A., Schlotterbeck, G., & Senn, H. (2006). Probabilistic quotient normalization as robust method to account for dilution of complex biological mixtures. Application in 1H NMR metabonomics. *Analytical Chemistry, 78*, 4281–4290.
16. Du, P., Kibbe, W. A., & Lin, S. M. (2006). Improved peak detection in mass spectrum by incorporating continuous wavelet transform-based pattern matching. *Bioinformatics, 22*, 2059–2065.
17. Du, P., Stolovitzky, G., Horvatovich, P., Bischoff, R., Lim, J., & Suits, F. (2008). A noise model for mass spectrometry based proteomics. *Bioinformatics, 24*, 1070–1077.

18. Fiedler, G. M., Leichtle, A. B., Kase, J., Baumann, S., Ceglarek, U., Felix, K., et al. (2009). Serum peptidome profiling revealed platelet factor 4 as a potential discriminating peptide associated with pancreatic cancer. *Clinical Cancer Research, 15*, 3812–3819.
19. Friedman, J. H. (1984). *A variable span smoother*. Technical Report, DTIC Document.
20. Gammerman, A., Nouretdinov, I., Burford, B., Chervonenkis, A., Vovk, V., & Luo, Z. (2008). Clinical mass spectrometry proteomic diagnosis by conformal predictors. *Statistical Applications in Genetics and Molecular Biology, 7*, 13.
21. Gibb, S., & Strimmer, K. (2012). MALDIquant: A versatile R package for the analysis of mass spectrometry data. *Bioinformatics, 28*, 2270–2271.
22. Gibb, S., & Strimmer, K. (2015). Differential protein expression and peak selection in mass spectrometry data by binary discriminant analysis. *Bioinformatics, 31*, 3156–3162.
23. Gil, J. Y., & Kimmel, R. (2002). Efficient dilation, erosion, opening, and closing algorithms. *IEEE Transactions on Pattern Analysis and Machine Intelligence, 24*, 1606–1617.
24. Gregori, J., Villarreal, L., Méndez, O., Sánchez, A., Baselga, J., & Villanueva, J. (2012). Batch effects correction improves the sensitivity of significance tests in spectral counting-based comparative discovery proteomics. *Journal of Proteomics, 75*(13), 3938–3951.
25. He, Q. P., Wang, J., Mobley, J. A., Richman, J., & Grizzle, W. E. (2011). Self-calibrated warping for mass spectra alignment. *Cancer Informatics, 10*, 65–82.
26. House, L. L., Clyde, M. A., & Wolpert, R. L. (2011). Bayesian nonparametric models for peak identification in MALDI-TOF mass spectroscopy. *The Annals of Applied Statistics, 5*, 1488–1511.
27. Hu, J., Coombes, K. R., Morris, J. S., & Baggerly, K. A. (2005). The importance of experimental design in proteomic mass spectrometry experiments: Some cautionary tales. *Briefings in Functional Genomics and Proteomics, 3*, 322–331.
28. Jeffries, N. (2005). Algorithms for alignment of mass spectrometry proteomic data. *Bioinformatics, 21*, 3066–3073.
29. Kim, S., Koo, I., Fang, A., & Zhang, X. (2011). Smith-Waterman peak alignment for comprehensive two-dimensional gas chromatography-mass spectrometry. *BMC Bioinformatics, 12*, 235.
30. Lange, E., Gröpl, C., Reinert, K., Kohlbacher, O., & Hildebrandt, A. (2006). High-accuracy peak picking of proteomics data using wavelet techniques. In *Pacific Symposium on Biocomputing* (Vol. 11, pp. 243–254).
31. Leek, J. T., Scharpf, R. B., Bravo, H. C., Simcha, D., Langmead, B., Johnson, W. E., et al. (2010). Tackling the widespread and critical impact of batch effects in high-throughput data. *Nature Reviews Genetics, 11*, 733–739.
32. Leichtle, A. B., Dufour, J.-F., & Fiedler, G. M. (2013). Potentials and pitfalls of clinical peptidomics and metabolomics. *Swiss Medical Weekly, 143*, w13801.
33. Li, X. (2005). PROcess: Ciphergen SELDI-TOF Processing. R package version 1.44.0.
34. Lilley, K. S., Deery, M. J., & Gatto, L. (2011). Challenges for proteomics core facilities. *Proteomics, 11*(6), 1017–1025.
35. Lin, S. M., Haney, R. P., Campa, M. J., Fitzgerald, M. C., & Patz, E. F. (2005). Characterising phase variations in MALDI-TOF data & correcting them by peak alignment. *Cancer Informatics, 1*, 32–40.
36. Liu, Q., Krishnapuram, B., Pratapa, P., Liao, X., Hartemink, A., & Carin, L. (2003). Identification of differentially expressed proteins using MALDI-TOF mass spectra. *Signals, Systems & Computers, 2003. Conference Record* (Vol. 2, pp. 1323–1327).
37. Liu, Q., Sung, A. H., Qiao, M., Chen, Z., Yang, J. Y., Yang, M. Q., et al. (2009). Comparison of feature selection and classification for MALDI-MS data. *BMC Genomics, 10*(Suppl 1), S3.
38. Liu, L. H., Shan, B. E., Tian, Z. Q., Sang, M. X., Ai, J., Zhang, Z. F., et al. (2010). Potential biomarkers for esophageal carcinoma detected by matrix-assisted laser desorption/ionization time-of-flight mass spectrometry. *Clinical Chemistry & Laboratory Medicine, 486*, 855–861.
39. Martens, L., Chambers, M., Sturm, M., Kessner, D., Levander, F., Shofstahl, J., et al. (2011). mzML–a community standard for mass spectrometry data. *Molecular & Cellular Proteomics, 10*, R110.000133.

40. Mertens, B. J. A., de Noo, M. E., Tollenaar, R. A. E. M., & Deelder, A. M. (2006). Mass spectrometry proteomic diagnosis: Enacting the double cross-validatory paradigm. *Journal of Computational Biology, 13*, 1591–1605.
41. Meuleman, W., Engwegen, J. Y., Gast, M.-C. W., Beijnen, J. H., Reinders, M. J., & Wessels, L. F. (2008). Comparison of normalisation methods for surface-enhanced laser desorption and ionisation (SELDI) time-of-flight (TOF) mass spectrometry data. *BMC Bioinformatics, 9*, 88.
42. Morháč, M. (2009). An algorithm for determination of peak regions and baseline elimination in spectroscopic data. *Nuclear Instruments and Methods in Physics Research Section A: Accelerators, Spectrometers, Detectors and Associated Equipment, 600*, 478–487.
43. Morris, J. S., Baggerly, K. A., Gutstein, H. B., & Coombes, K. R. (2010). Statistical contributions to proteomic research. *Methods in Molecular Biology, 641*, 143–166.
44. Morris, J. S., Coombes, K. R., Koomen, J., Baggerly, K. A., & Kobayashi, R. (2005). Feature extraction and quantification for mass spectrometry in biomedical applications using the mean spectrum. *Bioinformatics, 21*, 1764–1775.
45. Norris, J. L., Cornett, D. S., Mobley, J. A., Andersson, M., Seeley, E. H., Chaurand, P., et al. (2007). Processing MALDI mass spectra to improve mass spectral direct tissue analysis. *International Journal of Mass Spectrometry, 260*, 212–221.
46. Pedrioli, P. G. A., Eng, J. K., Hubley, R., Vogelzang, M., Deutsch, E. W., Raught, B., et al. (2004). A common open representation of mass spectrometry data and its application to proteomics research. *Nature Biotechnology, 22*, 1459–1466.
47. Purohit, P. V., & Rocke, D. M. (2003). Discriminant models for high-throughput proteomics mass spectrometer data. *Proteomics, 3*, 1699–1703.
48. R Core Team (2015). R: A language and environment for statistical computing. R Foundation for Statistical Computing, Vienna, Austria.
49. Robb, R. A., Hanson, D. P., Karwoski, R. A., Larson, A. G., Workman, E. L., & Stacy, M. C. (1989). Analyze: A comprehensive, operator-interactive software package for multidimensional medical image display and analysis. *Computerized Medical Imaging and Graphics, 13*, 433–454.
50. Ryan, C. G., Clayton, E., Griffin, W. L., Sie, S. H., & Cousens, D. R. (1988). SNIP, a statistics-sensitive background treatment for the quantitative analysis of PIXE spectra in geoscience applications. *Nuclear Instruments and Methods in Physics Research Section B: Beam Interactions with Materials and Atoms, 34*, 396–402.
51. Sakoe, H., & Chiba, S. (1978). Dynamic programming algorithm optimization for spoken word recognition. *IEEE Transactions on Acoustics, Speech and Signal Processing, 26*, 43–49.
52. Sauve, A. C., & Speed, T. P. (2004). Normalization, baseline correction and alignment of high-throughput mass spectrometry data. In *Proceedings of the Data Proceedings Gensips*.
53. Savitzky, A., & Golay, M. J. E. (1964). Smoothing and differentiation of data by simplified least squares procedures. *Analytical Chemistry, 36*, 1627–1639.
54. Schramm, T., Hester, A., Klinkert, I., Both, J.-P., Heeren, R. M. A., Brunelle, A., et al. (2012). imzML–a common data format for the flexible exchange and processing of mass spectrometry imaging data. *Journal of Proteomics, 75*, 5106–5110.
55. Shin, H., & Markey, M. K. (2006). A machine learning perspective on the development of clinical decision support systems utilizing mass spectra of blood samples. *Journal of Biomedical Informatics, 39*, 227–248.
56. Sköld, M., Rydén, T., Samuelsson, V., Bratt, C., Ekblad, L., Olsson, H., et al. (2007). Regression analysis and modelling of data acquisition for SELDI-TOF mass spectrometry. *Bioinformatics, 23*, 1401–1409.
57. Smith, C. A., Want, E. J., O'Maille, G., Abagyan, R., & Siuzdak, G. (2006). XCMS: Processing mass spectrometry data for metabolite profiling using nonlinear peak alignment, matching, and identification. *Analytical Chemistry, 78*, 779–787.
58. Smith, R., Ventura, D., & Prince, J. T. (2013). LC-MS alignment in theory and practice: A comprehensive algorithmic review. *Briefings in Bioinformatics, 16*(1), 104–117.
59. Strimmer, K. (2014). crossval: Generic functions for cross validation. R package version 1.0.1.

60. Tibshirani, R., Hastie, T., Narsimhan, B., & Chu, G. (2003). Class prediction by nearest shrunken centroids, with applications to DNA microarrays. *Statistical Science, 18*, 104–117.
61. Tibshirani, R., Hastie, T., Narasimhan, B., Soltys, S., Shi, G., Koong, A., et al. (2004). Sample classification from protein mass spectrometry, by 'peak probability contrasts'. *Bioinformatics, 20*, 3034–3044.
62. Toppoo, S., Roveri, A., Vitale, M. P., Zaccarin, M., Serain, E., Apostolidis, E., et al. (2008). MPA: A multiple peak alignment algorithm to perform multiple comparisons of liquid-phase proteomic profiles. *Proteomics, 8*, 250–253.
63. Torgrip, R. J. O., Åberg, M., Karlberg, B., & Jacobsson, S. P. (2003). Peak alignment using reduced set mapping. *Journal of Chemometrics, 17*, 573–582.
64. Tracy, M. B., Chen, H., Weaver, D. M., Malyarenko, D. I., Sasinowski, M., Cazares, L. H., et al. (2008). Precision enhancement of MALDI-TOF MS using high resolution peak detection and label-free alignment. *Proteomics, 8*, 1530–1538.
65. van Herk, M. (1992). A fast algorithm for local minimum and maximum filters on rectangular and octagonal kernels. *Pattern Recognition Letters, 13*, 517–521.
66. Veselkov, K. A., Lindon, J. C., Ebbels, T. M. D., Crockford, D., Volynkin, V. V., Holmes, E., et al. (2009). Recursive segment-wise peak alignment of biological (1)h NMR spectra for improved metabolic biomarker recovery. *Analytical Chemistry, 81*, 56–66.
67. Wang, B., Fang, A., Heim, J., Bogdanov, B., Pugh, S., Libardoni, M., et al. (2010). DISCO: distance and spectrum correlation optimization alignment for two-dimensional gas chromatography time-of-flight mass spectrometry-based metabolomics. *Analytical Chemistry, 82*, 5069–5081.
68. Wehrens, R., Bloemberg, T., & Eilers, P. (2015). Fast parametric time warping of peak lists. *Bioinformatics, 15*, 3063–3065.
69. Williams, B., Cornett, S., Dawant, B., Crecelius, A., Bodenheimer, B., & Caprioli, R. (2005). An algorithm for baseline correction of MALDI mass spectra. In *Proceedings of the 43rd Annual Southeast Regional Conference* (Vol. 1, pp. 137–142). ACM-SE 43.
70. Yasui, Y., McLerran, D., Adam, B., Winget, M., Thornquist, M., & Feng, Z. (2003). An automated peak-identification/calibration procedure for high-dimensional protein measures from mass spectrometers. *Journal of Biomedicine and Biotechnology, 4*, 242–248.
71. Yasui, Y., Pepe, M., Thompson, M. L., Adam, B.-L., Wright, G. L., Qu, Y., et al. (2003). A data-analytic strategy for protein biomarker discovery: profiling of high-dimensional proteomic data for cancer detection. *Biostatistics, 4*, 449–463.

Model-Based Analysis of Quantitative Proteomics Data with Data Independent Acquisition Mass Spectrometry

Gengbo Chen, Guo Shou Teo, Guo Ci Teo, and Hyungwon Choi

1 Introduction

Data independent acquisition (DIA) mode of mass spectrometry (MS) is a novel analysis strategy for comprehensive collection of MS/MS data and simultaneous quantification to overcome the limitation of the conventional data dependent acquisition (DDA) mode of analysis [21]. In DDA, precursor ions in an MS1 scan are sorted by intensity and, due to limited scan time between contiguous MS1 scans, the ions of greater intensity are preferentially isolated and fragmented for sequence identification (Fig. 1a). In each isolation window, mostly containing one peptide ion, subsequent MS/MS scans record mass-to-charge (m/z) ratio and intensity of fragment ions of the peptide, and this information provides clues to the most likely amino acid sequence via database search or de novo sequencing methods [18]. An identified peptide (with assigned sequence) is usually quantified by calculating the area under the elution profile consisting of peak heights across contiguous MS1 scans, often termed the extracted ion chromatogram (XIC). We refer to the peak area as "peptide intensity" hereafter. Hence DDA utilizes MS/MS data for peptide identification of selected precursor ions, not for quantification. Following peptide quantification, protein intensities are *inferred* using heuristic approaches such as summation of peptide intensities within each protein [17].

Although DDA has been the platform of choice in the past decade, the idea of DIA had emerged years before [13, 21]. In DIA, the mass range of interest (e.g., 400–900 m/z) is segmented into a set of sequential windows, and MS/MS spectra are acquired for *all* precursors in each window separately, yielding an

G. Chen • G.S. Teo • G.C. Teo • H. Choi (✉)
Saw Swee Hock School of Public Health, National University of Singapore,
Singapore, Singapore
e-mail: hyung_won_choi@nuhs.edu.sg

© Springer International Publishing Switzerland 2017
S. Datta, B.J.A. Mertens (eds.), *Statistical Analysis of Proteomics, Metabolomics,
and Lipidomics Data Using Mass Spectrometry*, Frontiers in Probability and the
Statistical Sciences, DOI 10.1007/978-3-319-45809-0_7

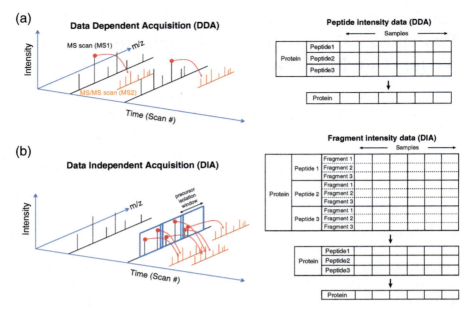

Fig. 1 Instrumental framework and data structure in DDA versus DIA. (**a**) In DDA, peptide ions with higher intensity in MS1 scans are prioritized for MS/MS fragmentation in order to assign a peptide sequence to each precursor ion. For ions with identified peptide sequences, XICs are traced from MS1 scans for peak area calculation (intensity), and the peptide intensities are further summarized into a protein intensity in each sample. (**b**) In DIA, all precursor ions in each isolation window are fragmented together, and thus more rich MS/MS data are produced without the abundance bias of MS1 scans. This allows XIC extraction for individual fragments of each peptide, allowing repeated quantification of peptides. The fragment intensity data can be rolled up to peptide intensities, and then to protein intensities by summation in existing pipelines

unbiased set of precursors. Initially constrained by the low scan speed and resolving power of old generation mass spectrometers, DIA implementation did not live up to its expectation of drastic improvement in the peptide coverage and quantitative accuracy. However, Gillet et al. [7] showed that DIA can be successfully performed for the analysis of complex protein mixtures on a Qq-TOF AB SCIEX instrument using a relatively large, 25 m/z-wide precursor isolation windows, termed SWATH-MS (Fig. 1b). Subsequently, others groups have also reported alternative DIA setups on Thermo Fisher Q Exactive mass spectrometers [6, 14].

Unlike DDA, note that MS/MS scans are performed for each isolation window without preferential selection of precursor ions from MS1 scans. This allows for unbiased MS/MS scans for all possible precursor ions, and elution profiles can be constructed for a large number of fragment ions (quantification of fragments). Bringing the source of quantification from MS data to MS/MS data is probably the most important innovation of DIA-MS for quantitative proteomics. On the downside, as elaborated in the next section, the specificity of fragments to their precursor peptide ion is lost in DIA, which has to be recovered computationally.

In hindsight, DDA can be thought of an extreme form of DIA: DDA is DIA with a very small isolation window, where a small number of windows are chosen for MS/MS fragmentation based on precursor intensity. As the scan speed and measurement accuracy of MS instrument improve, the two frameworks are nevertheless bound to converge to similar performance since preferential selection of precursors is no longer necessary with faster scan speed.

2   Challenges in the Analysis of DIA-MS Data

While the DIA analysis presents a significant leap towards comprehensive quantification of the proteome, an array of informatics challenges ironically stems from the way DIA collects MS/MS spectra: all precursor peptide ions within each MS1 isolation window are co-isolated and subjected to fragmentation to produce *multiplex* MS/MS spectra. Therefore, it becomes a computational problem to assign each fragment to their most likely mother peptide(s). In general, the size of an MS1 isolation window is inversely associated with the specificity of fragments to the peptide precursors. For example, wide windows (e.g., 25 m/z in the initial form of SWATH-MS [7]) tend to include co-eluting peptides and thus the complexity of MS/MS is high. In contrast, small windows (4 m/z in MSX approach [6]) have increased purity of MS/MS fragments to their parent peptide, but it requires far more MS/MS scans within the fixed cycle time (between two MS1 scans) and reduces scan time for MS/MS fragments in each isolation window, potentially compromising the quality of quantitative MS/MS data.

Another important issue restraining the coverage improvement is the fact that the current data extraction pipeline depends on existing spectral assay libraries [7, 15], which are the records of mass-to-charge ratio (m/z) and retention time coordinates of annotated MS/MS fragment peaks of known peptides [16]. Moreover, given the library and the data from DIA experiments, it requires specialized data extraction tools such as OpenSWATH [15], Skyline [11], or PeakView (AB Sciex) to obtain final quantification tables (Fig. 2a), each of which applies different quality scoring to ensure high-quality quantitative data. However, this not only means that spectral assay library must be generated by additional DDA experiments on the same or similar samples, but also that the quantification coverage of peptides in DIA experiments is inherently limited by that of the spectral library. This critical dependence has called for alternative ways to extract data without the dependence on spectral assay libraries. DIA-Umpire [20] is the first tool of this kind, which avoids library building exercise and builds its own internal library directly from DIA data and performs targeted extraction using the resulting library (Fig. 2b).

Whether data are extracted in a library-dependent or library-free manner, extraction tools typically report >100,000 fragments per sample from tens of thousands of peptides in the initially reported quantification table. However, not all fragment intensities are reliable and reproducible enough for downstream statistical analysis since many of them are not consistently observed across multiple samples.

(a)   Library-dependent extraction   (b)   Library-free extraction

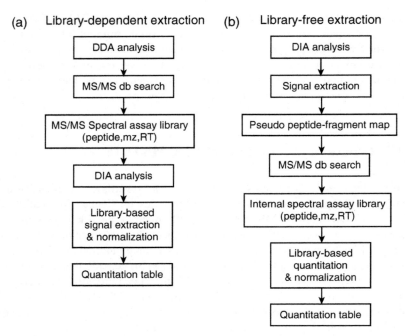

Fig. 2 (**a**) In the library-dependent extraction (OpenSWATH, Skyline), DDA experiments are performed to obtain a peptide-to-fragment map (spectral assay library) first. Next DIA experiments are performed for the samples of interest, and XIC curves are extracted for the MS/MS fragments from designated coordinates in the library (i.e., targeted extraction), and the final quantification table is constructed after data normalization and cross-sample alignment. (**b**) Library-free extraction does not require pre-existing spectral assay library (DDA experiments). Instead, it builds its own library internally using the pseudo peptide-to-fragment map and follows similar extraction steps afterwards. This process tends to produce a lot more fragments per peptide than the library-dependent extraction, since it typically includes putative peptide-fragment pairs in the pseudo map construction (prior to database search)

Moreover, even in the data sets extracted using a comprehensive spectral library, not all fragments assigned to the same precursor peptide are well correlated amongst themselves. These observations suggest the need for careful quality control steps such as outlier detection and selection of MS/MS fragments as well as careful handling of missing data in the downstream statistical analysis.

As a remedy to address these issues, we will first describe a comprehensive data analysis tool *mapDIA* [19] which implements crucial quality control procedures and performs model-based statistical significance analysis on the filtered data using a hierarchical Bayesian modeling approach [22]. We will then illustrate the method in the analysis of recently published SWATH-MS dataset [8] featuring renal cancer tumor tissues paired with healthy control tissues from six patients. Using this data, we will demonstrate that the analysis of fragment-level intensity data allows for improved sensitivity in detection of differentially expressed proteins compared to the analysis using the protein intensity data derived from fragment intensities. We will also compare the results between a library-dependent extraction tool (OpenSWATH [15]) and a library-free extraction tool (DIA-Umpire [20]).

3   mapDIA: Data Preprocessing and Differential Expression Analysis

Here we introduce mapDIA with brief description of its data preprocessing and statistical analysis method. mapDIA features three analysis modules: (1) intensity data normalization, (2) data filtering procedures, and (3) statistical analysis. In this work, we focus on detailed description of the statistical model and scoring method for the selection of differentially expressed proteins, omitting the data normalization and preprocessing modules. See Teo et al. [19] for the full details of the first two modules.

3.1   Fragment Filtering and Selection

In the data preprocessing module, mapDIA performs a series of fragment filtering/selection steps. The filtering removes outlying intensity observations and also discard fragments that are poorly correlated with the other fragments within each protein, ensuring fragment-to-fragment consistency of quantitative data within each peptide and protein. All calculations are performed in log2 scale with proper normalization steps applied to the initially extracted data.

We first define outlier fragment intensity as an observation that deviates more than k times the standard deviation away from the median fragment intensity in the same protein within the same sample. The default value of k is currently 2, yet this option is specified by the user and it should be set to a value large enough to avoid removing too many values. Once this step is done, we set the next cross-sample filtering criteria for individual fragments. For each fragment, the correlation is calculated between the fragment and all other fragments belonging to the same protein, and the fragment is retained for downstream analysis if the median correlation is above a user-specified threshold. This step is designed to remove a small proportion of fragments within each protein, since the majority of fragments are likely to be positively correlated in each protein if the underlying protein abundance sufficiently varied across the samples. For this reason, the default threshold is set to a moderately large value 0.2 (modifiable through options).

The last fragment selection step imposes the final inclusion criteria based on the number of fragments and peptides available for each protein. Since the benefit of DIA data comes from repeated measurements for each peptide, it is recommended to perform the statistical analysis with peptides with at least two fragments passing the outlier and median correlation requirements described above. On the other hand, including too many peptides and fragments can introduce false positives in the downstream statistical analysis. For this reason, mapDIA also allows the user to choose the maximum number of fragments per peptide and maximum number of peptides per protein. If there are more than a specified maximum number of peptides or fragments, the software selects the fragments scoring the highest median correlations. This measure prevents some proteins from having disproportionately many data points and being more disposed to the risk of false positive differential expression calls.

3.2 Statistical Model for Differential Expression Analysis

Using the filtered fragment intensity data, the current practice for statistical analysis is to aggregate the intensity data up to peptide or protein intensity data [4, 7, 8, 10]. There are different ways to quantify peptides and proteins using this data, and the most widely used approach is to sum the intensities over the top k most intense fragments per peptide and top ℓ most intense peptides per protein afterwards, following the logic of Silva et al. [17] used in the protein quantification in DDA experiments. As mentioned earlier, the model in mapDIA can be applied to any of the three levels of intensity data, and we will show that directly modeling fragment intensities provides the greatest statistical power to detect change (fold changes) of smaller magnitude in their performance comparison later.

The underlying probability model in mapDIA is an extension of the work by Wei and Li [22], a Bayesian latent variable model with Markov random field prior. We made two changes to their work: (1) the likelihood accounts for protein-peptide-fragment hierarchy of the data, with most model parameters specified at the peptide level; and (2) the Gamma likelihood with Gamma prior was replaced by Gaussian likelihood with Gaussian-Inverse Gamma prior for the log2 intensity data. Here we describe the model for fragment intensity data for comparison of two groups of subjects (group A and group B), without rolling up the intensity to peptide or protein level. Denoting the fragment intensity data as $\mathbf{Y}$, the latent variable model can be first written as

$$\pi(\mathbf{Y}|\mathbf{Z}) = \prod_{p=1}^{P} \pi(\mathbf{y}_p|z_p),$$

where $\mathbf{Z} = (z_1, \ldots, z_P)$ is the set of latent variables indicating differential expression status for all proteins, and $\mathbf{y}_p$ is the observed fragment intensity data for protein $p = 1, \ldots, P$. For each protein p, $z_p = 1$ if the protein is differentially expressed between the two comparison groups, or $z_p = 0$ otherwise. Under a set of likelihood and prior (see below), the objective is to find the optimal latent states $\mathbf{Z}_*$ maximizing the posterior distribution of $\mathbf{Z}$, i.e.,

$$\mathbf{Z}_* = \text{argmax } \pi(\mathbf{Z}|\mathbf{Y}).$$

Here the joint prior distribution of $\mathbf{Z}$ is approximated by the Markov random field model [1]

$$\pi(z_p = z) \propto \exp\left(\gamma_z - \beta \sum_{k \in \partial p} 1\{z_k \neq z\}\right) \quad \text{for } z \in \{0, 1\},$$

where ∂p denotes the set of network neighbors of protein p on a previously known biological network. Note that, if the network information is not utilized ($\beta = 0$), then the entire model becomes equivalent to a mixture model treating the latent states as independent binary random variables, which we assume in this illustration.

To specify the marginal conditional likelihood $\pi(\mathbf{y}_p | z_p = z)$, we first let $(\mathbf{y}_q^A, \mathbf{y}_q^B)$ and $(\mathbf{y}_f^A, \mathbf{y}_f^B)$ denote fragment intensity data for peptide q or a single fragment f in groups A and B, respectively. Then,

$$
\begin{aligned}
\pi(\mathbf{y}_p | z_p = z) &= \prod_{q \in \mathscr{I}_p} \pi(\mathbf{y}_q^A, \mathbf{y}_q^B | z_p = z) \\
&= \prod_{q \in \mathscr{I}_p} \int \varphi(\mathbf{y}_q^A, \mathbf{y}_q^B | z_p = z, \Theta_z) \pi(\Theta_z) d\Theta_z \\
&= \prod_{q \in \mathscr{I}_p} \int \prod_{f \in \mathscr{I}_{pq}} \varphi(\mathbf{y}_f^A, \mathbf{y}_f^B | z_p = z, \Theta_z) \pi(\Theta_z) d\Theta_z.
\end{aligned}
$$

where $\pi(\Theta_z)$ denotes the prior distribution of all model parameters for differential expression status z, $\mathscr{I}_p$ and $\mathscr{I}_{pq}$ denote the peptide index set for protein p and fragment index set for peptide q, respectively. Here $\varphi(\cdot)$ denotes the product of all element-wise Gaussian densities, i.e.,

$$\varphi(\mathbf{y}_f^A, \mathbf{y}_f^B | z_p = 1, \Theta_1) = \prod_{g \in \{A, B\}} \prod_{s \in \mathscr{S}_g} \frac{1}{\sigma_q \sqrt{2\pi}} \exp\left\{-\frac{(y_{fs} - \mu_{qg})^2}{2\sigma_q^2}\right\}$$

$$\varphi(\mathbf{y}_f^A, \mathbf{y}_f^B | z_p = 0, \Theta_0) = \prod_{s \in \{\mathscr{S}_A, \mathscr{S}_B\}} \frac{1}{\sigma_q \sqrt{2\pi}} \exp\left\{-\frac{(y_{fs} - \mu_q)^2}{2\sigma_q^2}\right\}$$

where fragment f is from peptide q, $\Theta_1 = \{(\mu_{qA}, \mu_{qB}, \sigma_q^2)\}$ and $\Theta_0 = \{(\mu_q, \sigma_q^2)\}$ in protein p.

The prior distribution for μ_{qg}, the mean intensities in peptide q group $g \in \{A, B\}$ and μ_q, or the mean intensities in peptide q group A and B, is conditional on the variance parameter σ_q^2 and is the Gaussian distribution with mean 0 and variance $(\sigma_q^2 \cdot V)$. The hyperparameter V is set large enough (default value 1000) to render this prior to be least subjective.

$$\mu_{qg}|\sigma_q^2 \sim \mathcal{N}(0, \sigma_q^2 \cdot V)$$

$$\mu_q|\sigma_q^2 \sim \mathcal{N}(0, \sigma_q^2 \cdot V).$$

The prior distribution for σ_q^2, the variance of the all intensities in peptide q group A and B, is the inverse gamma distribution with hyperparameters (a, b). The hyperparameters (a, b) are set to the method of moments estimates of the inverse gamma distribution based on the sample variance calculated assuming equal means across the two groups (i.e., assuming non-differential expression).

Putting all elements of the model together, we can now derive the full closed form expression of conditional marginal likelihoods for $z_p = 0$ and $z_p = 1$. For $z_p = 0$,

$$\pi(\mathbf{y}_q^A, \mathbf{y}_q^B|z_p = 0) = \int_0^\infty \int_{-\infty}^\infty \varphi(\mathbf{y}_q^A, \mathbf{y}_q^B|\mu_q, \sigma_q^2)\varphi(\mu_q|0, \sigma_q^2 V)\mathscr{IG}(\sigma_q^2|a, b)\,d\mu_q d\sigma_q^2$$

$$= \frac{1}{\sqrt{(n_A + n_B)V + 1}} \frac{\Gamma(a + (n_A + n_B)/2)}{\Gamma(a)} \frac{1}{(2\pi)^{(n_A+n_B)/2}}$$

$$\times \frac{b^a}{\left[b + \frac{1}{2}\left(\sum_{y \in \mathbf{y}_q^A, \mathbf{y}_q^B} y^2 - (\frac{1}{n_A+n_B+1/V})(\sum_{y \in \mathbf{y}_q^A, \mathbf{y}_q^B} y)^2\right)\right]^{a+(n_A+n_B)/2}}$$

where $\mathscr{IG}(\cdot)$ denotes the inverse gamma density function. Likewise, the closed form expression for the case $z_p = 1$,

$$\pi(\mathbf{y}_q^A, \mathbf{y}_q^B|z_p = 1)$$

$$= \int_0^\infty \int_{-\infty}^\infty \int_{-\infty}^\infty \varphi(\mathbf{y}_q^A|\mu_{qA}, \sigma_q^2)\varphi(\mathbf{y}_q^B|\mu_{qB}, \sigma_q^2)$$

$$\times \varphi(\mu_{qA}|0, \sigma_q^2 V)\varphi(\mu_{qB}|0, \sigma_q^2 V)\mathscr{IG}(\sigma_q^2|a, b)\,d\mu_{qA}d\mu_{qB}d\sigma_q^2$$

$$= \frac{1}{\sqrt{n_A V + 1}} \frac{1}{\sqrt{n_B V + 1}} \frac{\Gamma(a + (n_A + n_B)/2)}{\Gamma(a)} \frac{1}{(2\pi)^{(n_A+n_B)/2}}$$

$$\times \frac{b^a}{\left[b + \frac{1}{2}\left(SS_{qA} - (\frac{1}{n_A+1/V})(S_{qA})^2 + SS_{qB} - (\frac{1}{n_B+1/V})(S_{qB})^2\right)\right]^{a+(n_A+n_B)/2}},$$

where $SS_{qA} = \sum_{y \in \mathbf{y}_q^A} y^2$, $S_{qA} = \sum_{y \in \mathbf{y}_q^A} y$, $SS_{qB} = \sum_{y \in \mathbf{y}_q^B} y^2$, $S_{qB} = \sum_{y \in \mathbf{y}_q^B} y$, and $n_g = \sum_{y \in \mathbf{y}_q^g} I\{y \text{ observed}\}$ is the number of observed intensities in peptide q group g.

Notice that the probability model for the fragment intensity data can be directly applied to the peptide intensity data or the protein intensity data derived from the fragment intensities. For example, peptide intensity data can be considered as fragment intensity data with a single fragment. Protein intensity can likewise be considered as a data with each protein containing one peptide and one fragment only. All three types of data can be used to produce protein-specific probability score of differential expression, as will be illustrated in Sect. 4.

3.3 Estimation

The model parameters $\Phi = (\gamma, \beta)$ are estimated by the iterative conditional maximization (ICM) algorithm [1]:

1. Obtain an initial estimate $\hat{\mathbf{Z}}$ using simple two-sample t-tests.
2. Estimate Φ by the value $\hat{\Phi}$ maximizing the pseudo-likelihood $\prod_p \pi(z_p|z_{(\partial p)}, \Phi)$.
3. Carry out a single cycle of ICM based on the current values of $\hat{\mathbf{Z}}, \hat{\Theta}, \hat{\Phi}$, to obtain a new $\hat{\mathbf{Z}}$. For $p = 1, \ldots, P$, update z_p so as to maximize

$$\pi(z_p|\mathbf{y}, \hat{z}_{(\Omega/p)}) \propto \left[\prod_{q \in \mathscr{I}_p} \pi(\mathbf{y}_q^A, \mathbf{y}_q^B | z_p, \hat{\Theta}) \right] \pi(z_p | \hat{z}_{(\partial p)}, \hat{\Phi}).$$

4. Repeat step 3 until $\hat{\mathbf{Z}}$ converges.

This estimation is performed for all pairwise comparisons specified by the user and a single set of MRF coefficients is simultaneously applied to all comparisons.

3.4 Scoring of Differentially Expressed Proteins

From this model, the posterior probability of differential expression is calculated for each protein as the statistical significance score in comparing two groups, which is expressed as

$$\hat{s}_p = \pi(z_p = 1|\mathbf{y}, \hat{z}_{(\Omega/p)})$$

$$= \frac{e^{\hat{\gamma}_1 - \hat{\beta} \sum_{k \in \partial p}(1-\hat{z}_k)} \pi(\mathbf{y}_p | z_p = 1, \hat{z}_{(\Omega/p)})}{e^{\hat{\gamma}_1 - \hat{\beta} \sum_{k \in \partial p}(1-\hat{z}_k)} \pi(\mathbf{y}_p | z_p = 1, \hat{z}_{(\Omega/p)}) + e^{\hat{\gamma}_0 - \hat{\beta} \sum_{k \in \partial p} \hat{z}_k} \pi(\mathbf{y}_p | z_p = 0, \hat{z}_{(\Omega/p)})}.$$

If no gene neighbor information is utilized ($\beta = 0$), then

$$\hat{s}_p = \pi(z_p = 1|\mathbf{y}) = \frac{e^{\hat{\gamma}_1} \pi(\mathbf{y}_p | z_p = 1)}{e^{\hat{\gamma}_1} \pi(\mathbf{y}_p | z_p = 1) + e^{\hat{\gamma}_0} \pi(\mathbf{y}_p | z_p = 0)},$$

where $\mathbf{y}_p$ denotes the fragment intensity data for protein p. Using these probability scores, Newton et al. [12] allow computation of the Bayesian FDR associated with each threshold s^*

$$\text{BFDR}(s^*) = \frac{\sum_{\hat{s}_p > s^*} (1 - \hat{s}_p)}{\sum_{\hat{s}_p > s^*} 1}.$$

4 Results

Simulation studies for the evaluation of the proposed model-based method had already been conducted in the original mapDIA manuscript [19]. Here we illustrate the statistical analysis of a recently published renal cell carcinoma (RCC) SWATH-MS data set, in which tumor and non-tumor control biopsy samples were obtained from nine RCC patients. To avoid potential tumor heterogeneity across histomorphological types, we chose to work with six clear cell RCC samples. In this study, each tumor and control sample has been divided into two tubes and analyzed twice as biological replicates. We first obtained the raw fragment intensity data from the authors as processed by OpenSWATH [8], and also separately performed DIA-Umpire extraction using all default parameters. In both sets of data, we verified the correlation between replicates was 0.98 and above in all samples, and we then took average of the fragment intensities for each sample.

4.1 Quality Control of Extracted DIA Data

Before applying the fragment filtering steps in mapDIA, the quantification table from OpenSWATH consisted of 84,225 fragments from 12,346 peptides and 3088 proteins. A majority of peptides had six or fewer fragments due to the extraction parameter setting, and most proteins were quantified by one to twenty peptides. Based on our empirical observations, OpenSWATH with its default parameter setting tends to report fewer fragments than the alternatives such as DIA-Umpire, since the former tool applies scoring steps to select the "best" MS/MS fragments for peptides as many as the user requests.

In the fragment filtering step of mapDIA, we applied two standard deviation threshold for outlier removal and median correlation threshold of 0.2. Since the statistical test will be based on paired tumor-control samples from the same patients, we also required that the fragments be quantified in at least 4 of 6 samples in both conditions (tumor and control). This requirement alone reduced the number of qualifying fragments to 26,231 (4390 peptides, 1570 proteins), indicating serious lack of reproducible quantification across multiple DIA experiments. The outlier and median correlation filter also removed additional fragments, eventually filtering the data down to 11,146 fragments in 3915 peptides and 1422 proteins.

As such, it is necessary to take caution in reading the reported counts of proteins/peptides/fragments in DIA data sets: the tallies are usually before careful data processing steps are applied and minimal reproducibility of quantification is met.

4.2   Comparison of Fragment, Peptide, and Protein Intensity Data

The next question for the downstream analysis is how the intensity data will be used to derive statistical analysis of differential expression. As stated earlier, the common approach is to sum the intensity values from fragments to peptides, and then from peptides to proteins, to have a simple numerical summary of abundance for each protein in each sample. However, such an automated roll-up must be performed after careful selection of fragments. Teo et al. [19] have already shown that the common practice of choosing the top k intense fragments from the raw data is not the most robust way to derive protein-level intensity summary. The reason behind this observation is that individual peptide abundance can be relatively poorly correlated with the overall protein abundance and the most intense peptides tend to contribute to the discrepancy the most.

Figure 3a shows that the fold change estimates between tumors and controls (obtained by mapDIA after fragment selection) are not necessarily consistent when we use the three different intensity data. Here we derived peptide-level intensity values by summing up the intensities from five fragments with the highest intensity values in each sample, regardless of whether each fragment is reproducibly quantified in other samples or not. Similarly, we then derived protein-level intensity values by summing up the intensities from three peptides with the highest intensity values in each sample. As the figure shows, the protein fold change estimates using protein intensity data are poorly correlated with those estimated from the other two levels of data, whereas the fold change estimates from the peptide intensity data and those from the fragment intensity data are highly correlated. This suggests that directly rolling up the fragment intensities to protein-level intensity may incur overestimation of effect sizes.

When we compared the three data sets in terms of mapDIA analysis, the benefit of analyzing fragment-level intensity became even more evident: the presence of *repeated measures* (fragment intensities) of peptides allowed us to detect differentially expressed proteins with smaller magnitude of change (Fig. 3b). Note that this observation is partly due to the fact that the protein fold change is overestimated in protein intensity data. Selecting differentially expressed proteins at the probability threshold 0.94 associated with 1 % BFDR threshold, the analysis using the fragment-level data reported 855 differentially expressed proteins, whereas those from peptide-level and protein-level data reported 307 and 191 proteins, respectively, indicating that there is a significant loss in statistical information (e.g., statistical power). However, this does not necessarily mean that the fragment-level

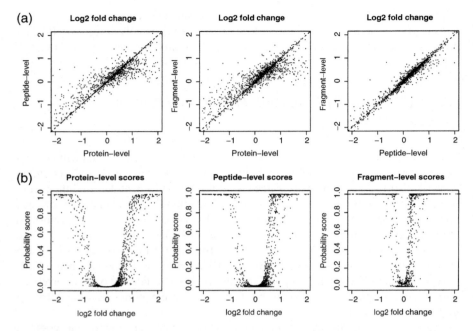

Fig. 3 (**a**) The correlation among fold changes between tumors and corresponding controls using the protein-level summary intensity, peptide-level summary intensity, and fragment-level intensity. The effect size estimates are more consistent between peptide-level summary intensities and fragment-level intensities. (**b**) Probability score against the log2 fold change as reported by mapDIA when using the three different levels of input data

data is always beneficial. As it was illustrated in Teo et al. [19], it is likely that the fragment-level analysis reports a high number of false positives if lenient criteria were applied in the data filtering stage. To prevent inclusion of such false positives, it will be useful to utilize the log posterior odds of differential expression reported by mapDIA, where high positive log odds (e.g., >log(100)) typically indicates reliable significance calls.

4.3 Library-Dependent Extraction Versus Library-Free Extraction

We have also performed library-free data extraction using the DIA-Umpire package on the same data and performed statistical analysis using mapDIA. We used X!Tandem search engine [5] with high mass accuracy parameter setting (MS1 tolerance of 10 ppm, MS2 tolerance of 0.02 Da) in the internal library building of DIA-Umpire, and quantified fragment intensities using all default parameters in the quantification. While DIA-Umpire also screens fragments using an array of quality scores, their default output reported far more fragments in its internal peptide-fragment map so that the fragment selection tool in mapDIA can play a more influential role in the filtering step. DIA-Umpire produced fragment intensity data consisting of 344,319 fragments from 16,064 peptides and 1921 proteins, much greater in volume than the data extracted from OpenSWATH.

Applying the same fragment selection criteria, the number of fragments used for the downstream statistical analysis also reduced significantly in the DIA-Umpire data. Among the total of 344,319 fragments, only 72,691 fragments in 5406 peptides and 1164 proteins met the criterion that the fragment intensity is present in at least four matched tumor-control pairs. The outlier and correlation filter also removed additional fragments from this step, resulting in a filtered list of 15,197 fragments in 5860 peptides and 1060 proteins. In comparison to the OpenSWATH output, this represents fewer proteins yet with a deeper coverage of each peptide.

Given the differences in the input data after filtering, we assumed that the outcome of mapDIA analysis may also differ. At the respective probability score thresholds associated with 1 % FDR, OpenSWATH and DIA-Umpire reported 828/3088 and 507/1921 proteins to be differentially expressed (Fig. 4a). When we limited the comparison to the proteins that passed the outlier and median correlation filter in both data sets, the list of differentially expressed proteins from DIA-Umpire data was almost nested within the list of proteins from OpenSWATH.

To investigate whether the discrepancy comes from the filtered data, we compared the fold change estimates from mapDIA between the two data sets. Interestingly, Fig. 4b shows that the magnitude of fold change tends to be greater in OpenSWATH than in DIA-Umpire, suggesting that the higher number of differentially expressed proteins in the former may come from different quantification method or the quality of library used in the extraction step. Given that the pattern between the posterior probabilities and the fold changes were similar in the mapDIA analysis of both data sets (Fig. 4c, d), we can infer that the fragment selection step was one of the major factors contributing to the differences between the two data extraction methods. However, since the gold standard proteins with validated differential expression between the tumor and the control cells are not known in this data, it is premature to conclude on the relative performance of each extraction method in this particular data set and leave the evaluation of quantification quality and their impact on the downstream statistical analysis to future studies.

5   Conclusion

In this work, we introduced mapDIA as one of the first comprehensive tools for statistical analysis of quantitative proteomics data generated in the DIA mode. We used two independent data extraction tools to obtain quantitative DIA data from a recently published renal cancer proteomics study. We illustrated that not all MS/MS signals reported by the data extraction tools are useful for statistical analysis since

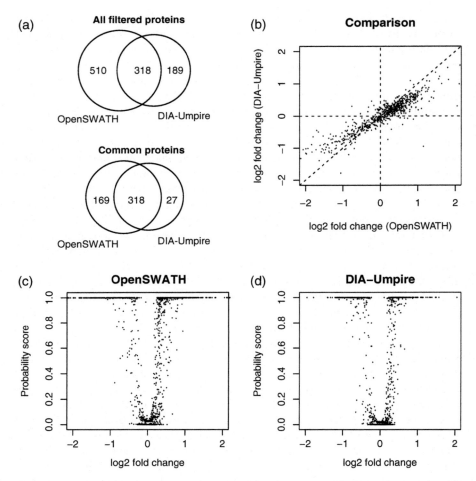

Fig. 4 (a) The number of differentially expressed proteins at 1 % FDR threshold, using all quantified proteins in each data set (*top*) and the proteins quantified in both data sets (*bottom*). (b) Comparison of fold change estimates from mapDIA between the two extraction methods. (c) The posterior probability score against the fold change estimate in mapDIA analysis using the OpenSWATH extracted data. (d) The posterior probability score against the fold change estimate in mapDIA analysis using the DIA-Umpire extracted data

the initial extraction does not guarantee reproducible quantification of the same fragments across the samples even after elaborate attempts of peak alignment. Therefore, the sensitivity of extraction methods, i.e., the ability to quantify a large number of MS/MS fragments, should not necessarily be the sole criterion in the evaluation of quantification methods, and more attention should be directed towards their impact on the downstream statistical analysis.

Related to this issue, we remark that the issue of missing data emerges as one of the most critical challenges to the DIA data analysis, which has been discussed in a number of previous works in the DDA setting [2, 3, 9]. In fragment intensity data, missing data can be addressed in a more systematic manner since the abundance of MS/MS fragments should be correlated within the same peptide and also across different peptides in the same protein. This possibility of pattern-based imputation using correlated fragments should also be considered together with other mechanisms of missing values such as co-eluting precursor ions, transition interference, or variable detection limits in the DIA. In addition, one can also re-mine potentially misaligned low intensity peaks from the raw MS/MS data, which must precede data imputation discussed above. To date, there is no systematic method to address the imputation accounting for the unique structure of DIA data, and we expect to see future development of such methods in this area.

Overall, constantly improving instrumentation for DIA analysis brings us closer towards quantitative proteomics with a genomics-like coverage of the proteome. However, development of rigorous informatics and statistical tools will be a critical prerequisite for reliable interpretation of the data and successful biological application of this technology. In the long run, we expect that the gap between library-dependent and library-free data extraction will diminish as more spectral assay libraries are accumulated and their qualities improve over time. At the moment, however, both approaches seem to be complimentary in this early stage of development, and future refinement of data extraction procedures should benefit from the advantages of competing methods.

References

1. Besag, J. (1986). On the statistical analysis of dirty pictures. *Journal of the Royal Statistical Society: Series B, 48*, 259–302.
2. Choi, H., Kim, S., Fermin, D., Tsou, C. C., & Nesvizhskii, A. I. (2015). QPROT: Statistical method for testing differential expression using protein-level intensity data in label-free quantitative proteomics. *Journal of Proteomics, 129*, 121–126.
3. Clough, T., Key, M., Ott, I., Ragg, S., Schadow, G., & Vitek, O. (2009). Protein quantification in label-free LC-MS experiments. *Journal of Proteome Research, 8*(11), 5275–5284.
4. Collins, B. C., Gillet, L. C., Rosenberger, G., Röst, H. L., Vichalkovski, A., Gstaiger, M., et al. (2013). Quantifying protein interaction dynamics by SWATH mass spectrometry: Application to the 14-3-3 system. *Nature Methods, 10*(12), 1246–1253.
5. Craig, R., & Beavis, R. C. (2003). A method for reducing the time required to match protein sequences with tandem mass spectra. *Rapid Communications in Mass Spectrometry, 17*, 2310–2316.
6. Egertson, J. D., Kuehn, A., Merrihew, G. E., Bateman, N. W., MacLean, B. X., Ting, Y. S., et al. (2013). Multiplexed MS/MS for improved data-independent acquisition. *Nature Methods, 10*, 744–746.
7. Gillet, L. C., Navarro, P., Tate, S., Röst, H. L., Selevsek, N., Reiter, L., et al. (2012). Targeted data extraction of the MS/MS spectra generated by data-independent acquisition: A new concept for consistent and accurate proteome analysis. *Molecular & Cellular Proteomics, 11*(6), O111.016717.

8. Guo, T., Kouvonen, P., Koh, C. C., Gillet, L. C., Wolski, W., Röst, H. L., et al. (2015). Rapid mass spectrometric conversion of tissue biopsy samples into permanent quantitative digital proteome maps. *Nature Medicine, 21*(4), 407–413.
9. Karpievitch, Y., Stanley, J., Taverner, T., Huang, J., Adkins, J. N., Ansong, C., et al. (2009). A statistical framework for protein quantitation in bottom-up MS-based proteomics. *Bioinformatics, 25*(16), 2028–2034.
10. Lambert, J. P., Ivosev, G., Couzens, A. L., Larsen, B., Taipale, M., Lin, Z. Y., et al. (2013). Mapping differential interactomes by affinity purification coupled with data-independent mass spectrometry acquisition. *Nature Methods, 10*(12), 1239–1245.
11. MacLean, B. X., Tomazela, D. M., Shulman, N., Chambers, M., Finney, G. L., Frewen, B., et al. (2010). Skyline: An open source document editor for creating and analyzing targeted proteomics experiments. *Bioinformatics, 26*(7), 966–968.
12. Newton, M. A., Noueiry, A., Sarkar, D., & Ahlquist, P. (2004). Detecting differential gene expression with a semiparametric hierarchical mixture method. *Biostatistics, 5*(2), 155–176.
13. Panchaud, A., Scherl, A., Shaffer, S. A., von Haller, P. D., Kulasekara, H. D., Miller, S. I., et al. (2009). PAcIFIC: How to dive deeper into the proteomics ocean. *Analytical Chemistry, 81*(15), 6481–6488.
14. Prakash, A., Peterman, S., Ahmad, S., Sarracino, D., Frewen, B., Vogelsang, M., et al. (2013). Hybrid data acquisition and processing strategies with increased throughput and selectivity: pSMART analysis for global qualitative and quantitative analysis. *Journal of Proteome Research, 12*, 5415–5430.
15. Röst, H. L., Rosenberger, G., Navarro, P., Gillet, L., Miladinović, S. M., Schubert, O. T., et al. (2014). Openswath enables automated, targeted analysis of data-independent acquisition MS data. *Nature Biotechnology, 32*(3), 219–223.
16. Schubert, O. T., Gillet, L. C., Collins, B. C., Navarro, P., Rosenberger, G., Wolski, W., et al. (2015). Building high-quality assay libraries for targeted analysis of SWATH MS data. *Nature Protocols, 10*(3), 426–441.
17. Silva, J. C., Gorenstein, M. V., Li, G. Z., Vissers, J. P. C., & Geromanos, S. J. (2006). Absolute quantification of proteins by LCMSE: A virtue of parallel ms acquisition. *Molecular & Cellular Proteomics, 5*, 144–156.
18. Steen, H., & Mann, M. (2004). The abc's (and xyz's) of peptide sequencing. *Nature Reviews Molecular Cell Biology, 5*(9), 699–711.
19. Teo, G. S., Kim, S., Tsou, C.-C., Gingras, A.-C., Nesvizhskii, A. I., & Choi, H. (2015). mapDIA: Preprocessing and statistical analysis of quantitative proteomics data from data independent acquisition mass spectrometry. *Journal of Proteomics, 129*, 108–120.
20. Tsou, C.-C., Avtonomov, D., Larsen, B., Tucholska, M., Choi, H., Gingras, A.-C., et al. (2015). DIA-Umpire: Comprehensive computational framework for data independent acquisition proteomics. *Nature Methods, 12*(3), 258–264.
21. Venable, J. D., Dong, M. Q., Wohlschlegel, J., Dillin, A., & Yatesm, J. R. (2004). Automated approach for quantitative analysis of complex peptide mixtures from tandem mass spectra. *Nature Methods, 1*(1), 39–45.
22. Wei, Z., & Li, H. (2007). A Markov random field model for network-based analysis of genomic data. *Bioinformatics, 23*(12), 1537–1544.

The Analysis of Human Serum Albumin Proteoforms Using Compositional Framework

Shripad Sinari, Dobrin Nedelkov, Peter Reaven, and Dean Billheimer

1 Introduction

A relatively small number of genes yield the enormous diversity in the eukaryotes by means of posttranslational modifications of proteins. These modified proteins, also called proteoforms, give rise to new functional capabilities and regulate the cellular environment [13, 27]. They have been implicated in diseases such as cancer [8] and age related dementia [22].

Identifying these proteoforms with sensitivity and specificity has been a challenge, especially when the abundance is low. Mass spectrometry immuno assay (MSIA) is an approach developed to address these challenges. MSIA combines the sensitivity of the immuno assay based approaches with the specificity of detection from mass spectroscopy. Moreover, MSIA allows the simultaneous detection of multiple proteoforms in a single assay. Nelson et al. [18] is a useful reference for the details of this approach.

S. Sinari
BIO5 Institute, The University of Arizona, Tucson, AZ, USA
e-mail: shripad@email.arizona.edu

D. Nedelkov
Molecular Biomarkers Laboratory, Biodesign Institute, Arizona State University, Tempe, AZ, USA
e-mail: dobrin.nedelkov@asu.edu

P. Reaven
Phoenix VA Health Care System, Phoenix, AZ, USA
e-mail: Peter.Reaven@va.gov

D. Billheimer (✉)
Epidemiology and Biostatistics, BIO5 Institute, The University of Arizona, Tucson, AZ, USA
e-mail: dean.billheimer@arizona.edu

© Springer International Publishing Switzerland 2017   141
S. Datta, B.J.A. Mertens (eds.), *Statistical Analysis of Proteomics, Metabolomics, and Lipidomics Data Using Mass Spectrometry*, Frontiers in Probability and the Statistical Sciences, DOI 10.1007/978-3-319-45809-0_8

The use of immuno based assays to enrich for proteoforms imposes a constraint on the measurement of their concentrations. Thus the resulting peak areas capture information on relative abundances of the proteoforms rather than their absolute concentration. Measurements of the *relative abundance* of multiple components is a characteristic of compositional data [1]. In this paper we demonstrate that the compositional data analysis framework is ideally suited to exploring and analyzing MSIA data. Particularly, the framework allows interpretation of complicated covariance structure, guarantees consistency between analyses of a part and the whole composition, and permits the use of standard multivariate statistical methods, all while respecting structural constraints inherent in the observed data.

We begin by exploring the compositional nature of MSIA data in our example of albumin proteoforms. We show how the compositional framework allows for reference free normalization of this data. We explore the association of glycosylated and cysteinylated albumin proteoforms with prognosis of chronic kidney disease (CKD) in patients with Type 2 diabetes mellitus. Finally, we discuss the role of compositional data in our application as well as in genomics and conclude with remarks on the exploratory analysis of albumin proteoforms.

2   MSIA and Albumin Proteoforms

Our MSIA measurements comprise albumin proteoforms from 283 patients with Type 2 diabetes mellitus. Glycosylation [26] and cysteinylation [17] of albumin are two important post translational modifications that have been associated with advanced CKD. Here we explore the association of these proteoforms with CKD.

Table 1 shows a small subset of the raw data. The first column is the sample identifier and the remaining nine columns represent the raw peak areas of the nine albumin post translational modifications.

The most abundant form is called wild type and is denoted by "wt". The cysteinylated proteoforms are annotated with ".cys" and the glycosylated proteoforms with ".gly". The proteoforms annotated with "des" are truncated forms of wild type protein. The data matrix consists of 283 rows and nine columns. Each row being a composition and thus a point in $\mathscr{S}^8$ which is the space of 9 part composition given by

$$\mathscr{S}^8 = \left\{ \mathbf{x} = (x_1, x_2, \ldots, x_9) : x_i > 0 (i = 1, 2, \ldots, 9), \sum_{i=1}^{9} x_i = 1 \right\} \quad (1)$$

This is an 8 dimensional simplex embedded in the 9 dimensional real vector space $\mathbb{R}^9$. Table 2 gives the compositions formed from the data subset shown in Table 1.

Typically additional information may be present that provides clinical status associated with each sample. In our data, we have two additional columns. One gives the CKD status of the patient and the other gives the value of the glomerular

Table 1 Table of raw peak areas of a small subset of the albumin data

ID	des.DA	des.D	des.DA.cys	wt	wt.cys	wt.gly	wt.cys.gly	wt.gly.gly	wt.cys.gly.gly
546101	4969.01	6021.65	3318.04	68552.31	55486.38	27058.15	15544.28	9834.52	4291.44
546103	7272.77	6704.79	8614.81	98730.16	134177.76	35674.16	28190.84	10905.35	5562.96
546104	6589.51	5673.29	8419.23	107413.43	104393.18	40453.52	33830.79	15787.60	10996.31
546105	7119.19	6802.57	8144.94	98650.74	90278.81	29793.74	17440.50	8504.92	3608.09
546106	5880.67	4774.71	6249.13	67389.77	89762.67	25233.69	23333.77	8539.34	5624.68

Table 2 Table of compositions of the small subset of the albumin data

ID	des.DA	des.D	des.DA.cys	wt	wt.cys	wt.gly	wt.cys.gly	wt.gly.gly	wt.cys.gly.gly
546101	0.03	0.03	0.02	0.35	0.28	0.14	0.08	0.05	0.02
546103	0.02	0.02	0.03	0.29	0.40	0.11	0.08	0.03	0.02
546104	0.02	0.02	0.03	0.32	0.31	0.12	0.10	0.05	0.03
546105	0.03	0.03	0.03	0.36	0.33	0.11	0.06	0.03	0.01
546106	0.02	0.02	0.03	0.28	0.38	0.11	0.10	0.04	0.02

Table 3 Table of centered log ratios of the small subset of the albumin data

ID	des.DA	des.D	des.DA.cys	wt	wt.cys	wt.gly	wt.cys.gly	wt.gly.gly	wt.cys.gly.gly
546101	−0.91	−0.72	−1.31	1.71	1.50	0.78	0.23	−0.23	−1.06
546103	−0.97	−1.05	−0.80	1.64	1.95	0.62	0.39	−0.56	−1.23
546104	−1.17	−1.31	−0.92	1.63	1.60	0.65	0.47	−0.29	−0.65
546105	−0.79	−0.83	−0.65	1.84	1.75	0.64	0.11	−0.61	−1.47
546106	−0.91	−1.12	−0.85	1.53	1.82	0.55	0.47	−0.54	−0.95

filtration rate (GFR) for the patient. GFR is used to determine the health of the kidney and classify the patient in one of the three CKD status (low, medium, or high).

The simplex $\mathscr{S}^8$ is a Hilbert space with a metric defined by

$$d(x, y) = \left[\sum_{i=1}^{9} \left(\log \left(\frac{x_i}{g_9(x)} \right) - \log \left(\frac{y_i}{g_9(y)} \right) \right)^2 \right]^{1/2} \tag{2}$$

Here $g_9(x)$ is the geometric mean of vector $x \in \mathscr{S}^8$. Multiple co-ordinate systems exist on this Hilbert space. Details of these co-ordinates as well as their equivalence can be found in Egozcue et al. [11]. In our analysis we will use the centered log ratio transformation which is given by

$$\mathscr{L} : \mathscr{S}^8 \to \mathbb{R}^9$$

$$x \longmapsto \left(\log \left(\frac{x_1}{g_9(x)} \right), \log \left(\frac{x_2}{g_9(x)} \right), \ldots, \log \left(\frac{x_9}{g_9(x)} \right) \right) \tag{3}$$

The centered log ratio allows us to look at all the proteoforms and gives a covariance matrix that is more interpretable than the original composition, and is suitable for exploration using the principal component analysis (PCA). Table 3 gives the centered log ratios of compositions from Table 2. Note that centered log ratios now sum to zero for each sample.

For some proteins a naturally occurring native or highly abundant form exists. In applications where there is such a highly abundant form, it is the ratio with this form that is often of most interest. In such cases, an alternate co-ordinate system named the additive log ratio transform [Eq. (9) in Appendix] may be more useful. Additive log ratio transform may also be useful in parameter estimation in linear models when all proteoforms are included as a multivariate outcome. Standard multivariate methods, as well as multiple regression techniques, can now be applied to these appropriately transformed data.

2.1   Normalization of Proteomic Measurements as Compositions

In addition to providing a convenient structure to apply the usual multivariate analyses methods, the co-ordinate transformations in the compositional setting also perform a normalization of the data. Figure 1 shows that the variability as well as skewness in each variable is reduced by the centered log ratio transform.

When a convenient reference standard exists, it may be included in the MSIA assay [18, 25]. The reference standard is used to determine the *absolute* concentrations of the proteoforms from a calibration curve. Typically, it is a modified version of the protein with a known mass-to-charge ratio. If poorly matched to the

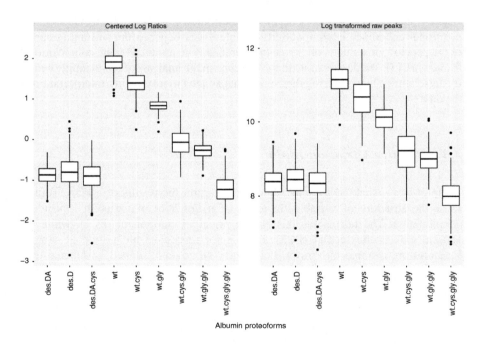

Fig. 1 Box plots of the log transformed raw values and the centered log ratios of albumin proteoforms. The box plots for each proteoform indicate a decrease in overall variability and skewness in the centered log ratios

target protein, however, the reference standard may increase the variability in the calibrated data. For our dataset, no reference standard was used. In this situation, taking compositional nature of the data into account and applying appropriate transformations provides a good normalization scheme with reduction in the total variability of measurements.

2.2 Interpretation of Multivariate Proteoform Analysis

Aitchison (see [2]) discusses the difficulty of interpretation of the PCA on the raw data. In particular, issues arise due to lack of spherically symmetric distributions. This is dealt by the use of logistic normal distribution [Eq. (10) in Appendix]. The centered log ratio transformed composition gives an isotropic invariant covariance structure from which a measure of total variability of a composition can be expressed as

$$\sum_{i=1}^{9} \text{var}\left[\log\left(\frac{x_i}{g_9(x)}\right)\right] \tag{4}$$

and the principal components become orthonormal log linear contrasts. Subcompositional coherence is compatibility of inferences between the full and a subset of the proteoforms. Such coherence in inference is guaranteed by the compositional framework [3]. We demonstrate this coherence in the analysis of our example below. It is also shown how ignoring the constraint can lead to misleading interpretation of the data.

2.3 Relative Variation Biplot

A biplot is a visual aid to understand and interpret the results of a PCA. Biplots show the structure of variables in terms of major axes of variability (principal components). The horizontal axis is first principal component (PC1), while the vertical axis is the second (PC2). Each point represents an individual sample. Variables are denoted by arrows. Points may be colored, shaped, or labeled by a classification or for identification.

A biplot resulting from the PCA of a covariance matrix of variables is called a *relative variation biplot*. Any biplot can be displayed in two forms. One is called the covariance biplot where distances between the variables are approximations of the standard deviations of the corresponding log ratios and angle cosines between links estimate the correlations between log ratios. Links are the difference vectors connecting the tips of the arrows representing the variables. The second is called a form biplot where distance between points are approximation of the distances given by the metric in Eq. (2).

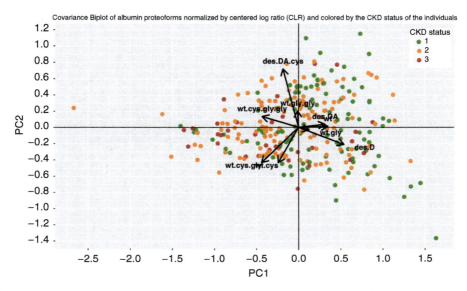

Fig. 2 Covariance biplot of individuals with chronic kidney disease (CKD) as classified from the measurement of their glomerular filtration rate (GFR). The albumin proteoform values are centered log ratio (CLR). The axes are the first two principal components with first component along the *horizontal* and the second along the *vertical axis*. The *points* represent individuals (about 283) whose samples contributed to the MSIA measurements and whose CKD status were known at the start of the study

Figures 2 and 3 are the covariance and the form relative variation biplots, respectively, for the albumin dataset. Points are colored by the CKD status of the individual from whom the sample was obtained. The biplots indicate that the relative proportions of the albumin proteoforms in the sample can distinguish between the higher CKD status (encoded as 3 and color coded as red) and the lowest CKD status (encoded as 1 and color coded as green). The samples with lower CKD status are mostly to the right and those with higher CKD status mostly to the left (see Table 4). The plots also show that higher proportion of wild type albumin is associated with lower CKD status as one would expect. Higher proportions of the cysteinylated versions of albumin proteoforms are associated with poor CKD status. The proportion of variance explained by the covariance and the form biplots are 73 % and 70 %, respectively.

These plots provide an approximation to the covariance structure of the albumin proteoforms. An example is that of the link between the proteoforms wt and des.D in the covariance biplot (Fig. 2). The length of the link is approximately 0.293 whereas the actual standard deviation of the log ratio is 0.283.

Aitchison [4] provides a good introduction and insights into numerous useful properties of the relative variation biplots and proves the equivalence of the biplots under various co-ordinate systems.

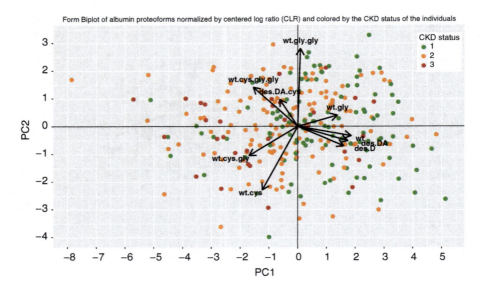

Fig. 3 Form biplot of individuals with chronic kidney disease (CKD) as classified from the measurement of their glomerular filtration rate (GFR). The albumin proteoform values are centered log ratio (CLR). The axes are the first two principal components with first component along the *horizontal* and the second along the *vertical axis*. The *points* represent individuals (about 283) whose samples contributed to the MSIA measurements and whose CKD status were known at the start of the study

Table 4 Table of patients with negative or non-negative loading on first principal component (PC1) by their CKD status

CKD status	PC1 values		
	Negative	Non-negative	Total
1	35 (32 %)	75 (68 %)	110 (39 %)
2	83 (59 %)	58 (41 %)	141 (50 %)
3	22 (69 %)	10 (31 %)	32 (11 %)
Total	140 (49 %)	143 (51 %)	283

The values correspond to Form Biplot (Fig. 3)

The associations seen in the biplots can also be confirmed in the linear regression of the proteoforms with the continuous measurement of CKD status, GFR. Each row of Table 5 gives the coefficients of the linear regression of the proteoform with GFR.

2.4 Results Without the Unit Sum Constraint

We now look at a similar analysis with log transformed raw peaks of the proteoforms. Figure 4 is a form biplot and Table 6 contains the results of regression of the log transformed raw peak areas with GFR.

Table 5 Table of coefficients of linear regression with GFR. The model uses the centered log ratios of the proteoform. Each individual proteoform was regressed against GFR. These are nine separate simple linear regression models each with a single proteoform as the explanatory variable

	Estimate	Std. error	t value	Pr(> |t|)
des.DA	27.34	5.65	4.84	0.00
des.D	14.78	3.45	4.29	0.00
des.DA.cys	−10.41	3.80	−2.74	0.01
wt	27.91	5.26	5.30	0.00
wt.cys	−17.95	4.77	−3.77	0.00
wt.gly	34.51	9.93	3.48	0.00
wt.cys.gly	−16.07	3.94	−4.08	0.00
wt.gly.gly	3.41	6.72	0.51	0.61
wt.cys.gly.gly	−12.77	3.79	−3.37	0.00

Table 6 Table of coefficients of linear regression with GFR. The model uses the log transformed raw peak areas of the proteoform. Each individual proteoform was regressed against GFR. These are nine separate simple linear regression models each with a single proteoform as the explanatory variable

	Estimate	Std. error	t value	Pr(> |t|)
des.DA	10.33	3.52	2.93	0.00
des.D	9.66	2.80	3.45	0.00
des.DA.cys	−6.39	3.03	−2.11	0.04
wt	11.34	3.42	3.31	0.00
wt.cys	−4.53	2.46	−1.84	0.07
wt.gly	5.07	3.79	1.34	0.18
wt.cys.gly	−4.96	2.25	−2.20	0.03
wt.gly.gly	1.53	4.09	0.37	0.71
wt.cys.gly.gly	−6.77	2.81	−2.41	0.02

Although the regression tables indicate similar relationships between each proteoform and GFR, the strength of the association is reduced or deemed insignificant. The picture in the biplots, however, is less clear. In Fig. 4 all proteoforms point to the left indicating association of the total signal of albumin with CKD status. However the association of cysteinylated proteoforms with poor CKD status is lost.

Consistency between the part, i.e., the univariate analysis (Table 5), and the whole composition, i.e., the PCA using all proteoforms (Fig. 3), is evident for analysis done with centered log ratios. The proteoform "wt.gly.gly" does not carry information about the health of the kidney of the patient. This consistency is absent between the univariate analysis (Table 6) and the corresponding PCA (Fig. 4) done with the log transformed raw peaks of the proteoforms.

As seen in this example, consideration of compositional structure brings added insights into the covariance structure of the components and vindicates the use of standard analytical tools. This is consistent with observation in Lovell et al. [15], that use of compositional framework may not lead to dramatically different results across the board but the application of Aitchison distance [Eq. (2)] provides more meaningful insights.

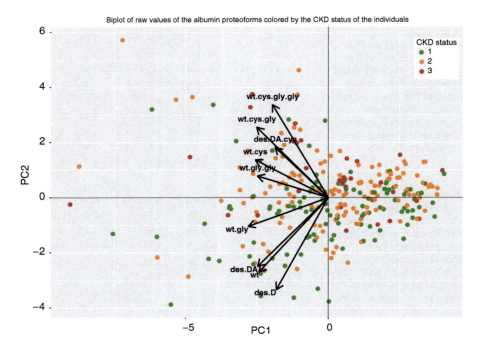

Fig. 4 Form biplot of individuals with chronic kidney disease (CKD) as classified from the measurement of their glomerular filtration rate (GFR). The albumin proteoform values are log transformed raw peak areas. The axes are the first two principal components with first component along the *horizontal* and the second along the *vertical axis*. The *points* represent individuals (about 283) whose samples contributed to the MSIA measurements and whose CKD status were known at the start of the study

3   Discussion

The multivariate exploration of albumin proteoforms highlights the importance of the cysteinylated proteoforms of albumin in the prognosis of diabetic patients with CKD in our data. Such insights are absent from the analysis that does not take the compositional constraint into account. Recently, Borges et al. [7] have shown that cysteinylation of albumin can result from sample storage or handling. In such cases, the consideration of compositional framework can reflect on the quality of the data. Thus such analysis brings about better understanding of the roles of cysteinylated versus glycosylated proteoforms of albumin in the prognosis of CKD or serves to provide a quality check on samples. The compositional framework also provides a convenient and interpretable normalization scheme. The normalization constant is the mean of log transformed components. Hence the name centered log ratio. In the albumin proteoforms this means that the variability such as batch effects due to antibody used for immuno affinity capture is normalized. In general, all non-proteoform specific variability is reduced.

Many widely used high throughput technologies produce similar data that are inherently compositional. Two prominent ones being quantification of gene expression using RNA-Seq and metagenomics, which is the study of the composition of microbial genomes in a sample. In RNA-Seq the counts of mRNA are often reported as a composition. The extraction of mRNA from a fixed volume of starting material puts the unit sum constraint on the counts observed in the experiment. Differential expression of genes between conditions is the question about change in relative abundance with respect to a reference, called a housekeeping gene. It is important to realize that information on absolute abundance is lost in these measurements. Two genes may have the same level of mRNA production but one might be differentially expressed while the other is not. However correlation induced due to the sum constraint can lead to misleading interpretation in case of such data, if the compositional nature is not taken into account. See Lovell et al. [15] for detailed exposition on this issue.

Similarly, applications in metagenomics involve comparison of the compositions within or between different conditions of genetically diverse microorganisms. The compositional structure of microbial community here is more evident. Community composition analysis is employed in diverse applications such as exploring the biodiversity of habitat [23], common pathogens in clinical settings [19] or classification of the microbes into genus [10], and phylogeography [9]. Cell fractionation techniques or size selection similar to proteomics is often used in sample collection to create homogeneous populations of cells and enrichment of the target DNA [24]. These methods impose a compositional constraint. Statistical analysis of such data can benefit from the use of compositional framework [14].

One limitation of compositional approach is worth mentioning. This is the problem of *essential zeros*. Essential zeros arise when zero is valid value for some parts of the composition. This is distinct from the inability to detect a signal due to the signal being lower than the limit of detection. Such below the detection limit zeros are called *rounded zeros* in the compositional literature. An example of essential zero arises in a compositional data consisting of family budgets. Some families may not consume alcohol and hence the money allocated to this expenditure may be zero.

In proteomic applications, zeros are often treated as rounded zeros (e.g., below detection limit). Thus rounded zeros are often replaced by multiplicative strategy. In this strategy, the zeros in a composition are replaced by small non-zero values. To maintain unit sum constraint, the non-zero components are multiplied by a suitable value. In our data set, we replaced the zero values in raw peak areas with half of the lowest non-zero terms for that proteoform, before computing the centered log ratios or the log transformations. A detailed discussion on zeros as well as the several methods of dealing with rounded zeros can be found in Martín-Fernández et al. [16]. An important point to note is that, samples with a part value as zero lie on the edge of the simplex which is excluded from our definition of composition for mathematical convenience. Problems arise in extending the metric as well as the full logistic normal distribution to the edges.

4 Conclusions

The results of the exploratory analysis of albumin data using compositional data framework show that changes in the proportions of the cysteinylated albumin proteoforms can reveal information about the status of the chronic CRD in an individual, or indicate issues with data storage and handling ex vivo. This analysis implies that MSIA assays can be used to explore the clinical role of posttranslational modifications of a protein. Compositional framework is essential in inference related to such relative proportions data. The framework provides for normalization of data and also validate the application of conventional multivariate analysis techniques. It provides for consistency between analysis of the part and the whole composition through the principle of subcompositional coherence. Ignoring the limitation imposed by the summation constraint in these relative proportions data, as is often the case, can result in loss of valuable insights or worse, lead to misleading conclusions.

Acknowledgements The authors would like to thank Dr. Borges for guidance on various mechanisms that can lead to higher proportions of cysteinylated albumin proteoforms in MSIA data. Acknowledgment is also due to National Institutes of Health (NIH) for supporting the work presented in this chapter under awards numbered R24DK090958-01A1 and P30ES006694.

Appendix

A Synopsis on Compositional Framework

Compositional data describe the proportion that each of D components contributes to the whole. Coherence of inference between subset of components and the whole composition, also called subcompositional coherence, is an important feature of the analysis. Scale invariance is essential for such analysis.

A D part composition is defined as an element of the d-dimensional positive simplex

$$\mathscr{S}^d = \left\{ \mathbf{x} = (x_1, x_2, \ldots, x_D) : x_i > 0 (i = 1, 2, \ldots, D), \sum_{i=1}^{D} x_i = 1 \right\} \quad (5)$$

where $d = D - 1$. We will use the convention $d = D - 1$ in the rest of Appendix. Thus the sample space is a subset of the space $\mathscr{S}^d$.

Let C denote the *closure* operator on $\mathbb{R}^D$ which normalizes the vector to a unit sum. That is, for $z \in \mathbb{R}^D$,

$$C(z) = \left(\frac{z_1}{\sum_{i=1}^{D} z_i}, \frac{z_2}{\sum_{i=1}^{D} z_i}, \ldots, \frac{z_D}{\sum_{i=1}^{D} z_i} \right) \in \mathscr{S}^d \quad (6)$$

Also, we define two additional operations. For any two elements $x = (x_1, x_2, \ldots, x_D)$ and $y = (y_1, y_2, \ldots, y_D) \in \mathscr{S}^d$ and for $\alpha \in \mathbb{R}$, we define

$$x \oplus y = C((x_1 \cdot y_1, x_2 \cdot y_2, \ldots, x_D \cdot y_D)) \tag{7}$$

$$\alpha \odot x = C((x_1^\alpha, x_2^\alpha, \ldots, x_D^\alpha)). \tag{8}$$

The operations in Eqs. (7) and (8) are called the perturbation and power operators, respectively. With perturbation operator as addition and power operator as the scalar multiplication, $\mathscr{S}^d$ acquires the structure of a d-dimensional Hilbert space [6, 21] with a metric given by (2).

The additive log ratio transform defined as:

$$\phi : \mathscr{S}^d \to \mathbb{R}^d$$

$$x \longmapsto \left(\log\left(\frac{x_1}{x_D}\right), \log\left(\frac{x_2}{x_D}\right), \ldots, \log\left(\frac{x_d}{x_D}\right) \right) \tag{9}$$

and the centered log ratio (3) are alternative co-ordinate systems on this space.

The dependence structure induced by the unit sum (compositional) constraint is often addressed by using the class of logistic normal distributions as appropriate models of the data. For illustration, consider an element $x = (x_1, x_2, \ldots, x_D) \in \mathscr{S}^d$. Following Aitchison [1] the pullback of the multivariate normal distribution using ϕ from Eq. (9) is the logistic normal density function given by:

$$f(x|\mu, \Sigma) = (2\pi)^{d/2} |\Sigma|^{-1/2} \left(\frac{1}{\prod_{i=1}^k x_i} \right) \exp\left[-\frac{1}{2} (\phi(x) - \mu)' \Sigma^{-1} (\phi(x) - \mu) \right] \tag{10}$$

where μ is the location parameter in $\mathbb{R}^d$ and Σ is the $d \times d$ variance–covariance matrix, $(\prod_{i=1}^D x_i)^{-1}$ is the Jacobian of the transformation. In the following, we will denote this d-dimensional logistic normal distribution by $\mathscr{LN}_d$ and A' will denote the transpose if A is a matrix.

The part S composition is orthogonal projection of the full composition, with respect to the inner product on $\mathscr{S}^d$ [12]. The class preserving property of logistic normal distributions, i.e., if $x \in \mathscr{LN}_D(\mu, \Sigma)$ and A is an $n \times D$ matrix, then $Ax \in \mathscr{LN}_n(A\mu, A\Sigma A')$ [5], ensures that the orthogonal projections satisfy the property of subcompositional coherence.

A comprehensive review of analytical techniques and applications of compositional framework is available in the book Pawlowsky-Glahn et al. [20].

References

1. Aitchison, J. (1982). The statistical analysis of compositional data. *Journal of the Royal Statistical Society. Series B (Methodological)*, 139–177. http://www.jstor.org/stable/10.2307/2345821.
2. Aitchison, J. (1983). Principal component analysis of compositional data. *Biometrika, 70*(1), 57–65. http://biomet.oxfordjournals.org/content/70/1/57.short.
3. Aitchison, J. (2001). Simplicial inference. In M. A. G. Viana & D. S. P. Richards (Eds.), *Algebraic methods in statistics and probability*. Contemporary mathematics (Vol. 287). Providence, RI: American Mathematical Society. doi:10.1090/conm/287. http://www.ams.org/conm/287/.
4. Aitchison, J., & Greenacre, M. (2002). Biplots of compositional data. *Applied Statistics, 51*(4), 375–392. doi:10.1111/1467-9876.00275. http://dx.doi.org/10.1111/1467-9876.00275.
5. Aitchison, J., & Shen, S. M. (1980). Logistic-normal distributions: Some properties and uses. *Biometrika, 67*(2), 261–272. http://biomet.oxfordjournals.org/content/67/2/261.short.
6. Billheimer, D., Guttorp, P., & Fagan, W. F. (2001). Statistical interpretation of species composition. *Journal of the American Statistical Association, 96*(456), 1205–1214. doi:10.1198/016214501753381850. http://www.tandfonline.com/doi/abs/10.1198/016214501753381850.
7. Borges, C. R., Rehder, D. S., Jensen, S., Schaab, M. R., Sherma, N. D., Yassine, H., et al. (2014). Elevated plasma albumin and apolipoprotein A-I oxidation under suboptimal specimen storage conditions. *Molecular & Cellular Proteomics: MCP, 13*(7), 1890–1899. doi:10.1074/mcp.M114.038455. http://www.mcponline.org/cgi/doi/10.1074/mcp.M114.038455.
8. Chammas, R., Sonnenburg, J. L., Watson, N. E., Tai, T., Farquhar, M. G., Varki, N. M., et al. (1999). De-N-acetyl-gangliosides in humans: Unusual subcellular distribution of a novel tumor antigen. *Cancer Research, 59*(6), 1337–1346. http://cancerres.aacrjournals.org/content/59/6/1337.full.
9. Chanturia, G., Birdsell, D. N., Kekelidze, M., Zhgenti, E., Babuadze, G., Tsertsvadze, N., et al. (2011). Phylogeography of Francisella tularensis subspecies holarctica from the country of Georgia. *BMC Microbiology, 11*(1), 139. doi:10.1186/1471-2180-11-139. http://eutils.ncbi.nlm.nih.gov/entrez/eutils/elink.fcgi?dbfrom=pubmed&id=21682874&retmode=ref&cmd=prlinks.
10. Consortium, T. H. M. P. (2012). A framework for human microbiome research. *Nature, 486*(7402), 215–221. doi:10.1038/nature11209. http://dx.doi.org/10.1038/nature11209.
11. Egozcue, J. J., & Barcelo-Vidal, C. (2011). Elements of simplicial linear algebra and geometry. In V. Pawlowsky-Glahn & A. Buccianti (Eds.), *Compositional data analysis* (pp. 141–156). New York: Wiley. doi:10.1002/9781119976462.ch4. http://dx.doi.org/10.1002/9781119976462.ch4.
12. Egozcue, J. J., & Pawlowsky-Glahn, V. (2006). Simplicial geometry for compositional data. *Geological Society, London, Special Publications, 264*(1), 145–159. http://sp.lyellcollection.org/content/264/1/145.short.
13. Karve, T. M., & Cheema, A. K. (2011). Small changes huge impact: The role of protein posttranslational modifications in cellular homeostasis and disease. *Journal of Amino Acids, 2011*(2), 1–13. doi:10.4061/2011/207691. http://www.hindawi.com/journals/jaa/2011/207691/.
14. Li, H. (2015). Microbiome, metagenomics, and high-dimensional compositional data analysis. *Annual Review of Statistics and Its Application, 2*(1), 73–94. doi:10.1146/annurev-statistics-010814-020351. http://dx.doi.org/10.1146/annurev-statistics-010814-020351.
15. Lovell, D., Müller, W., Taylor, J., & Zwart, A. (2011). Proportions, percentages, ppm: Do the molecular biosciences treat compositional data right. In: V. Pawlowsky-Glahn & A. Buccianti (Eds.), *Compositional data analysis*. New York: Wiley. http://books.google.com/books?hl=en&lr=&id=Ggpj3QeDoKQC&oi=fnd&pg=PT215&dq=Proportions+Percentages+PPM+Do+the+Molecular+BiosciencesTreat+Compositional+Data+Right&ots=cII3kxnfSb&sig=icwOFojg2zPXj2WPUj9IQ2K4MCk.

16. Martín-Fernández, J. A., Palarea-Albaladejo, J., & Olea, R. A. (2011). Dealing with zeros. In: V. Pawlowsky-Glahn & A. Buccianti (Eds.), *Compositional data analysis* (pp. 43–58). New York: Wiley. doi:10.1002/9781119976462.ch4. http://dx.doi.org/10.1002/9781119976462. ch4.
17. Nagumo, K., Tanaka, M., Chuang, V. T. G., Setoyama, H., Watanabe, H., Yamada, N., et al. (2014). Cys34-cysteinylated human serum albumin is a sensitive plasma marker in oxidative stress-related chronic diseases. *PloS One, 9*(1), e85,216–9. doi:10.1371/journal.pone.0085216. http://dx.plos.org/10.1371/journal.pone.0085216.
18. Nelson, R. W., Krone, J. R., Bieber, A. L., & Williams, P. (1995). Mass-spectrometric immunoassay. *Analytical Chemistry, 67*(7), 1153–1158. doi:10.1021/ac00103a003. http:// pubs.acs.org/doi/abs/10.1021/ac00103a003.
19. Pallen, M. J. (2014). Diagnostic metagenomics: Potential applications to bacterial, viral and parasitic infections. *Parasitology, 141*(14), 1856–1862. doi:10.1017/S0031182014000134. http://www.journals.cambridge.org/abstract_S0031182014000134.
20. Pawlowsky-Glahn, V., & Buccianti, A. (2011). *Compositional data analysis*. Theory and Applications. New York: Wiley. http://books.google.com/books?id=Ggpj3QeDoKQC& printsec=frontcover&dq=intitle:Compositional+Data+Analysis+Theory+and+Applications& hl=&cd=1&source=gbs_api.
21. Pawlowsky-Glahn, V., & Egozcue, J. J. (2001). Geometric approach to statistical analysis on the simplex. *Stochastic Environmental Research and Risk Assessment, 15*(5), 384–398. doi:10.1007/s004770100077. http://link.springer.com/10.1007/s004770100077.
22. Peleg, S., Sananbenesi, F., Zovoilis, A., Burkhardt, S., Bahari-Javan, S., Agis-Balboa, R. C., et al. (2010). Altered histone acetylation is associated with age-dependent memory impairment in mice. *Science, 328*(5979), 753–756. doi:10.1126/science.1186088. http://www.sciencemag. org/content/328/5979/753.full.
23. Teeling, H., & Glockner, F. O. (2012). Current opportunities and challenges in micro-bial metagenome analysis–a bioinformatic perspective. *Briefings in Bioinformatics, 13*(6), 728–742. doi:10.1093/bib/bbs039. http://bib.oxfordjournals.org/cgi/doi/10.1093/bib/bbs039.
24. Thomas, T., Gilbert, J., & Meyer, F. (2012). Metagenomics - a guide from sampling to data analysis. *Microbial Informatics and Experimentation, 2*(1), 3. doi:10.1186/2042-5783-2-3. http://www.microbialinformaticsj.com/content/2/1/3.
25. Trenchevska, O., Schaab, M. R., Nelson, R. W., & Nedelkov, D. (2015). Development of multiplex mass spectrometric immunoassay for detection and quantification of apolipoproteins C-I, C-II, C-III and their proteoforms. *Methods*, 1–7. doi:10.1016/j.ymeth.2015.02.020. http:// dx.doi.org/10.1016/j.ymeth.2015.02.020.
26. Vos, F. E., Schollum, J. B., & Walker, R. J. (2011). Glycated albumin is the preferred marker for assessing glycaemic control in advanced chronic kidney disease. *Clinical Kidney Journal, 4*(6), 368–375. doi:10.1093/ndtplus/sfr140. http://ckj.oxfordjournals.org/cgi/doi/10.1093/ndtplus/ sfr140.
27. Walsh, C. T., & Tsodikova, S. G. (2005). Protein posttranslational modifications: The chemistry of proteome diversifications. *Angewandte Chemie International Edition in English*. http:// onlinelibrary.wiley.com/doi/10.1002/anie.200501023/full.

Variability Assessment of Label-Free LC-MS Experiments for Difference Detection

Yi Zhao, Tsung-Heng Tsai, Cristina Di Poto, Lewis K. Pannell, Mahlet G. Tadesse, and Habtom W. Ressom

1 Introduction

Mass spectrometry is an analytical technique that allows the measurement of the mass-to-charge ratio of ions present in a sample. It is often coupled to chromatographic techniques, such as liquid or gas (LC-MS or GC-MS), resulting into effective approaches for the measurement of biomolecular abundance changes, across biological samples [11]. In this chapter, we focus on variability assessment of label-free experiments conducted by LC-MS to improve the detection of true differences in ion abundance. Label-free LC-MS methods measure relative abundance of biomolecules without the use of stable isotope labeling [13]. However, they require a rigorous workflow to detect differential abundances. In addition to analytical considerations, crucial steps include: (1) an experimental design that reduces bias during data acquisition and enables effective utilization of available resources [12]; (2) a data preprocessing pipeline that extracts meaningful features [9]; and (3) a statistical test that identifies significant changes taking into account the experimental design [2, 10]. Specifically, we focus on experimental design and statistical tests. Good experimental design provides an opportunity to process and compare samples in an unbiased manner. This benefit can diminish if the data analyst fails to conduct

Y. Zhao • T.-H. Tsai • C. Di Poto • H.W. Ressom (✉)
Lombardi Comprehensive Cancer Center, Georgetown University, Washington, DC, USA
e-mail: hwr@georgetown.edu

L.K. Pannell
Mitchell Cancer Institute, University of South Alabama, Mobile, AL, USA
e-mail: lpannell@health.southalabama.edu

M.G. Tadesse
Department of Mathematics and Statistics, Georgetown University, Washington, DC, USA
e-mail: mgt26@georgetown.edu

© Springer International Publishing Switzerland 2017
S. Datta, B.J.A. Mertens (eds.), *Statistical Analysis of Proteomics, Metabolomics, and Lipidomics Data Using Mass Spectrometry*, Frontiers in Probability and the Statistical Sciences, DOI 10.1007/978-3-319-45809-0_9

the subsequent statistical tests in accordance with the experimental design. For example, t-test is commonly applied for detecting differences between groups; however, the independence assumption is not valid in a study using multiple analytical and/or technical replicates and could lead to false positives.

Variability assessment is a key component in both the design of experiments and the evaluation of hypothesis tests. It provides guidelines for replication assignment, sample size calculation, and identification of significant differences in statistical tests. In this chapter, we investigate the sources of variability arising in the analysis of LC-MS data derived from label-free experiments focusing on peak-level assessment of variability. Mixed effects models can accurately reflect the variability in peak intensities of LC-MS data, and overcome the drawbacks that the t-test suffers from, due to its failure to account for the dependence structure.

The remaining part of this chapter is organized as follows. In Sect. 2, two commonly used peak-level methods for difference detection, i.e., a marginal t-test and a mixed effects model, are introduced. The partition of the total variance into different sources of variability under the mixed effects model is discussed. Section 3 explains the experimental design issues including replication assignment and sample size calculation. Data from a pilot label-free LC-MS experiment are presented in Sect. 4 to compare the performance of the marginal t-test against the mixed effects model and to demonstrate the design of experiment based on the results of the pilot study. LC-MS data from a biomarker discovery study are analyzed using the introduced peak-level mixed effects model and the marginal t-test. The results demonstrate the benefits of conducting a pilot experiment before a large-scale study for appropriate assignment of replicates and calculation of the required sample size. Other statistical methods for biomarker candidate discovery are discussed in Sect. 5. Section 6 concludes the chapter with some discussion.

2   Statistical Methods for Difference Detection

In an LC-MS experiment for discovery of candidate biomarkers, an important step is to identify significant differences in peak intensities between biological populations. In a case-control design, for example, the differential analysis is performed by assessing the evidence in the data against the null hypothesis that the population means of the pth peak are the same. (i.e., $H_0^{(p)} : \mu_1^{(p)} = \mu_2^{(p)}$, where $\mu_i^{(p)}$ is the pth peak population mean of group i, $i = 1, 2$. To keep the following discussion uncluttered, we drop the peak index p hereafter.) Since the tests are performed one at a time for multiple peaks, multiple testing correction needs to be applied to control the false discovery rate (FDR) [1].

Given samples from different biological conditions/groups (biological replicates), multiple aliquots of the same sample (analytical replicates), and multiple injections of the same aliquot into the LC-MS (technical replicates) may be considered in order to estimate the various sources of measurement uncertainty and adequately evaluate the biological differences. A logarithm transformation is usually applied to the peak intensities to ensure that the normality assumption is satisfied [12]. In general, let y_{ijkl} denote the log-transformed intensity of a peak for the lth technical replicate and kth analytical replicate in the jth biological sample of group i.

2.1 A Marginal t-Test Approach

A naive way of performing difference detection is to apply a marginal t-test. For each biological sample, the average over analytical and technical replicates is calculated first, i.e.,

$$\bar{y}_{ij\cdot\cdot} = \frac{1}{bc} \sum_{k=1}^{b} \sum_{l=1}^{c} y_{ijkl}, \quad i = 1, 2, \quad j = 1, \ldots, a_i, \tag{1}$$

where a_i is the number of biological samples in group i, b is the number of analytical replicates of each biological sample, and c is the number of technical replicates of each analytical replicate. Then, a two-sample t-test with unequal variance is performed to compare the two biological groups using the a_1 and a_2 averaged values from each group.

This marginal t-test approach is easy to implement. However, it is less robust compared to the mixed effects models introduced in Sect. 2.2, which takes into account the dependence structure. In large-scale experiments, samples are usually divided into multiple batches or sub-experiments. A marginal t-test approach can only be applied within each batch/sub-experiment, potentially leading to a loss of statistical power since smaller sample sizes are considered.

2.2 Mixed Effects Models

For an experimental design with multiple analytical and/or technical replicates, the peak intensities of the same biological sample are correlated. The variance of the peak intensities is a combination of the deviation due to biological differences, analytical variations, and technical variations. A peak-level mixed effects model with these three variance components is given by

$$y_{ijkl} = \mu_i + \epsilon_{j(i)} + \epsilon_{k(ij)} + \epsilon_{l(ijk)}, \tag{2}$$

where $\mu_i = \mu + G_i$, μ is the average of the peak over all populations, G_i ($i = 1, \ldots, g$) are the group effects, such that $\sum_i a_i G_i = 0$, a_i is the number of subjects in group i; $\epsilon_{j(i)}$, $\epsilon_{k(ij)}$, and $\epsilon_{l(ijk)}$ are assumed to be normally distributed with zero mean and variances σ_α^2, σ_β^2, and σ_γ^2, respectively. All the random effects are assumed to be mutually independent.

Under model (2), the measurements from different biological subjects j and j' ($j \neq j'$) are independent and the correlation between measurements y_{ijkl} and $y_{ij'kl}$ is 0,

$$\rho(y_{ijkl}, y_{ij'kl}) = \frac{\text{cov}(y_{ijkl}, y_{ij'kl})}{\sqrt{\text{var}(y_{ijkl}) \cdot \text{var}(y_{ij'kl})}} = 0. \tag{3}$$

For different analytical replicates k and k' from the same biological subject j, the measurements are dependent with correlation coefficient

$$\rho(y_{ijkl}, y_{ijk'l}) = \frac{\text{cov}(y_{ijkl}, y_{ijk'l})}{\sqrt{\text{var}(y_{ijkl}) \cdot \text{var}(y_{ijk'l})}} = \frac{\sigma_\alpha^2}{\sigma_\alpha^2 + \sigma_\beta^2 + \sigma_\gamma^2}. \tag{4}$$

For different technical replicates l and l' from the same kth analytical replicate of subject j, the correlation is stronger and is given by

$$\rho(y_{ijkl}, y_{ijkl'}) = \frac{\text{cov}(y_{ijkl}, y_{ijkl'})}{\sqrt{\text{var}(y_{ijkl}) \cdot \text{var}(y_{ijkl'})}} = \frac{\sigma_\alpha^2 + \sigma_\beta^2}{\sigma_\alpha^2 + \sigma_\beta^2 + \sigma_\gamma^2}. \tag{5}$$

Therefore, the mixed effects model (2) captures the dependence between analytical and technical replicates, and explicitly reflects the sources of variation in the data.

Under the mixed effects model, when the number of analytical replicates b and the number of technical replicates c are the same for all biological samples, the variance of the mean difference between the two biological groups is

$$\text{var}(\bar{y}_{1\ldots} - \bar{y}_{2\ldots}) = \sum_{i=1}^{2} \left(\frac{\sigma_\alpha^2}{a_i} + \frac{\sigma_\beta^2}{a_i b} + \frac{\sigma_\gamma^2}{a_i bc} \right). \tag{6}$$

From Eq. (6), we see that as the number of replicates increases, the variance gets smaller and the difference between the two groups may be detected more easily. We should also note that the sample size a_i for each biological group has the most influence on the variance. If we can estimate the individual components of the variance, i.e., σ_α^2, σ_β^2, and σ_γ^2, we can design the allocation of replicates more effectively according to the estimated variances and levels of sensitivity and specificity we wish to reach. We elaborate these considerations when discussing experimental design in Sect. 3.

The mixed effects model (2) does not require the number of analytical and/or technical replicates to be identical for all the subjects. When only one analytical and technical replicate is generated ($b = c = 1$), the mixed effects model (2) is equivalent to an analysis of variance (ANOVA) model, since the variance components cannot be decomposed with a single observation. For studies with multiple batches or sub-experiments, a mixed effects model with fixed batch effects can be considered. For more complex designs, the mixed effects model can be easily modified accordingly [3]. Model (2) is a peak-level model, but we can also utilize the ion annotation information and instead of averaging the peaks with identical identification, consider a mixed effects model with one more hierarchy that captures the within-peak variability.

3 Replication Assignment and Sample Size Calculation

An important application of variability assessment is to offer better experimental design for follow-up experiments. One important consideration in experimental design is the estimation of the optimal number of biological, as well as analytical and/or technical replications to achieve a balance between the desired statistical power while controlling for multiple testing and the cost and time of running assays. A pilot experiment is often conducted to obtain estimates for effect size, the number of peaks considered, as well as the variation in biological heterogeneity, sample preparation, and instrument measurement. Since multiple testing issues arise in the analysis of LC-MS data, the required sample sizes are estimated controlling the FDR. It has been shown that controlling the FDR at level q amounts to controlling the average type I error rate α^* over all features in the experiment at

$$\alpha^* \le (1 - \beta^*) \frac{q}{1 + (1 - q)m_0/m_1},\tag{7}$$

where $(1 - \beta^*)$ is the average power over all features, m_0 and m_1 are the numbers of truly non-different and different features, respectively [1]. The number of biological samples in each group under a balanced design can be estimated by

$$a = 2 \left(\frac{z_{1-\beta^*} + z_{1-\alpha^*/2}}{\delta \sqrt{\theta}} \right)^2,\tag{8}$$

where z_α is the αth percentile of the standard normal distributions; $\delta = |\mu_1 - \mu_2|/\sqrt{\sigma_\alpha^2 + \sigma_\beta^2 + \sigma_\gamma^2}$ is the absolute value of effect size; $\theta = (\sigma_\alpha^2 + \sigma_\beta^2 + \sigma_\gamma^2)/(\sigma_\alpha^2 + \sigma_\beta^2/b + \sigma_\gamma^2/bc)$ with σ_α^2, σ_β^2, and σ_γ^2 the variances of the random effects in model (2). When $b = c = 1$, i.e., only one measurement for each biological subject, this sample size calculation is equivalent to the commonly used two-sample t-test sample size calculation.

4 Real Data Example

In this section, we first apply the mixed effects model (2) to an in-house pilot LC-MS dataset and obtain estimates for the variance components. Experimental design is discussed and the performances of the marginal t-test and the mixed effects model are compared through simulation studies based on the estimates from the pilot study. Finally, we illustrate an application on LC-MS data from a hepatocellular carcinoma (HCC) biomarker discovery study.

4.1 In-House Pilot LC-MS Dataset

The in-house pilot LC-MS experiment is designed to evaluate the sample preparation protocol, where high abundance proteins in human serum were depleted using MARS Hu-14 column (Agilent Technologies) and a mixture of peptides from five non-human proteins were spiked-in [20]. LC-MS/MS data were acquired from five biological samples with three analytical and two technical replicates using an Agilent 1200 nano-LC coupled to a ThermoFisher LTQ-Orbitrap mass spectrometer. We used DifProWare (in-house computational tool developed by Dr. Pannell at the University of South Alabama, Mobile, USA) to perform LC-MS data preprocessing including deisotoping of mass spectra, peak detection, and charge state deconvolution. Each LC-MS run was preprocessed separately. Peak detection was performed on the basis of LC-MS data without using the MS/MS spectra.

The mixed effects model (2) is applied to estimate the variabilities introduced by sample heterogeneity, sample preparation, and instrument measurement error, where a single biological group is considered (i.e., $g = 1$), $\mu_i = \mu$ is the overall mean value, and $a_1 = 5, b = 3, c = 2$. Figure 1 shows the estimated variance components for all the 1082 detected peaks in the log-transformed data and Table 1 provides the related summary statistics. Restricted maximum likelihood (REML) estimation is

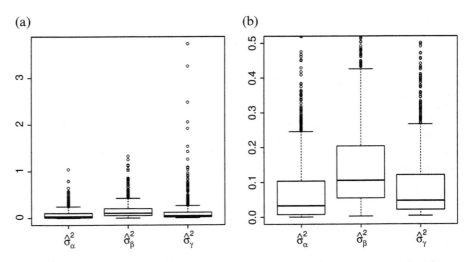

Fig. 1 Boxplot of variability assessment for the pilot LC-MS dataset. (**a**) Estimated variance components. (**b**) Estimated variance components zooming in range $(0, 0.5)$

Table 1 Variability assessment for the pilot LC-MS dataset

	Median	Q1	Q3	IQR
$\hat{\sigma}_\alpha^2$	0.0331	0.0076	0.1038	0.0962
$\hat{\sigma}_\beta^2$	0.1057	0.0551	0.2037	0.1486
$\hat{\sigma}_\gamma^2$	0.0479	0.0219	0.1212	0.0994

"Q1", "Q3" and "IQR" represent the first, third quartiles and interquartile range, respectively

used to ensure that the estimates are both non-negative and unbiased. We note that σ_β^2 has the highest median estimate with a value of 0.1057 and shows the greatest variability. The median estimate of σ_γ^2 (0.0479) is slightly higher than that of σ_α^2 (0.0331), while the IQRs are very close (shown in Table 1). The estimation results indicate that the sample preparation introduces the most variability in this pilot dataset. Here, we want to emphasize that the proportions of the variance components may vary in different LC-MS experiments. For example, in our in-house dataset, the largest proportion of total variation is due to the analytical variance, while in the LC-MS dataset from Oberg and Vitek's study [12], the technical variance accounts for the largest proportion of variability.

4.2 Comparison of Methods for Difference Detection

In this section, we compare through simulation studies the performance in detecting differences using the t-test on the averages over analytical and/or technical replicates versus the mixed effects model. The model parameters are based on the results from the pilot LC-MS study (Sect. 4.1). The sample vectors $y_i = (y_{i111}, \ldots, y_{i11c}, \ldots, y_{i1bc}, \ldots, y_{ia_ibc})^\top$ ($i = 1, 2$) are generated from a normal distribution with mean vector μ_i and variance covariance matrix $\Sigma_i = (\sigma_{st})$, where $\sigma_{st} = \sigma_\alpha^2 + \sigma_\beta^2 + \sigma_\gamma^2$ if $s = t$; $\sigma_{st} = \sigma_\alpha^2 + \sigma_\beta^2$ if the sth and tth components in y_i are from the same analytical replicate of the same biological subject; $\sigma_{st} = \sigma_\alpha^2$ if the sth and tth components in y_i are from the same biological subject but different analytical replicates; and $\sigma_{st} = 0$ if the sth and tth components in y_i are from different biological samples. The variance component values are set to their corresponding median estimates in the in-house pilot study. The effect sizes (δ's) are generated from a sequence in the range from zero to 1.5, and the absolute difference between the group means is calculated as $|\mu_1 - \mu_2| = \delta \sqrt{\sigma_\alpha^2 + \sigma_\beta^2 + \sigma_\gamma^2}$.

Single Peak Difference Detection

We first compare the two methods in identifying the difference between two biological populations under the single peak setting. The simulation is repeated

20,000 times. The achieved Type I error rate and power for the test statistics under each of the two models are compared at a nominal significance level of 0.05 using different replication assignments (Table 2).

When the number of biological samples in each group is small, the marginal t-test is slightly conservative with Type I error rate lower than the designated significance level. The mixed effects model, on the other hand, can effectively control the Type I error rate around the specified level of 0.05, even for sample size as small as five. In addition, the mixed effects model attains higher power compared to the marginal t-test. As the number of biological subjects increases, the performance of the marginal t-test gets closer to that of the mixed effects model in terms of both Type I error and statistical power. When comparing different analytical and/or technical replication assignments, increasing the number of analytical replicates leads to a greater gain in statistical power as the sample preparation accounts for more variability than the instrumental variations in our pilot LC-MS study.

Difference Detection with Multiple Testing

To investigate the performances of the marginal t-test and the mixed effects model in the context of multiple testing, the same number of peaks (1082 peaks) as in the pilot LC-MS dataset is generated. Their variances are set to the median estimates from the pilot study. Among these 1082 peaks, 108 peaks are considered as differentially expressed across biological groups with the same effect sizes. The simulation is repeated 200 times. The FDR achieved by the two methods and their receiver operating characteristic (ROC) curves are compared.

Figure 2 compares the ROC curves for an effect size of 0.5. For both the marginal t-test and the mixed effects model, with the same number of biological subjects, increasing the number of analytical replicates provides better performances compared to increasing the number of technical replicates (Fig. 2a, b). We also notice that the improvement from two analytical replicates to three is not as significant as that from one analytical replicate to two, suggesting that with a similar variability structure, running two analytical replicates could be sufficient in practice. This provides a practical guideline for experimental design in how to balance the number of injections to achieve a desired statistical power while controlling the cost and time of running samples. Figure 2c shows that the most efficient way to achieve better performance in identifying differentially distributed ions is to increase the number of biological subjects in each group. Similar to the conclusion in single peak difference detection (Section "Single Peak Difference Detection"), the marginal t-test gives slightly lower area under the curve for small sample size, while this difference diminishes as the number of biological subjects increases.

The achieved FDR for designated levels q for an effect size of 0.5 are compared in Fig. 3. With five biological subjects in each group, the FDR is well controlled under each replication assignment setting for the mixed effects model. The marginal t-test,

Table 2 Attained Type I error rate and power for single peak difference detection at a nominal significance level of 0.05 under different replication assignments

	Model	$\alpha = 0.05$ Type I error	Effect size				
			0.5	0.6	0.8	1.0	1.5
			Power				
5B, 1A, 1T	t-test	0.045	0.096	0.118	0.180	0.262	0.520
	ME	0.051	0.106	0.131	0.196	0.283	0.547
5B, 1A, 2T	t-test	0.042	0.099	0.130	0.202	0.296	0.579
	ME	0.048	0.111	0.142	0.220	0.318	0.607
5B, 1A, 3T	t-test	0.042	0.106	0.131	0.210	0.309	0.598
	ME	0.048	0.115	0.144	0.231	0.332	0.626
5B, 2A, 1T	t-test	0.044	0.136	0.177	0.284	0.413	0.751
	ME	0.049	0.149	0.192	0.305	0.440	0.773
5B, 2A, 2T	t-test	0.045	0.153	0.200	0.316	0.457	0.793
	ME	0.052	0.167	0.218	0.339	0.484	0.816
5B, 2A, 3T	t-test	0.045	0.154	0.202	0.323	0.466	0.808
	ME	0.052	0.168	0.220	0.348	0.405	0.829
5B, 3A, 1T	t-test	0.044	0.168	0.223	0.360	0.521	0.854
	ME	0.050	0.183	0.242	0.386	0.550	0.872
5B, 3A, 2T	t-test	0.045	0.179	0.239	0.389	0.558	0.883
	ME	0.050	0.195	0.257	0.417	0.585	0.899
5B, 3A, 3T	t-test	0.042	0.179	0.244	0.401	0.568	0.894
	ME	0.048	0.197	0.265	0.426	0.597	0.908
10B, 1A, 1T	t-test	0.049	0.180	0.239	0.392	0.560	0.888
	ME	0.050	0.183	0.243	0.397	0.565	0.891
10B, 1A, 2T	t-test	0.046	0.203	0.270	0.435	0.614	0.918
	ME	0.047	0.206	0.275	0.440	0.619	0.921
10B, 2A, 1T	t-test	0.051	0.273	0.370	0.583	0.778	0.984
	ME	0.052	0.277	0.375	0.588	0.783	0.984
10B, 2A, 2T	t-test	0.048	0.300	0.405	0.634	0.827	0.992
	ME	0.050	0.305	0.412	0.639	0.830	0.993
10B, 3A, 3T	t-test	0.051	0.390	0.522	0.766	0.916	0.999
	ME	0.053	0.396	0.528	0.770	0.918	0.999
50B, 1A, 1T	t-test	0.049	0.694	0.844	0.976	0.998	1.000
	ME	0.049	0.694	0.844	0.976	0.998	1.000
50B, 1A, 2T	t-test	0.050	0.751	0.886	0.990	1.000	1.000
	ME	0.050	0.751	0.886	0.990	1.000	1.000
50B, 2A, 1T	t-test	0.050	0.894	0.970	0.999	1.000	1.000
	ME	0.050	0.894	0.971	0.999	1.000	1.000
50B, 2A, 2T	t-test	0.049	0.927	0.984	0.999	1.000	1.000
	ME	0.049	0.927	0.984	0.999	1.000	1.000

"xB, yA, zT" represents x biological samples, each with y analytical replicates, and z technical replicates per analytical replicate. "t-test" reports the result from the marginal t-test, and ME from the mixed effects model (2)

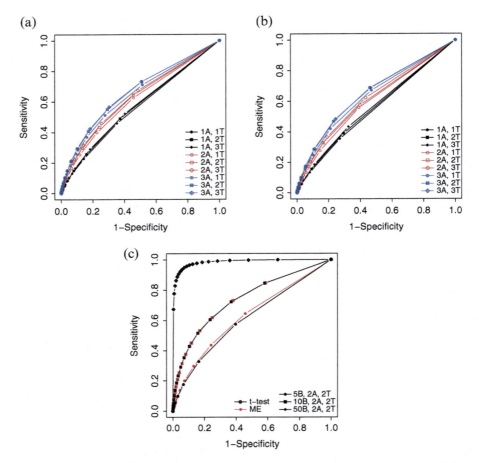

Fig. 2 Receiver operating characteristic (ROC) curves in multiple testing with effect size 0.5 ("xB, yA, zT" represents x biological samples, each with y analytical replicates, and z technical replicates per analytical replicate). (**a**) Mixed effects model with $a_1 = a_2 = 5$. (**b**) Marginal t-test with $a_1 = a_2 = 5$. (**c**) Model comparison with $b = c = 2$

on the other hand, yields much lower FDR than the nominal level q, indicating that it is too conservative for small sample size. For larger number of biological subjects, there is no significant difference between the two methods in terms of controlling false discoveries. Figure 4 shows the FDR for different effect sizes under a nominal FDR level of 0.05. From Fig. 4a, b, we observe that all the replication assignments considered yield similar FDR for both the mixed effects model and the marginal t-test. When $a_1 = a_2 = 5$, the FDR attained by the marginal t-test is much lower than the nominal level regardless of the effect size. Figure 4c demonstrates that the best way to improve the performance in controlling FDR is to increase the number of biological samples.

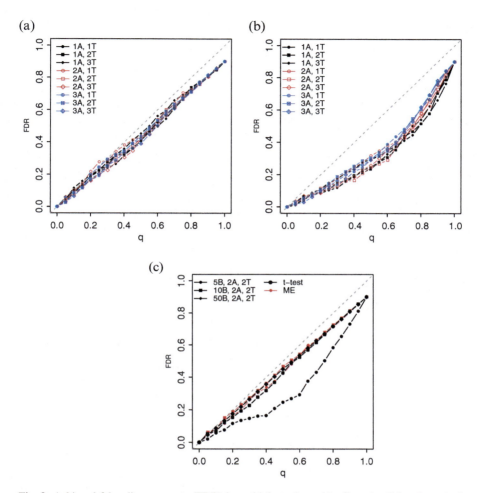

Fig. 3 Achieved false discovery rate (FDR) in multiple testing with effect size 0.5 and nominal FDR level q ("xB, yA, zT" represents x biological samples, each with y analytical replicates, and z technical replicates per analytical replicate). (**a**) Mixed effects model with $a_1 = a_2 = 5$. (**b**) Marginal t-test with $a_1 = a_2 = 5$. (**c**) Model comparison with $b = c = 2$

4.3 Replication Assignment and Sample Size Calculation

According to the simulation studies in sections "Single Peak Difference Detection" and "Difference Detection with Multiple Testing," the optimal way of improving the performances of both the marginal t-test and the mixed effects model is to increase the number of biological subjects in each group. However, in practice, recruiting more biological subjects may not be possible or easy, especially for rare diseases. One option to increase the statistical power is to generate multiple replications for

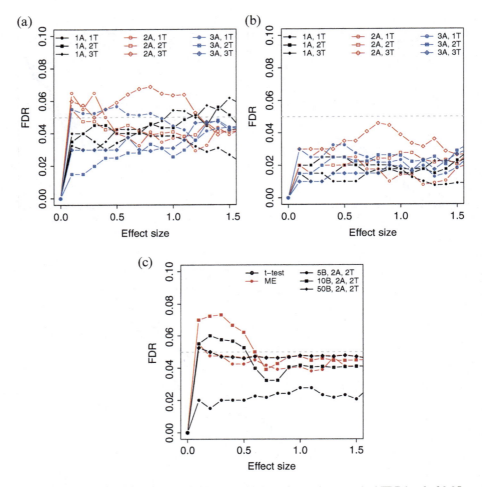

Fig. 4 Achieved false discovery rate (FDR) in multiple testing under a nominal FDR level of 0.05 ("xB, yA, zT" represents x biological samples, each with y analytical replicates, and z technical replicates per analytical replicate). (**a**) Mixed effects model with $a_1 = a_2 = 5$. (**b**) Marginal t-test with $a_1 = a_2 = 5$. (**c**) Model comparison with $b = c = 2$

each subject. In this section, we discuss the issues of replication assignment and sample size calculation based on the estimates from our pilot study.

Table 3 presents the required sample size in each biological group under the mixed effects model (2) controlling FDR at levels 0.05 and 0.1, assuming an average power of 80 and 90 % across all peaks, respectively. The variances are assumed to be the median estimates in the pilot study (Sect. 4.1). We observe that fewer biological samples are required when including analytical or technical replicates. In the current design, the numbers of analytical and technical replicates are the same, i.e., $b = c$. Increasing either b or c by one unit changes the value of σ_γ^2/bc

Table 3 Required biological samples in each group by controlling FDR at levels 0.05 and 0.1

q	$1 - \beta^*$	Effect size	$m_0/(m_0 + m_1) = 0.9$					$m_0/(m_0 + m_1) = 0.95$				
			0.5	0.6	0.8	1.0	1.5	0.5	0.6	0.8	1.0	1.5
0.05	0.8	1A, 1T	104	72	40	26	11	121	84	47	30	13
		1A, 2T	81	56	31	20	9	94	65	37	23	10
		2A, 1T	59	41	23	14	6	68	47	26	17	7
		2A, 2T	47	33	18	11	5	55	38	21	13	6
		3A, 3T	34	23	13	8	3	39	27	15	9	4
	0.9	1A, 1T	154	107	50	38	17	173	120	67	43	19
		1A, 2T	121	84	47	30	13	136	94	53	34	15
		2A, 1T	87	60	34	21	9	98	68	37	24	10
		2A, 2T	71	49	27	17	7	79	55	31	19	8
		3A, 3T	50	35	19	12	5	56	39	22	14	6
0.1	0.8	1A, 1T	102	71	40	25	11	119	83	46	29	13
		1A, 2T	80	55	31	20	8	93	65	36	23	10
		2A, 1T	58	40	22	14	6	68	47	26	17	7
		2A, 2T	47	32	18	11	5	55	38	21	13	6
		3A, 3T	33	23	13	8	3	39	27	15	9	4
	0.9	1A, 1T	153	106	59	38	17	172	119	67	43	19
		1A, 2T	120	83	46	30	13	135	93	52	33	15
		2A, 1T	87	60	33	21	9	97	67	38	24	10
		2A, 2T	70	48	27	17	7	79	55	30	19	8
		3A, 3T	50	34	19	12	5	56	39	22	14	6

"yA, zT" represents y analytical replicates with z technical replicates for each

in Eq. (8). However, increasing the number of analytical replicates b decreases the value of σ_β^2/b, thus, reducing the number of subjects required in each group. In our pilot study, where sample preparation accounted for most of the variability in the data, the inclusion of analytical replicates gives a more substantial reduction of the number of biological subjects. It should be emphasized, however, that the need for more biological samples cannot always be circumvented by increasing the number of analytical or technical replicates. It is noticed that relaxing the FDR level from 0.05 to 0.1 does not change the number of required biological samples by much, and essentially has no effect for larger effect sizes. As an example, when the number of truly non-different peaks is 90 %, using the variance components estimated from our pilot dataset, with one analytical and two technical replicates for each biological subject, 55 biological samples in each group are sufficient to detect an effect size of 0.6 at 80 % power controlling the FDR at level of 0.1, while 40 biological samples are necessary with two analytical and one technical replicates per subject.

4.4 LC-MS Based Biomarker Identification for Hepatocellular Carcinoma

We illustrate the application of the mixed effects model on LC-MS data from our hepatocellular carcinoma (HCC) biomarker discovery study. The study was conducted with sera samples from a US cohort [14]. The US participants were recruited from the hepatology clinics at MedStar Georgetown University Hospital. Serum samples were collected from 60 HCC cases and 129 patients with liver cirrhosis. Mass spectrometric analysis was performed on a Q-TOF premier (Waters) operating in positive mode. Among these samples, five HCC cases and five cirrhotic controls were chosen to constitute a second experiment. In both experiments, two technical replicates for each subject were considered with identical sample preparation protocols by the same technician.

Data Preprocessing

UPLC-QTOF data were acquired in centroid mode from 50 to 850 mass-to-charge ratio (m/z) in MS scanning using the MassLynx software (Waters). The XCMS package [16] (Scripps Center for Metabolomics, La Jolla, CA) was used to preprocess each LC-MS run separately. The peak matching algorithm in XCMS takes into account the two-dimensional anisotropic nature of the data. The first step consists of detecting the peaks cutting the UPLC-QTOF data into slices that are a fraction of a mass unit wide then applying a model peak matched filter on the individual slices over the chromatographic time domain. After detecting peaks in the individual samples, the peaks are matched across samples to allow calculation of retention time (rt) deviations and relative ion intensity comparison. This is accomplished using a method based on kernel density estimation that groups peaks in the mass domain [16]. These groups are then used to identify and correct drifts in rt from run to run. We detected 1586 ions in the first experiment and 612 in the second. Following data preprocessing, ion annotation is conducted to identify derivative ions such as isotopes, adducts, and insource fragments by using the R package CAMERA [17]. This ion annotation information was used to convert the m/z values of annotated isotopes/adducts/neutral-loss fragment ions to the corresponding neutral monoisotopic masses before searching them against databases to find the putative identifications [14, 23].

Variance Components Estimation

The ion intensities were first log-transformed to satisfy the normality assumption. Since there is only one analytical replicate for each subject, σ_β^2 cannot be isolated in the estimation of the random effects variances of the proposed model (2).

Figure 5 shows the boxplot of the estimated variance components for both experiments and Table 4 provides the corresponding summary statistics. The median estimate of $(\sigma_\alpha^2 + \sigma_\beta^2)$ and σ_γ^2 in Experiment 1 are significantly higher than those

(a)   (b)

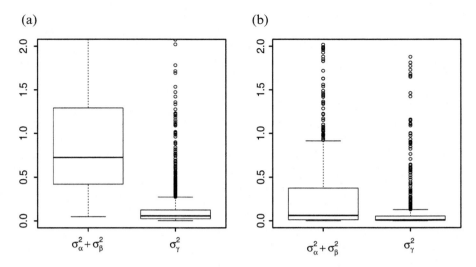

Fig. 5 Boxplot of variability assessment for LC-MS data from the HCC biomarker discovery study. (**a**) Experiment 1. (**b**) Experiment 2

Table 4 Variability assessment for the LC-MS data from the HCC biomarker discovery study

	Experiment	Median	Q1	Q3	IQR
$\sigma_\alpha^2 + \sigma_\beta^2$	Experiment 1	0.7253	0.4207	1.2910	0.8703
	Experiment 2	0.0607	0.0117	0.3741	0.3624
σ_γ^2	Experiment 1	0.0578	0.0244	0.1243	0.0999
	Experiment 2	0.0091	0.0036	0.0542	0.0507

in Experiment 2 with greater variation in the variance components estimates. This might be due to the underestimation of the subject variation in the second experiment, as the number of biological samples is much smaller compared to the first one. On the other hand, having more subjects requires longer machine time to run the samples, which is possibly the reason that the technical variation in the first experiment is higher as well.

Difference Detection

To identify differentially abundant peaks among HCC cases versus cirrhotic controls, both the mixed effects model and the marginal t-test are conducted. For multiple testing consideration, the p-values are adjusted following the Benjamini–Hochberg procedure and the threshold for significance is set to 0.1.

In the first experiment, we identified 11 significant ions by applying the mixed effects model, while no significant ion was identified by the marginal t-test. In the second experiment, 59 significant ions were identified by the mixed effects model and 52 by the marginal t-test. For the significant ions discovered in Experiment 1, the effect sizes are all between 0.50 and 0.60. The effect size of the significant peaks in Experiment 2 are in the range of 2.0–3.5. As the variance components could be possibly underestimated in Experiment 2, these effect sizes may not be reliable and the identified peaks can be false positives.

Putative identifications of the significant monoisotopic masses were found using MetaboSearch [27] by searching against four databases: the Human Metabolite DataBase (HMDB) [22], Metlin [15], Madison Metabolomics Consortium Database (MMCD) [4], and LIPID MAPS [7]. Table 5 lists the selected significant peaks with putative identification, which are discovered in both of the experiments using the mixed effects model. Both glycodeoxycholic acid (GDCA) and its fragments, as well as fragment of glycocholic acid (GCA), were verified by comparing their MS/MS fragmentation patterns and retention time with those of authentic standard compounds [14, 23]. Figure 6 shows the individual value plots of fragment of GDCA and GDCA in both experiments. These figures illustrate that the intensities in the two patient groups have different means. The intensities of GDCA and its fragment are discriminately distributed between the two groups in the second experiment. With small effect size, the mixed effects model has slightly higher power to reliably detect candidate biomarkers for further targeted quantitation than the marginal t-test procedure.

Table 5 Metabolites corresponding to m/z values with putative identifications among peaks identified to be differentially abundant in the LC-MS-based HCC biomarker discovery study

	Metabolite ID	FDR-adjusted p-value		Raw p-value rank		Effect size
		Marginal t-test	Mixed effects	Marginal t-test	Mixed effects	
Experiment 1	Fragment of GDCA Adduct: $[-H_2O]$ m/z = 414.300 rt = 200.744	0.122	0.086	1	1	0.589
	Fragment of GDCA Adduct: $[-2H_2O]$ m/z = 432.309 rt = 200.948	0.212	0.086	2	5	0.572
	GDCA Adduct: $[+H]$ m/z = 450.320 rt = 201.166	0.212	0.087	5	9	0.539
Experiment 2	Fragment of GDCA Adduct: $[-H_2O]$ m/z = 414.302 rt = 225.705	0.051	0.043	31	34	2.742
	Fragment of GDCA Adduct: $[-2H_2O]$ m/z = 432.312 rt = 225.705	0.051	0.043	25	30	2.767
	Fragment of GCA Adduct: $[-2H_2O]$ m/z = 432.310 rt = 197.764	0.095	0.057	46	47	2.473
	GDCA Adduct: $[+H]$ m/z = 450.322 rt = 225.761	0.070	0.052	39	44	2.532

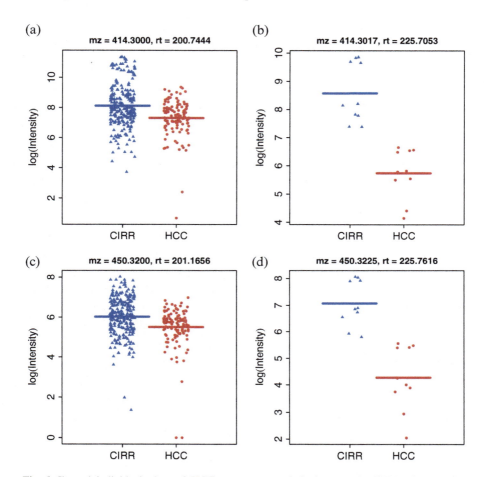

Fig. 6 Jittered individual plots of HCC cases versus cirrhotic controls (CIRR) for selected differentially expressed peaks with putative identification. The *horizontal bar* in each plot corresponds to the sample mean. (**a**) Fragment of GDCA from Experiment 1. (**b**) Fragment of GDCA from Experiment 2. (**c**) GDCA from Experiment 1. (**d**) GDCA from Experiment 2

5   Other Methods

In Sect. 2, we mainly discussed a marginal t-test and a mixed effects model approaches to discover the differentially abundant metabolites. Both are univariate approaches, in which the peak intensities are considered as random variables and the means of different biological populations are compared. Recent nonparametric rank tests for clustered data may also be applicable to this problem [5, 6].

An alternative approach is to treat the biological grouping as a binary outcome and identify the peaks that are significantly associated with the risk of being a "case" [24]. For studies with only one measurement per subject, this can be achieved through a logistic regression model

$$\log \left(\frac{P(G = 1)}{1 - P(G = 1)} \right) = \beta_0 + \beta_1 y, \tag{9}$$

where G is the group label with value one for "cases," and y is the peak intensity. If each subject has replicates, a generalized linear mixed effects model can be considered. The significant peaks can be identified by testing the hypothesis $H_0 : \beta_1 = 0$. The p-values also need to be adjusted to control either the FDR or the family-wise error rate (FWER).

In most mass spectrometry (MS) studies, the number of features (p) is much greater than the number of samples (n), which is referred to as a "large p, small n" problem. Multivariable models, such as multivariable mixed effects models, cannot be applied directly without resorting to dimension reduction or variable selection methods. A lasso [18] method can be used to achieve variable selection and coefficient estimation simultaneously by including an ℓ_1 penalty to the residual sum of squares minimization problem. Under appropriate assumptions, lasso can be shown to recover the exact support of a sparse model from data generated by this model if the covariates are not highly correlated [21, 26]. However, the detected peaks in an MS dataset can be highly correlated since they can be the isotopes/fragment ions of the same metabolite. Elastic net [29] may help improve the performance when there exists strong correlation between the predictors. By including an ℓ_2 penalty, it encourages a grouping effect among the covariates. Motivated by a SELDI MS study, Tibshirani et al. [19] introduced the fused lasso method, which imposes smoothness among adjacent peaks by encouraging sparsity. Some other regularization techniques, for example, adaptive lasso [28], SCAD [8] and MCP [25], can also be applied in MS studies.

6 Discussion

In this chapter, we presented two peak-level methods for the analysis of LC-MS data, a marginal t-test approach and a mixed effects model. We also discussed experimental design issues. The mixed effects model takes into account the correlation within analytical and technical replicates by explicitly estimating the different sources of variability in the data.

Based on simulation studies, the commonly used t-test applied after averaging the analytical and technical replicate measurements within each biological sample performs slightly worse compared to the mixed effects model for small sample size. This difference diminishes with sufficiently large samples size. Using the variance estimates from a pilot label-free LC-MS experiment, where sample preparation accounts for a larger proportion of the variability in the data, increasing the number of analytical replicates leads to more gain in statistical power compared to increasing the number of technical replicates. Experimental design issues including replication assignment and sample size calculation are discussed based on the variance estimates from the pilot dataset. Performing a pilot study before large-scale experiments is essential to determine more effectively the required sample size and the assignment of replicates. In sum, a better understanding of the sources of variability in LC-MS data is needed to decide on the most appropriate experimental design and statistical methods to identify true differences in ion abundance. We also briefly mentioned other statistical methods for high-dimensional data analysis, which could be applied to MS based proteomics data analysis.

Acknowledgements This work is in part supported by the National Institutes of Health Grants U01CA185188 and R01GM086746 awarded to HWR.

References

1. Benjamini, Y., & Hochberg, Y. (1995). Controlling the false discovery rate: A practical and powerful approach to multiple testing. *Journal of Royal Statistical Society: Series B (Methodological), 57*, 289–300.
2. Clough, T., Key, M., Ott, I., Ragg, S., Schadow, G., & Vitek, O. (2009). Protein quantification in label-free LC-MS experiments. *Journal of Proteome Research, 8*, 5275–5284.
3. Clough, T., Thaminy, S., Ragg, S., Aebersold, R., & Vitek, O. (2012). Statistical protein quantification and significance analysis in label-free LC-MS experiments with complex designs. *BMC Bioinformatics, 13*(Suppl 16), S6.
4. Cui, Q., Lewis, I. A., Hegeman, A. D., Anerson, M. E., Li, J., Schulte, C., et al. (2008). Metabolite identification via the Madison Metabolomics Consortium Database. *Nature Biotechnology, 26*, 162–164.
5. Datta, S., & Glen, A. S. (2005). Rank-sum tests for clustered data. *Journal of the American Statistical Association, 100*, 908–915.
6. Dutta, S., & Datta, S. (2016). A rank-sum test for clustered data when the number of subjects in a group within a cluster is informative. *Biometrics, 72*(2), 432–440.
7. Fahy, E., Sud, M., Cotter, D., & Subramaniam, S. (2007). LIPID MAPS online tools for lipid search. *Nucleic Acids Research, 35*, W606–W612.
8. Fan, J., & Li, R. (2001). Variable selection via nonconcave penalized likelihood and its oracle properties. *Journal of the American Statistical Association, 96*, 1348–1360.
9. Karpievitch, Y. V., Polpitiya, A. D., Anderson G. A., Smith, R. D., & Dabney, A. R. (2010). Liquid chromatography mass spectrometry-based proteomics: Biological and technological aspects. *The Annals of Applied Statistics, 4*, 1797–1823.
10. Karpievitch, Y. V., Stanley, J., Taverner, T., Huang, J., Adkins, J. N., Ansong, C., et al. (2009). A statistical framework for protein quantitation in bottom-up MS-based proteomics. *Bioinformatics, 25*, 2028–2034.
11. Nilsson, T., Mann, M., Aebersold, R., Yates III, J. R., Bairoch, A., & Bergeron, J. J. (2010). Mass spectrometry in high-throughput proteomics: Ready for the big time. *Nature Methods, 7*, 681–685.
12. Oberg, A. L., & Vitek, O. (2009). Statistical design of quantitative mass spectrometry-based proteomic experiments. *Journal of Proteome Research, 8*, 2144–2156.

13. Patel, V. J., Thalassinos, K., Slade, S. E., Connolly, J. B., Crombie, A., Murrell, J. C., et al. (2009). A comparison of labeling and label-free mass spectrometry-based proteomics approaches. *Journal of Proteome Research, 8,* 3752–3759.
14. Ressom, H. W., Xiao J. F., Tuli, L., Varghese, R. S., Zhou, B., Tsai, T., et al. (2012). Utilization of metabolomics to identify serum biomarkers for hepatocellular carcinoma in patients with liver cirrhosis. *Analytica Chimica Acta, 743,* 90–100.
15. Smith, C. A., O'Maille, G., Want, E. J., Qin, C., Trauger, S. A., Brandon, T. R., et al. (2005). METLIN: A metabolite mass spectral database. *Therapeutic Drug Monitoring, 27,* 747–751.
16. Smith, C. A., Want, E. J., O'Maille, G., Abagyan, R., & Siuzdak, G. (2006). XCMS: Processing mass spectrometry data for metabolite profiling using nonlinear peak alignment, matching, and identification. *Analytical Chemistry, 78,* 779–787.
17. Tautenhahn, R., Bottcher, C., & Neumann, S. (2007). Annotation of LC/ESI-MS mass signals. In *Proceedings of the First International Conference on Bioinformatics Research and Development* (pp. 371–380).
18. Tibshirani, R. (1996). Regression shrinkage and selection via the lasso. *Journal of the Royal Statistical Society: Series B (Methodological), 58*(1), 267–288.
19. Tibshirani, R., Saunders, M., Rosset, S., Zhu, J., & Knight, K. (2005). Sparsity and smoothness via the fused lasso. *Journal of the Royal Statistical Society: Series B (Statistical Methodology), 67,* 91–108.
20. Tsai, T. H., Tadesse, M. G., Di Poto, C., Pannel, L. K., Mechref, Y., Wang, Y., et al. (2013). Multi-profile Bayesian alignment model for LC-MS data analysis with integration of internal standards. *Bioinformatics, 29,* 2274–2280.
21. Wainwright, M. (2009). Sharp thresholds for noisy and high-dimensional recovery of sparsity using ℓ_1-constrained quadratic programming (lasso). *IEEE Transactions on Information Theory, 55,* 2183–2202.
22. Whishart, D. S., Tzur, D., Knox, C., Eisner, R., Guo, A. C., Young, N., et al. (2007). HMDB: The human metabolome database. *Nucleic Acids Research, 35,* D521–D526.
23. Xiao, J. F., Varghese, R. S., Zhou, B., Ranjbar, M. R., Zhao, Y., Tsai, T. H., et al. (2012). LC-MS based serum metabolomics for identification of hepatocellular carcinoma biomarkers in Egyptian cohort. *Journal of Proteome Research, 11,* 5914–5923.
24. Xiao, J. F., Zhao, Y., Varghese, R. S., Zhou, B., Di Poto, C., Zhang, L., et al. (2014). Evaluation of metabolite biomarkers for hepatocellular carcinoma through stratified analysis by gender, race and alcoholic cirrhosis. *Cancer Epidemiology, Biomarkers & Prevention, 23,* 64–72.
25. Zhang, C.-H. (2010). Nearly unbiased variable selection under minimax concave penalty. *The Annals of Statistics, 38*(2), 894–942.
26. Zhao, P., & Yu, B. (2006). On model selection consistency of lasso. *The Journal of Machine Learning Research, 7,* 2541–2563.
27. Zhou, B., Wang, J., & Ressom, H. W. (2012). MataboSearch: Tool for mass-based metabolite identification using multiple databases. *PLoS One, 7,* e40096.
28. Zou, H. (2006). The adaptive lasso and its oracle properties. *Journal of the American Statistical Association, 101,* 1418–1429.
29. Zou, H., & Hastie, T. (2005). Regularization and variable selection via the elastic net. *Journal of the Royal Statistical Society: Series B (Methodological), 67,* 301–320.

Statistical Approach for Biomarker Discovery Using Label-Free LC-MS Data: An Overview

Caroline Truntzer and Patrick Ducoroy

1 Introduction

The identification of new diagnostic, prognostic, or theranostics biomarkers is one of the main aims of clinical research. Technologies like mass spectrometry (MS) focus on the discovery of proteins as biomarkers and are commonly being used for this purpose. Mass spectrometry consists in the separation by gas of charged molecules, based on their mass-over-charge. Liquid chromatography coupled to tandem mass spectrometry (LC-MS/MS) first involves a separation by liquid chromatography (LC) followed by mass spectrometry in the MS and MS/MS modes. It allows the identification and quantification of peptides and proteins in complex biological mixtures like plasma, cell extracts, cell cultures, tissues, etc. In this way it delivers the characterization and the quantitative comparison of the full proteome of samples presenting specific biological status and is thus one method of choice for the discovery of new biomarkers.

The process works as follows. Proteins extracted from the biological mixtures are first subjected to enzymatic digestion and the resulting peptides are continuously separated through liquid chromatography. Depending on their retention time, peptides then enter the mass spectrometer where they are ionized. Inside the mass spectrometer, in the MS step, the peptides are quantified in a relative way according to the intensity of the resulting features, and in the MS/MS step peptides are identified by determining the amino-acid sequence of the features (Fig. 1).

C. Truntzer (✉) • P. Ducoroy
CLIPP (Clinical Innovation Proteomic Platform), Pôle de Recherche
Université de Bourgogne, F- 21000 Dijon, France
e-mail: caroline.truntzer@clipproteomic.fr

© Springer International Publishing Switzerland 2017
S. Datta, B.J.A. Mertens (eds.), *Statistical Analysis of Proteomics, Metabolomics, and Lipidomics Data Using Mass Spectrometry*, Frontiers in Probability and the Statistical Sciences, DOI 10.1007/978-3-319-45809-0_10

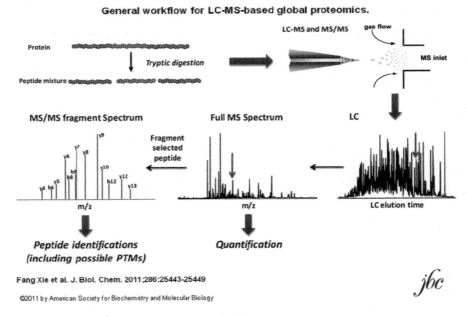

Fig. 1 Label-free LC-MS/MS workflow for raw data generation

The raw data generated from one LC-MS/MS experiment provides 3-dimensional information: one dimension for the retention time, one dimension for the mass-over-charge value (m/z), and one dimension for the quantitation of the abundances of the peptides.

This procedure leads to the identification and relative quantification of thousands of peptides, and therefore the corresponding proteins. These results are then used to find biomarkers by comparing the abundance of peptides/proteins in samples that represent different biological conditions. Sandin et al. [1] recently discussed the challenge of label-free MS–MS in the context of biomarker discovery. They believe that this technique is a suitable solution when dealing with complex biological samples and is suitable for the problem of biomarker discovery.

In this chapter we focus on a particular use of this technology, called label-free LC-MS/MS. The relative quantification of peptides and proteins is performed in a non-targeted way, in contrast with targeted techniques that use isotopic labeling for quantification and which are not the topic here.

The structure of the chapter follows the different steps encountered in a study using label-free LC-MS/MS for biomarker discovery: considerations about experimental design, description of the statistical pre-processing strategies to get peptide and protein quantification from raw data, and finally the description of statistical methods proposed to select potential biomarkers among the quantified proteins. The different steps are illustrated in Fig. 2.

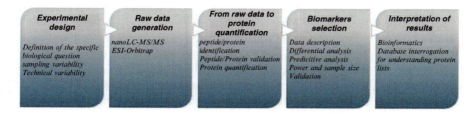

Fig. 2 Steps of the analysis of label-free LC-MS/MS data

The goal of this chapter is to give an overview of the statistical questions raised by the analysis of label-free LC-MS/MS data. We do not claim to make a complete review but we will highlight the key points of the analysis.

For ease of comprehension, only the selection of biomarkers in case of between-group comparisons has been considered in this chapter.

2 Raw Data Generation and Experimental Design

More precisely, label-free LC-MS/MS workflow for raw data generation works as follows.

Proteins contained in complex biological samples are first cut into small fragments, namely the peptides, through an enzymatic digestion step. Those peptides are then continuously separated through liquid chromatography by gradient elution. Peptides arrive sequentially, given their retention time, in the ion source of the mass spectrometer so that they are ionized. Once ionized, peptides are analyzed through two consecutive analyzers.

1. First analyzer—MS step: it determines the intensity and the mass-over-charge (m/z) of each ionized peptide, and it transmits the most intense peptides (precursor ions) to the second analyzer. The relation between the m/z ratio and the intensity is given through a spectrum with m/z in the x-axis and intensity in the y-axis.
2. Second analyzer—MS/MS step: after the precursor ions are fractionated into several ions, the amino-acid sequencing can be completed for the corresponding peptide.

Thereafter the flow of one sample through the mass spectrometer is called an injection or a run.

The resulting data are three-dimensional as each peptide is described by its retention time, its m/z, and its intensity, proportional to its abundance in the analyzed sample.

The mass spectrometer generates one raw file per run (about 500 Mb per file). Each file contains relevant information about MS and MS/MS acquisitions and is exported in proprietary data format. The MS acquisitions will be used to quantify peptides, themselves identified thanks to MS/MS acquisitions.

It is essential to define a specific question before generating the data because it is this definition that will define the way data must be generated and then analyzed: what comparisons, to what end, on what population, with how many and what type of replicates, etc. In particular, this will orientate the experimental design.

The data are affected by different sources of variability. The most interesting source of variability is the one generated by the biological condition of interest, for example, the variability of proteomic contents in samples from patients undergoing different treatments, or between diseased and healthy tissues.

Other sources of variability arise from sampling variability and technical variability of the procedures during the pre-analytical steps.

To formalize this point, total variability observed in a dataset can be divided into between-group and within-group variability. Between-group variability corresponds to the variability raised by the factor of interest; for example, variability observed between two diagnostic groups constituted, respectively, by healthy and diseased subjects. Within-group variability corresponds to biological variation between the different samples in each group of interest, for example, the variability observed between the healthy subjects. Within-group variability also includes technical variability in cases of technical replicates.

2.1   Sampling Variability

For obvious reasons, whole populations cannot be sampled. To ensure conclusions that generalize to the population, and thus make inference, sampling must be as representative as possible of the whole population from which it was taken.

Biological variability, that is to say within-group variability, must be evaluated for subjects/entities constituting this sample.

Biological replicates are used to assess the variability of the biomarkers of interest that is inherent to the whole population. Figure 3, inspired by Oberg et al. [2], illustrates the importance of these replicates. Let us suppose that the goal of the study is to find differential biomarkers between healthy and diseased subjects. On the left, the distribution of the abundance of one marker in a population of healthy (green) and diseased (red) subjects is represented. On the right, two different sampling strategies are represented. In the first one, no replicates are used. The observed difference of means between the two subject groups (vertical segments on Fig. 3) are different according to sampling situation Ex1 or Ex2. Results are not robust. In the second strategy, several subjects are drawn from each population. Estimated means are now better estimations of the true means in each of the sub-populations. Multiple biological replicates also reduce the variance of the calculated statistics.

Moreover, for the discovery of robust biomarkers, between-group variability should be larger than the within-group variability in the population. If there is only a single biological replicate per condition, between-group and within-group variability may be confounded.

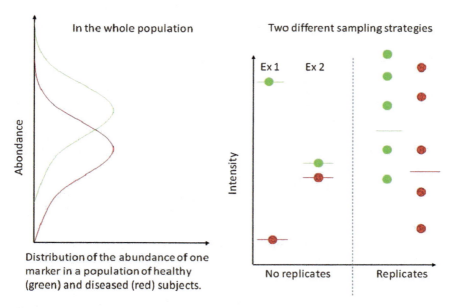

Fig. 3 Examples of sampling strategies

Note that from a biological point of view, this sampling variability may depend on the nature of the sample (tissue, cell extracts, etc.).

2.2 Technical Variability

Besides biological replicates, technical variability reflects all that is not related to biological variability. This technical variability may come from sample preparation, instrument calibration, peptide detection, etc. Technical variability should ideally be smaller than biological variability and must not be confounded with biological variability, hence the importance of technical replicates. Moreover, these technical replicates are essential to assess reproducibility and thus evaluate the quality of the experiment and, by consequence, of the whole study. We advise three technical replicates (called injections/runs thereafter) per sample, which seems to be a good compromise between economic concerns and robustness requirements. Three replicates provide protection in case of one bad quality run and ensures reproducibility. These technical replicates make the quantification of the variability induced by the different pre-analytical steps possible. If there is a lack of technical replicates, technical variability will be confounded with biological variability.

2.3   Experimental Design

Experimental design must answer the biological question and take into account the different sources of variability and in this way control and reduce technical variability as much as possible.

It ensures that both technical and biological sources of variability can be estimated. Different strategies, including replications, can be used for this purpose. The first strategy is randomization, that is to say samples are acquired in a random order. The aim is to avoid bias due to uncontrolled sources of variability. In case of LC-MS/MS data, these technical sources of variability mainly come from the chromatography column: aging or dirt due to incomplete washes may lead to a modification of the retention time for the same peptide, or to a decrease of separation capacity. Sample material batches of columns may also introduce variation when columns must be replaced.

The second is blocking, which diminishes the risk of bias due to known sources of variability. Imagine a longitudinal study with samples measured over time for each of the patients. Data from the same patient must be acquired with the same column and at the same time period.

If these recommendations are not observed it will not be possible to distinguish technical variability from the variability of interest. A complete review of these notions can be found in Oberg et al. [2].

Sampling variability and technical variability should be as low as possible so that they do not hide the principal source of variability, the one linked to the biological factor of interest. Technical variability can be controlled through framed procedures. Sampling variability is determined by the population, and the sampling strategy should cover the population as much as possible.

A preliminary study should be conducted to quantify the different sources of variability and to define the number of replications needed. The question of the calculation of sample size will be developed further in Sect. 4.4.

3   From Raw Data to Protein Quantification

Raw data are exported in proprietary data format. Proprietary data formats may be difficult to handle with non-commercial software, mostly preferred by the scientific community because of their transparency. Thus, the first step when dealing with such data is to convert the raw files from raw proprietary file format to open format like mzXML or mzML, which are common file formats for proteomics mass spectrometric data. These files contain relevant information about MS and MS/MS spectra corresponding to the acquisitions. MS/MS spectra will be used to identify peptides. MS spectra will be used to quantify peptides.

Once raw data is transformed, the following steps are followed:

- Peptide identification and validation
- Protein identification and validation
- Peptide and protein quantification

We describe here one possible workflow, but point out that it is one among other possibilities.

3.1 Peptide Identification and Validation

Peptides are identified from MS/MS information by interrogating databases through sequence search engines like Sequest, Comet, X! Tandem, or Mascot. The objective of these algorithms is to match the spectra with peptide sequences referenced in specific databases like UniProt. These search engines can also be combined, as described later.

The principle of search engines is to match the MS/MS spectra with corresponding peptide sequences from the protein sequence database. The quality of the peptide-spectrum matches (PSM) is evaluated through a scoring function whose calculation and parameters depend on the search engine. The evidence of correct identifications is based on these scores. Rather than using these scores directly, the PeptideProphet [3] algorithm provides an estimation of the confidence in the identifications. This algorithm is part of the open-source Trans-Proteomic Pipeline (TPP) software [4]. Based on a mixture-modeling approach (one distribution, respectively, for correct and incorrect identifications), the algorithm assigns a probability of correctness to each PSM and also gives a corresponding global False Discovery Rate (FDR). It takes other parameters, like mass deviation, number of missed cleavages, etc., into account when estimating the probabilities.

Peptides are now identified and the identifications are validated. Further peptide-level probabilities can be computed through the iProphet algorithm, which is an integral part of the TPP. This tool makes it possible to combine PSM scores from multiple search engines. It also improves probability estimates given by Peptide-Prophet by using additional information like the number of sibling search results (identifications based on the output of multiple search engines for the same set of spectra), the number of replicate spectra, and the number of sibling modifications (peptides that are identified with different mass modifications), etc. [5].

3.2 Protein Identification and Validation

Protein inference can then be performed, using, for example, another TPP tool, namely ProteinProphet [6]. Protein inference consists in assembling peptides experimentally identified from tandem mass spectra to deduce the proteins that were originally contained in the sample. This makes it possible to estimate the probability that one protein is present in one specific sample by adjusting the probabilities given by PeptideProphet. The probability associated with one protein corresponds to the probability that it is truly contained in the sample, and the probability that at least one PSM related to the considered protein is correct. The final list of proteins is the simplest one that can explain all of the peptides observed in the dataset.

3.3 Variability at the Identification Level

The list of peptides/proteins given at this stage constitutes the set among which potential biomarkers will be selected. These lists differ from one search engine to another because corresponding algorithms are not the same. This may disturb the biologist, who expects only one list, however it is obtained. The interest of combining several search engines, as proposed by the iProphet tool, becomes apparent in this context, as it reinforces the identification results.

Samples are often peptide rich. Among all of the peptides contained in the samples, only the most abundant peptides are retained for the MS/MS step and further identification. The same peptide is thus not necessarily identified in all of the injections, even if present because only the most intense peaks (the "top-20," for example, are commonly used) are fragmented for each MS scan. A particular peptide may have a different measured intensity depending on the intensity of the other peptides with which it is eluted. As a consequence, the same peptide is not necessarily identified in all of the injections, even if present. On the other hand, the more confident the identification, the more likely the peptide will be present in the other runs. By confident we mean that the peptide is identified with a high probability of correctness, or in other words a high likelihood that the peptide-spectrum-match is correct. If a peptide is identified with little confidence in one run, quantifying it in other runs may lead to useless information.

3.4 Missing Values

In the framework of label-free LC-MS/MS data analysis, missing values correspond to peptides that are observed in some runs, but unobserved in other runs, for different reasons partially described above. The origins of these missing values are not well understood because there are many. It may be a chance result related to the experiment or a true biological information and the peptide is unobserved because of post-translational modification, sequence variations, enzymatic cleavage, etc. Several explanations may thus be proposed:

- the peptide is missing in the sample
- the peptide is present in a detectable abundance but is in fact not detected or poorly identified

- the peptide is present but at a too low abundance to be detected by the instrument (limit of detection).

The question is then: must a peptide be quantified in a run where it was not identified, and if yes, in what manner? Different solutions can be found to deal with this question [7]:

- Withdrawal of peptides that are not present in all of the runs. This is based on the hypothesis that these peptides were subjected to measurement error in some runs; they are thus considered as poor quality features and removed from the dataset.
- Imputation of the missing value, which can be done in several ways

  - Use of the background by imputing the minimum of the abundance observed on the other runs from the same condition. This is based on the hypothesis that the peptide is missing because it is under the limit of detection.
  - The method of the k-nearest neighbors (knn) can be used for this purpose
  - One limit of this approach is that it modifies the mean and the variance structure of the data.

- Modeling of the quantity of the peptide using other runs and under the normality assumption.
- Transformation into a presence/absence information without further quantification.

Filters can also be set up upstream to limit the number of missing values by eliminating the least represented peptides. This filter can be based on the frequency of occurrence of the peptides. For example, only peptides detected in 2 of 3 of the runs for a given sample are conserved.

Each run is described by a list of peptides each characterized by its sequence, the ID of the protein it is related to, its mass-over-charge, and its retention time.

Note that proteins identified on the basis of one peptide only may not be reliable and need further confirmation.

3.5 Alignment Procedures

The objective of this step is to match the peptides identified in the different runs. A particular peptide can be observed at different retention times depending on the run it is identified because of uncertainties in peptides separation inherent to the LC technology and resulting in changes in flow rate; this corresponds to elution time drifts between the runs. In simplified terms, the same peptide may take different times to cross the column. It will thus be analyzed by the mass spectrometer at different times. The aim of the alignment procedure is to make identical peptides correspond across the samples even if observed at slightly different times. Retention times must thus be aligned so that each peptide has the same retention time over all the runs within which it is identified. Note that m/z drifts are minor compared with retention times. This alignment step is essential to compare peptide quantification in the different runs. As described by Lai et al. [8], several elution patterns can be observed depending on the peptides.

Several methods are proposed to align runs. Lange et al. proposed a review of the different strategies [9]. More recently, Smith et al. [10] referenced as many as 50 algorithms. Therefore, we will not make a complete review of all available algorithms, but limit ourselves to the major principles. Smith et al. detailed each of them. The idea is to estimate a linear or non-linear function that corrects the distortion observed from one run to another. A warping function is estimated by minimizing an objective function. The objective of this "warping function" is to modify the spectrum signal corresponding to peptides by monotonically shifting, stretching, or squeezing them to find consensus features.

This can be done as follows:

- by using a reference run or not
- by successive paired alignments or simultaneous alignment of all the spectra
- by using the complete profiles or only specific peptides or landmarks in the spectra.

Another approach corresponds to direct-match approaches. In this case, no warping function is applied but a correspondence is achieved from run to run [10].

Each run is described by a unique list of peptides that must now be quantified.

3.6 Quantification

LC-MS/MS is used as a relative quantification tool. Quantification corresponds to estimation of the abundance of one particular peptide in the sample it is observed. However, mass spectrometry technology is not designed to give the exact amount of a peptide in a sample; this is not possible because of differences in the ionization efficiency and/or detectability of the many peptides in a given sample. As a consequence, the technology cannot provide absolute—exact—quantification. However, abundance of a single peptide is comparable from one run to another and by extension from one condition to another. Mass spectrometry is thus used to reflect relative differences in abundance of peptides from one condition to the other. For that reason we speak about "relative" quantification. Most often biologists or clinicians are interested in conclusions at the protein level. Protein quantification may be derived from the quantification of peptides related to them.

Two different approaches are available for peptide quantification:

- Spectral counting, based on the number of identifications at the MS/MS level related to a same peptide. The number of MS/MS spectra that match peptides to a particular protein are then counted and this count is used to quantify the abundance of the protein in the biological sample.

- Abundances of peptides are quantified at the MS level. It is based on the intensity of ions of one peptide given by the chromatogram.

In the following we will not discuss spectral counting.

Peptide quantification is usually based on the signal from eXtracted or Selected Ion Chromatogram (XIC or SIC). This spectrum signal is first baseline- and noise-corrected and peaks, corresponding to peptides, are then detected through different algorithms [11, 12]. Once the peaks have been detected, peptide abundance is estimated through peak width, height, or area, for example.

The quantification of the peptides then serves at basis for the quantification of the proteins they are related to; this is the inference step. Three main strategies can be used to infer protein quantification.

Additive

In this case, the abundances of peptides related to a single protein are combined in an additive way. This can be done by taking the (standardized) sum from all or only a subset of the most abundant peptides.

Reference

In this case, a specific peptide is chosen as a reference to standardize the abundance of other peptides of the protein. The reference peptide can, for example, be chosen as the one with the fewest missing values; peptides are normalized on this reference and the median of the normalized peptides is used to quantify the protein. This solution is, for example, implemented in DAnTE software [13].

Linear Model

In this case, the abundance of proteins is modeled. Peptides are regarded as repeated measures of the protein and one linear model is fitted for each feature. The linear model can be additive with or without interaction [14].

Without interaction the model can be written as :

$$\bar{y}_{i.k} = \mu + gp_i + \text{error}_{ik},$$

where $\bar{y}_{i.k}$ is the average abundance over all peptides representing the considered protein, observed for sample k coming from disease group i, μ is the overall mean abundance, gp_i is the deviation due to disease group i, and error_{ik} the error term.

With interaction it can be written as :

$$y_{ijk} = \mu + gp_i + pep_j + (pep * gp)_{ij} + S_{k(i)} + \text{error}_{ijk},$$

where y_{ijk} is the abundance of peptide j observed in disease group i for sample k, $(pep * gp)_{ij}$ is the interaction term, $S_{k(i)}$ is the deviation of sample k from the overall group mean abundance, and $error_{ijk}$ the error term. $S_{k(i)}$ can be modeled as a random or a fixed effect depending on whether subjects are considered as a random selection from larger underlying population or not.

Protein abundance is then given by $\bar{y}_{i.k}$. This approach is implemented, for example, in R library MSstats [14, 15].

For more details about these models please refer to Clough et al. [14]

Summary

Matzke et al. [16] proposed a comparison of these different strategies. They concluded that the choice of the quantification method has less impact than the missing values imputation and the peptide filtering steps. Not all peptides should be used for protein quantification and a filter on peptides can help by removing poorly quantified peptides. For example, peptides can be filtered on the basis of their coefficient of variation over technical replicates as proposed by Lai et al. [8]

Matzke et al. also concluded that the best method is the one that is the most tightly controlled and easiest to set up.

Another question raised by the quantification step is the level at which quantification is applied: at the peptide or at the protein level. The answer depends largely on the biological question. Quantification of proteins can be seen as more precise than that of peptides because it is obtained from many entities representing the same protein. But for a particular protein, fold-changes may vary depending on the peptide for true biological reasons. This can come about for different reasons: some peptides are more variable than others in the same chromatographic conditions, post-translational modifications, isoforms, and peptides shared by several proteins. Biologists may be interested in accessing this level of information. For example, a particular disease can be characterized by a variant or post-translational modification of one particular peptide. In this context, peptides with extreme quantification values can in reality have a biological meaning and therefore must not be discarded.

3.7 Normalization

Whatever the quantification method, raw abundances are commonly transformed in the base 2 logarithm scale. The logarithmic transformation aims to remove excess skewness and to obtain approximately normally distributed intensities. Base 2 is used because further analysis uses fold-changes and interpretations are easier in this basis. Normalization allows for the (partial) compensation of some instrument fluctuations and takes into account potential differences in samples amounts. The objective of this step is to retrieve systematic biases from the data, which is any non-biological signal that affects the measure. These systematic errors can be due, for example, to sample processing conditions, instrument calibration, chromatographic columns, or temperature variations. They manifest themselves in variations of peaks intensities or retention time, measure precision, etc. [7, 17].

Several approaches are proposed.

Global Normalization

It is based on the hypothesis that the majority of the peptides do not vary; the distribution of peptide abundance is thus assumed to be similar from one sample to another. For this reason, abundance distribution is constrained to be centered around a constant (mean or median, for example). This can be written as

$$y'_{jk} = y_{jk} - \mu_k,$$

where y_{jk} is the observed abundance for peptide j in sample k and μ_k the mean (or another chosen constant) of all peptides quantified in sample k.

This type of normalization may correct for differences in runs.

Quantile Normalization

It is also based on the hypothesis that the distribution of the peptide abundance should be the same from one sample to another. In this case, distributions are constrained to be exactly identical from one sample to another [18].

Using a Reference Spectrum

Normalization can also be performed by using a reference spectrum, either by linear regression or by local regression. It is based on the "MA plot," which plots M values corresponding to the difference of log-intensities between the reference spectrum and the current one on the Y axis, against A, the mean of the log-intensities, on the X axis.

Linear regression is based on the hypothesis that systematic bias depends linearly on the order of magnitude of peptide abundance. This approach is useful to retrieve systematic bias-related carry-over in the LC column. Linear regression models the linear relationship between M and A. The value of M for a peptide j is given by m_j. The corrected normalized value of M for peptide j (m'_j) is thus given by

$$m'_j = m_j - m^*_j,$$

where m^*_j corresponds to the estimation of m_j given by the linear regression model.

Local regression is based on the hypothesis that systematic bias depends non-linearly on the order of magnitude of peptide abundance. This approach is useful to retrieve systematic bias due to peptides that are measured close to the saturation point or to background measurement. The concept is the same as for linear regression, except that the regression is performed in a piecewise manner.

Using Singular Value Decomposition

Another type of approach was proposed by Karpievitch et al. [19, 20]. It is based on singular value decomposition, as first proposed by Leek et al. in the context of DNA microarrays [21]. There is no need to identify the origin of the bias. In a nutshell, the effect of the factor of interest is estimated in the dataset through a specific model linking intensities for the peptide and the experimental factors of interest. The part of the data that was not explained by the model is decomposed into singular values and retrieved from the original dataset. This approach is proposed in the DAnTE software previously cited.

Summary

Each software suggests another approach. The relevance of one approach compared with another may also depend on the dataset. It is thus of interest to use different approaches and compare results. Chawade et al. proposed a tool for the simultaneous evaluation of several normalization methods [22]. This evaluation is based on quantitative and qualitative plots and guides the choice of the optimal normalization method. Lai et al. [23] advocate skipping the normalization step rather than normalizing a dataset under an incorrect hypothesis. For the authors, the appropriate filtering of peptides is more important than the normalization step. As a remark, Karpievitch et al. [7] recommended normalizing the data before imputing missing values.

Some authors have proposed introducing proteins in known quantities into samples and then using these as a normalization benchmark. We believe that these added proteins may interfere with the other proteins of interest and prefer not to use this strategy.

3.8   Conclusion

Numerous methods are available for this first part of the analysis. In a recent study, Chawade et al. [24] showed that the choice of the data processing strategy had an important impact on the final outcome. But at present, there is no consensus, whatever the step considered. It may be of interest to combine different software programs to optimize the whole workflow. The choice of method may also be adapted to each specific study and different methods can be compared.

In general, the wisest thing to do is to choose the most tightly controlled method.

In terms of software choice, the use of manufacturer software presents some disadvantages: there may be a lack of precise documentation and of traceability/reproducibility in cases of version changes as some parameters may not be changeable. One advantage is that it may be an easy way for the biologist to get preliminary results. Open-source software programs present all the advantages and allow the combination of modular blocks to obtain a controlled "in house" consensus.

Concerning formats, open-source formats allow data to be shared between different software programs and different laboratories, which seems essential.

4 Biomarker Selection

Once peptides and proteins are quantified and corresponding intensities suitably normalized, each sample is described by the abundance of a list of features that can be peptides and/or corresponding proteins. The choice between peptides or proteins as the statistical unit can be considered, as discussed in Sect. 3.6. In the following section, we will therefore use the term "feature" and specify "peptide" or "protein" if needed.

The abundances of these features are related to a factor of interest for which biomarkers should be found. The next step is thus the choice of a suitable method of analysis. The choice of the method must be dictated by the biological question defined previously at the experimental design step.

Three main questions can be identified.

1. Data description. Non-supervised analysis approaches will be used for this purpose.
2. Selection of biomarkers characteristic of one particular factor of interest. In this case, differential analysis approaches will be used.
3. Selection of prognostic biomarkers able to predict the factor of interest. In this case, supervised analysis approaches will be used.

The biological question defines the choice of the method, and not the opposite.

4.1 Data Description: Non-supervised Methods

Data description is the first step of the statistical analysis. The objective of this preliminary step is to visualize the global structure of the data. The aim of these methods is to cluster entities (resp. observations or features) in the space of resp. features or observations. These are exploratory methods, with no a priori information as input, hence the term "non-supervised" methods. We will describe two of the most frequently used non-supervised methods, namely Principal Component

Analysis (PCA) and Hierarchical Ascending Clustering. The objective is to give an overview of these two methods without going into detail so as to enable the interpretation of the corresponding results.

Hierarchical Ascending Clustering

The objective is to cluster similar entities into the same class. It results in the constitution of a hierarchical tree called dendrogram. This dendrogram is constructed iteratively: at each iteration the two most similar entities are gathered into the same class. The algorithm begins with each entity (feature or observation) forming a class. The algorithm ends when all the entities are clustered into a single unique class. The notion of similarity is specified by the choice of the metric, which is a measurement of similarity or dissimilarity. In cases of dissimilarity, the distance between two entities is minimized (e.g., Euclidean distance) whereas in cases of similarity it is maximized (e.g., correlation coefficient). Once this choice has been made, the similarity between two classes must also be defined. For example, in the centroid method, the distance between two clusters is defined by the distance between their centroids. For complete and single linkage, the distance between two clusters is defined, respectively, by the maximum and the minimum of the distances between any two points, one from each of the clusters.

The choice of the metric depends on the questions raised by the dataset.

Principal Components Analysis

The objective of this method is twofold: 1-visualize the different sources of variability that can be observed in the dataset. 2-reduce the dimension of the dataset.

The idea behind this approach is that observations cannot be visualized in the space of the features, because they are too numerous. The idea is hence to project observations into a lower space, this space being chosen as the one that maximizes the projected variability of the data. The corresponding variables, called components, are a linear combination of the original variables.

To go further, some authors have proposed a modification of PCA through sparse PCA. In classical PCA, all of the variables of origin are conserved in the components. In sparse PCA, the idea is to remove variables whose coefficients in the linear combination are below a defined threshold from the components. In this way, noisy variables are excluded from the PCA model, making the interpretation easier [25].

4.2 Differential Analysis

Multiple Testing

The objective here is to identify biomarkers related to the factor of interest. This is a univariate approach, in which features are tested one by one through parametric or non-parametric tests. It involves defining one null hypothesis $H0$ and its alternative hypothesis $H1$ and leads to the definition of two types of error: Type I and Type II. The first, denoted as α, corresponds to the incorrect rejection of a true null hypothesis and is related to the p-value and to the number of tested hypothesis. The second, denoted as β, corresponds to the failure to reject a false null hypothesis and is related to the power of the test defined as $(1 - \beta)$.

The testing of multiple hypotheses leads to four situations: True Positives which correspond to truly differential selected variables, False Positives which correspond to non-differential variables wrongly selected and are related to type I error, False Negatives which correspond to truly differential selected variables not selected by the test, and True Negatives which correspond to variables truly considered non-differential. To take into account this context, the simultaneous testing of a large number of hypotheses involves controlling type I error by adjusting raw p-values. Two strategies have been proposed for this purpose, namely the Family Wise Error Rate (FWER) and the False Discovery Rate (FDR). The first controls the probability of at least one False Positive, while the second controls the expected proportion of False Positives. A 5 % control of the FWER gives the confidence at 95 % that there are no False Positives among selected features. A 5 % control of the FDR gives the confidence that, on average, there are less than 5 % of False Positives. This control is less conservative than the FWER and, for this reason, is preferred and commonly used, through, for example, the Benjamini–Hochberg procedure [26].

4.3 Predictive Analysis: Supervised Methods

The objective of this approach is to classify a new observation into known classes. As for differential analysis, classical methods must be adapted to the high-dimensional setting. The high number of variables relative to the number of observations produces multi-collinearity and optimism. Different strategies have been proposed to deal with this issue.

1. Dimension reduction through selection or extraction of variables.
2. Specific methods that directly handle high-dimensions, for example, regularization methods, or k-nearest neighbors.

4.3.1 Dimension Reduction

Variable Selection Variable selection consists in selecting variables in a univariate way (*t*-test, ANOVA, etc.). Selection is based on ranking of the variables according to their relevance with regard to the factor of interest.

Variable Extraction Variable extraction consists in building new variables, called components, which are a linear combination of the original variables; in this way, observations are projected into a lower dimension space. This can be achieved through Principal Components Analysis or Partial Least Squares (PLS) [27, 28], for example. The first builds the components by maximizing the projected variance of the data, whereas the second builds the components so as to maximize the covariance between the response and the components. The major difference between the two methods is that PLS components are optimized so as to be predictive of the factor of interest.

The components are then considered the new variables that can be integrated into classical models.

4.3.2 Regularization Methods

Regularization methods, like regularized linear or logistic regression, are based on a penalized maximization of the likelihood to avoid overfitting. Regularization can be performed through $L1$ (LASSO [29]) or $L2$ (ridge regression [30]) penalty. The first leads to sparsity, thus variable selection, by shrinking the coefficients of non-informative variables toward 0. The second only shrinks the coefficients but keeps all of them in the model. By combining both penalties, the elastic net gives the possibility of a compromise between the two approaches [31].

These methods can be combined with PLS to introduce sparsity into the components, leading to sparse PLS; the combination of variable extraction and selection makes it easier to interpret the components of the PLS [32, 33].

4.4 Power and Sample Size Calculation

Power is defined as the capacity of a test to highlight a differential effect when it truly exists in the dataset. It corresponds to the control of type II risk. Power is directly linked to the number of observations included in the study.

More precisely, the required number of samples to get robust results depends on: the fold-change (FC) of interest δ, the desired power of the test $(1 - \beta)$, the FDR value (q), the ratio between the expected number of differential and non-differential proteins (m_0/m_1), the minimum number of peptides per protein, the number of technical replicates L and biological replicates K, the number of compared conditions J, and technical variability σ_{Error}^2. All these parameters can then be linked by the following equation [34]:

$$\delta = \frac{\hat{\sigma}^2_{\text{Error}}}{IKL}(t_{(1-\beta,df)} + t_{(\sigma/2,df)})^2,$$

with

$$\alpha = (1 - \beta) * \frac{q}{1 + (1 - q).m_0/m_1},$$

and

$$df = IJKL(L - 1) + (I - 1)J(K - 1),$$

This sample calculation is proposed in the R library MSstats [15].

In practice, the number of replicates is often limited by financial constraints. In any case, if the number of replicates is fixed, biological replicates are to be preferred over technical replicates; they will make it possible to detect smaller FC. However, if the number of biological replicates is limited, technical replicates can help reduce the detectable FC to a certain extent [2]. Note that increasing technical replicates cannot compensate for lack of robustness of the experimental design and will be useless in case of bad coefficient of repeatability between technical replicates.

As an example, Table 1 shows the evolution of the number of biological samples needed to detect differential proteins given different FC, FDR (q), and power ($1-\beta$). This was calculated for a study with 252 proteins represented by a minimum of 1 peptide, 3 technical replicates per sample, and $m_0/m_1 = 0.99$. Calculations were based on a preliminary study with two diagnostic groups of 10 and 17 patients, respectively. This table shows the influence of the order of magnitude of the parameters power FC and FDR. The higher the power, the more samples are needed. The same occurs for the FDR. Finally, selecting biomarkers with weak FC requires more samples.

Table 1 Number of required samples for varying FC, FDR (q), and power ($1 - \beta$)

	FC=1.25			FC=1.5			FC=2		
	β=0.1	β=0.2	β=0.3	β=0.1	β=0.2	β=0.3	β=0.1	β=0.2	β=0.3
$q = 0.01$	519	440	129	157	118	112	53	45	40
$q = 0.05$	439	367	321	133	128	97	45	38	33
$q = 0.1$	402	334	290	122	101	88	41	34	30

4.5   Importance of Validation

Models (and resulting candidates as biomarkers) must be internally and externally validated. Internal validation consists in validating the model using the same dataset whereas external validation consists in validating the model using a new dataset that has never "seen" the model. External validation should be preferred but is not always possible. In this case, internal validation must be performed by cross-validation by splitting the dataset into a train and a test set.

The model must be robust, that is to say have good predictive properties for new observations. The difference between the quality of adjustment and the quality of prediction raises the question of optimism.

The discovery of new biomarkers should be performed into two steps:

1. Identification studies designed to select a list of candidates as markers to be tested among a large number of features. These studies give a first estimation of the strength of association between each feature and the factor of interest.
2. Validation studies designed to confirm the previously selected candidates as biomarkers. These studies aim to re-estimate the strength of association of the previously selected candidates on independent datasets, and thus confirm (or invalidate) the relevance of candidates as markers.

Identification studies are based on sampling from the population. The selection of relevant markers—here proteins—then relies on estimating the strength of association between each feature and the factor of interest. Only features with a significantly high enough strength of association are selected; variables are selected as candidates because of their extreme strength of association. This leads to a selection bias, which is all the greater when the study is small; the mean estimated strength of association of selected features is greater than the "true" strength of association.

Validation studies use independent datasets to re-estimate the strength of association and to confirm the relevance of the selection of candidate markers, in this case proteins. They show optimism if identification studies were wrongly calibrated. The mechanism of variable selection induces optimism.

Truntzer et al. showed that this mechanism of selection was related to regression toward the mean [35]. Optimism decreases when the size of the identification study increases. It can be corrected by designing identification studies larger than validation studies. This is shown in Fig. 4, from Truntzer et al. [35]. In this work, the authors compared the estimated strength of association of variables selected as potential biomarkers in the identification study, with the strength of association of the same selected variables, but this time estimated on validation data. This work was done on simulations in the context of survival data: each subject is described by survival status and "omic" variables with known effects on survival. "Omic" variables were divided between active and non-active variables, where active variables correspond to variables that were predictive of survival. For more details about the simulation process, please refer to Truntzer et al. [35].

Simulated datasets were divided into identification and validation datasets and the two following steps were repeated 200 times:

1. Step 1—on identification datasets: The strength of association (SA) between each variable and the survival status was estimated through the regression coefficient of univariate Cox survival models. Variables with significant regression coefficients were selected as potential marker candidates. Those variables may be true or false positives according that they were simulated as active variables or not.
2. Step 2—on validation datasets: Univariate Cox models were estimated only for candidate markers selected at the previous step to confirm on independent datasets the effect on survival of the corresponding candidate biomarkers.

Figure 4 represents the distribution of the estimated SA on the identification (horizontal hatching) and validation (diagonal hatching) datasets for candidate markers selected at step 1. The vertical dotted line indicates the mean of the distribution of the estimated SA computed on the validation simulated datasets. The vertical continuous line indicates the value of the true simulated SA of active variables. The upper panel of the figure shows the results obtained with identification and validation datasets of, respectively, 100 and 1000 individuals. In this case, for the identification sets, variables are selected in both extremes of the estimates distribution, leading to a bimodal distribution. Many variables were wrongly selected, and had far lower estimates on validation datasets. In fact, the right mode mostly consists of the estimates of true positives, whereas the left mode mostly consists of estimates of false positives. Increasing the sample size of the identification data sets (1000 patients, bottom panel) SA are correctly estimated on the identification sets, and thus, confirmed on the validation sets, even though they are of smaller size. Selected variables are mostly true positives.

The identification step was improved by using larger sample sizes in the identification step. Increasing the size of validation sets cannot improve the first estimation obtained during the identification step.

5   Interpretation of Results

On completion of the statistical analysis, the statistician returns a list of proteins that is not particularly informative if the proteins are not considered in the context of the biological processes in which they are involved. Tools can thus be used to 1-integrate and visualize the results obtained by setting them in different biological networks. 2-visualize the interaction between networks involved and visualize signaling/metabolic pathways. This gives cohesion to the results and may allow the biologist/clinician to open up new research perspectives. We will not go more into detail here but want to point out that this part of the study is inseparable from the statistical analysis. This implies bioinformatics tools that are not the subject of this chapter. We can, however, cite, for example, the open-source software platform Cytoscape (http://www.cytoscape.org/) [36].

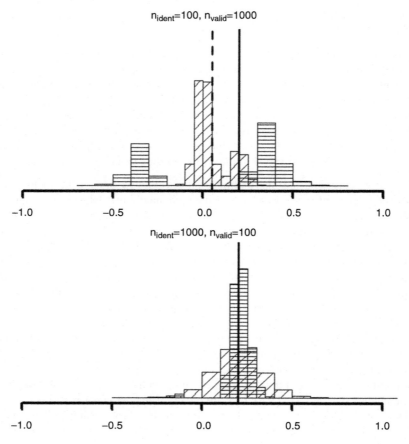

Fig. 4 Distribution of the strength of association estimated on identification (*horizontal hatching*) and validation (*diagonal hatching*) studies. The *vertical dotted line* indicates the mean of the distribution of the estimated strength of association computed on the validation simulated datasets. The *vertical continuous line* indicates the value of the true simulated strength of association of active variables. The *upper panel* shows the results obtained with identification and validation datasets of, respectively, $n_{ident} = 100$ and $n_{valid} = 1000$ individuals. The *bottom panel* shows the results obtained with identification and validation datasets of, respectively, $n_{ident} = 1000$ and $n_{ident} = 100$ individuals

6 Conclusion

Label-free LC-MS/MS is widely used for high-throughput quantitative proteomics. It enables the relative quantification and comparison of the abundance of peptides and/or proteins in complex biological mixtures representative of different biological or clinical conditions of interest. Sample preparation and data acquisition are straightforward and can be applied to sufficient number of samples to get robust studies. It is thus a method of choice for the discovery of new biomarkers. Candidate biomarkers should then be validated as relevant biomarkers through other technologies like electro-spray ionization mass spectrometer (ESI-MS) operating in multiple reactions monitoring (MRM) mode, for example, which makes it possible to have targeted absolute quantification for targeted proteins/peptides of interest.

To simplify the discussion, this chapter focused on between-group comparisons when there are two groups. Once pre-processed, label-free LC-MS/MS data corresponds to a matrix of intensities with runs in rows and peptides or proteins in columns. At this stage, all statistical methods developed in the context of high-dimensional data can be used to deal with other biological questions, such as continuous or survival outcome.

A large panel of statistical methods is proposed for the analysis of the datasets generated by label-free LC-MS/MS. Each manufacturer sells their own mass spectrometer with their own related analysis software. Each of these software solutions was developed independently with the associated algorithms more or less precisely described. In parallel, the scientific community develops open-source tools for each of the different pre-processing and processing steps. These tools may be embedded in open-source pipelines and it allows the user to pick among different methods for each step. No consensus was obtained because of the heterogeneity of the data, which does not lead to a "one size fits all" solution. As there is no consensus, the objective was not to give a list of immutable solutions but to point out the different questions raised by each stage and the major approaches that can be used to tackle them. The most important task is to control the tools that are employed and to use them to their best advantage by setting them up in appropriate and personalized workflows.

In the field of label-free LC-MS biomarker discovery, the equipment and corresponding pre-analytical methods are in constant evolution. Different mass spectrometers dedicated to LC-MS/MS exist. They differ in their resolution and sensitivity. We think that LC-MS/MS is a long-term approach for the discovery of proteomic biomarkers. The biggest issue is the reproducibility of the results over time because of the liquid chromatography step, but we imagine that future technological improvements will solve these problems. The mass spectrometer will also be more resolutive as well as faster in the analysis time.

So we guess that this technology will still have potential in the future.

Acknowledgements We wish to thank the members of the CLIPP Platform (University of Burgundy) for their contribution, and most particularly Géraldine Lucchi and Delphine Pecqueur, who read the article thoroughly. We also wish to thank the "Centre de Langues" (University of Burgundy) for editing the manuscript.

References

1. Sandin, M., Chawade, A., & Levander, F. (2015). Is label-free lc-ms/ms ready for biomarker discovery? *Proteomics Clinical Applications, 9*, 289–294.
2. Oberg, A. L., & Vitek, O. (2009). Statistical design of quantitative mass spectrometry-based proteomic experiments. *Journal of Proteome Research, 8*(5), 2144–2156.
3. Ma, K., Vitek, O., & Nesvizhskii, A. I. (2012). A statistical model-building perspective to identification of ms/ms spectra with peptideprophet. *BMC Bioinformatics, 13*(Suppl 16), S1.
4. Deutsch, E. W., Mendoza, L., Shteynberg, D., Farrah, T., Lam, H., Tasman, N., et al. (2010). A guided tour of the trans-proteomic pipeline. *Proteomics, 10*(6), 1150–1159.
5. Shteynberg, D., Deutsch, E. W., Lam, H., Eng, J. K., Sun, Z., Tasman, N., et al. (2011) iProphet: Multi-level integrative analysis of shotgun proteomic data improves peptide and protein identification rates and error estimates. *Molecular and Cellular Proteomics, 10*, M111.007690.
6. Nesvizhskii, A. I., Keller, A., Kolker, E., & Aebersold, R. (2003). A statistical model for identifying proteins by tandem mass spectrometry. *Analytical Chemistry, 75*(17), 4646–4658.
7. Karpievitch, Y., Dabney, A., & Smith, R. (2012). Normalization and missing value imputation for label-free lc-ms analysis. *BMC Bioinformatics, 13*(Suppl 16), S5.
8. Lai, X., Wang, L., Tang, H., & Witzmann, F. A. (2011). A novel alignment method and multiple filters for exclusion of unqualified peptides to enhance label-free quantification using peptide intensity in lc-ms/ms. *Journal of Proteome Research, 10*(10), 4799–4812.
9. Lange, E., Tautenhahn, R., Neumann, S., & Gröpl, C. (2008). Critical assessment of alignment procedures for lc-ms proteomics and metabolomics measurements. *BMC Bioinformatics, 9*, 375.
10. Smith, R., Ventura, D., & Prince, J. T. (2015). Lc-ms alignment in theory and practice: A comprehensive algorithmic review. *Briefings in Bioinformatics, 16*(1), 104–117.
11. Monroe, M. E., Shaw, J. L., Daly, D. S., Adkins, J. N., & Smith, R. D. (2008). Masic: A software program for fast quantitation and flexible visualization of chromatographic profiles from detected lc-ms(/ms) features. *Computational Biology and Chemistry, 32*(3), 215–217.
12. Valot, B., Langella, O., Nano, E., & Zivy, M. (2011). Masschroq: A versatile tool for mass spectrometry quantification. *Proteomics, 11*(17), 3572–3577.
13. Polpitiya, A. D., Qian, W.-J., & Jaitly, N. (2008). Dante: A statistical tool for quantitative analysis of - omics data. *Bioinformatics, 24*, 1556–1558.
14. Clough, T., Key, M., Ott, I., Ragg, S., Schadow, G., & Vitek, O. (2009). Protein quantification in label-free lc-ms experiments. *Journal of Proteome Research, 8*(11), 5275–5284.
15. Choi, M., Chang, C.-Y., & Vitek, O. (2014). *MSstats: Protein Significance Analysis in DDA, SRM and DIA for Label-free or Label-based Proteomics Experiments.* R package version 2.4.0.
16. Matzke, M., Brown, J. N., Gritsenko, M. A., Metz, T. O., Pounds, J. G., Rodland, K. D., et al. (2013). A comparative analysis of computational approaches to relative protein quantification using peptide peak intensities in label-free lc-ms proteomics experiments. *Proteomics, 13*(3–4), 493–503.
17. Callister, S. J., Barry, R. C., Adkins, J. N., Johnson, E. T., Qian, W.-J., Webb-Robertson, B.-J.M., et al. (2006) Normalization approaches for removing systematic biases associated with mass spectrometry and label-free proteomics. *Journal of Proteome Research, 5*(2), 277–286.
18. Bolstad, B. M., Irizarry, R. A., Astrand, M., & Speed, T. P. (2003). A comparison of normalization methods for high density oligonucleotide array data based on variance and bias. *Bioinformatics, 19*(2), 185–193.
19. Karpievitch, Y. V., Taverner, T., Adkins, J. N., Callister, S. J., Anderson, G. A., Smith, R. D., et al. (2009) Normalization of peak intensities in bottom-up ms-based proteomics using singular value decomposition. *Bioinformatics, 25*(19), 2573–2580.
20. Karpievitch, Y. V., Nikolic, S. B., Wilson, R., Sharman, J. E., & Edwards, L. M. (2014) Metabolomics data normalization with eigenms. *PLoS One, 9*(12), e116221.
21. Leek, J. T., & Storey, J. D. (2007). Capturing heterogeneity in gene expression studies by surrogate variable analysis. *PLoS Genet, 3*(9), e161.

22. Chawade, A., Alexandersson, E., & Levander, F. (2014). Normalyzer: A tool for rapid evaluation of normalization methods for omics data sets. *Journal of Proteome Research, 13*(6), 3114–3120.
23. Lai, X., Wang, L., & Witzmann, F.A. (2013). Issues and applications in label-free quantitative mass spectrometry. *International Journal of Proteomics, 2013,* Article ID 756039.
24. Chawade, A., Sandin, M., Teleman, J. N., Malmströ, J., & Levander, F. (2015). Data processing has major impact on the outcome of quantitative label-free lc-ms analysis. *Journal of Proteome Research, 14*(2), 676–687.
25. Zou, H., Hastie, T., & Tibshirani, R. (2004). Sparse principal component analysis. *Journal of Computational and Graphical Statistics, 15,* 2006.
26. Benjamini, Y., & Hochberg, Y. (1995) Controlling the false discovery rate: A practical and powerful approach to multiple testing. *Journal of the Royal Statistical Society, Series B, 57*(1), 289–300.
27. Wold, H. (2005). In S. Kots & N.L. Johnson (Eds.), *Partial least squares.* New York: Wiley.
28. Boulesteix, A.-L., & Strimmer, K. (2007). Partial least squares: A versatile tool for the analysis of high-dimensional genomic data. *Briefings in Bioinformatics, 8*(1), 32–44.
29. Tibshirani, R. (1996). Regression shrinkage, selection via the lasso. *Journal of the Royal Statistical Society, Series B, 58,* 267–288.
30. Hoerl, A., & Kennard, W. (1970). Ridge regression: Applications to nonorthogonal problems. *Technometrics, 12*(1), 69–82.
31. Zou, H., & Hastie, T. (2005). Regularization and variable selection via the elastic net. *Journal of the Royal Statistical Society, Series B: Statistical Methodology, 67*(2), 301–320.
32. Truntzer, C., Mostacci, E., Jeannin, A., Petit, J. M., Ducoroy, P., & Cardot, H. (2014). Comparison of classification methods that combine clinical data and high-dimensional mass spectrometry data. *BMC Bioinformatics, 15*(385), 1–12.
33. Chun, H., & Keles, S. (2010). Sparse partial least squares regression for simultaneous dimension reduction and variable selection. *Journal of the Royal Statistical Society, Series B: Statistical Methodology, 72,* 3–25.
34. Clough, T., Thaminy, S., Ragg, S., Aebersold, R., & Vitek, O. (2012). Statistical protein quantification and significance analysis in label-free lc-ms experiments with complex designs. *BMC Bioinformatics, 13*(Suppl 16), S6.
35. Truntzer, C., Maucort-Boulch, D., & Roy, P. (2013). Impact of the selection mechanism in the identification and validation of new "omic" biomarkers. *Journal of Proteomics and Bioinformatics, 6*(8), 164–170.
36. Lopes, C. T., Franz, M., Kazi, F., Donaldson, S. L., Morris, Q., & Bader, G. D. (2010) Cytoscape web: an interactive web-based network browser. *Bioinformatics, 26*(18), 2347–2348.

Bayesian Posterior Integration for Classification of Mass Spectrometry Data

Bobbie-Jo M. Webb-Robertson, Thomas O. Metz, Katrina M. Waters, Qibin Zhang, and Marian Rewers

1 Introduction

Mass spectrometry (MS) can be used to study multiple biomolecules and is routinely used to measure proteins, metabolites, and lipids [6, 10, 13, 14, 17, 26]. In addition, it has become more common recently to collect disparate clinical measurements and multiple omics data in the context of a single experiment. Building predictive models from these datasets is challenging since often the data is heterogeneous with different levels of variability and may have issues with missing values [29]. Thus, standard analysis and machine learning methods are often not appropriate or capable of analyzing these collections of datasets. Thus, robust statistical integration approaches are needed to identify the key panel of biomarkers related to the phenotype of interest associated with the study (e.g., cancer, diabetes).

2 Classification via Maximum Posterior Probability

The development of classification algorithms is based on a training dataset comprised of P features, $X = \{x_1, \ldots, x_P\}$, for which a pre-defined number of categories

B.M. Webb-Robertson (✉) • T.O. Metz • K.M. Waters
Pacific Northwest National Laboratory, Richland, WA, USA
e-mail: bj@pnnl.gov

Q. Zhang
University of North Carolina Greensboro, Greensboro, NC, USA

M. Rewers
School of Medicine, Barbara Davis Center for Diabetes, University of Colorado, Boulder, CO, USA

© Springer International Publishing Switzerland 2017
S. Datta, B.J.A. Mertens (eds.), *Statistical Analysis of Proteomics, Metabolomics, and Lipidomics Data Using Mass Spectrometry*, Frontiers in Probability and the Statistical Sciences, DOI 10.1007/978-3-319-45809-0_11

are set for each sample, $C = \{c_1, \ldots, c_J\}$. There are many algorithms that are available to generate classification functions to identify the most likely category for a new test sample, x, for example, Naïve Bayes (NB), logistic regression, linear discriminant analysis (LDA), support vector machines, K-nearest neighbors, and random forests based on classification trees (RT). Any of these classifiers that can produce a quantitative score associated with the likelihood that the sample x is in each class can generate a posterior probability summarizing the probability associated with each of the J classes [8, 16, 24]. The sample x can then be assigned to the group with the maximum posterior probability. NB classifiers are both a powerful classification methodology and simple to explain with respect to posterior probability. NB classifiers assume features are independent given the categories being classified:

$$P\left(X|c_j\right) = \prod_i P\left(x_i|c_j\right).$$

The probability of a given class given the observed data is the posterior probability as given by Bayes Theorem:

$$P\left(c_j|X\right) = \frac{P\left(c_j\right) \prod_i P\left(x_i|c_j\right)}{\sum_j P\left(c_j\right) \prod_i P\left(x_i|c_j\right)}. \tag{1}$$

The predicted class is associated with the maximum of $P\left(c_j|X\right)$. Since each variable is treated independently, NB can be used on large collections of features, continuous and discrete and has been shown to work well across various domains [11, 19, 27, 28].

3   Parallel Posterior Probability Product Integration

One common strategy to integrate disparate data is to generate classification models in parallel on independent data stream and then integrate at the posterior probability level [1, 12, 28]. Therefore, specific datasets, such as targeted proteomics and metabolomics, which may yield optimal prediction accuracies with different algorithms, such as LDA and RT, respectively, can be used and integrated via straightforward multiplication of the posteriors associated with each dataset for the data associated with the test sample from each dataset j, x^j.

$$P\left(c_j|x^1, \cdots, x^K\right) \approx \prod_i P\left(c_j|x^j\right). \tag{2}$$

There are many other posterior probability combination rules that can be utilized, such as averages, sums, maximums, and so forth [12], however in this circumstance the benefits of classifier integration are demonstrated via a Posterior Product Probability (P3).

4 Feature Selection

One key component of the development of predictive models using large numbers of predictors is feature selection. There are a multitude of feature selection algorithms that have been proposed and evaluated in biological applications [4, 21, 23]. Here a common algorithm, recursive feature elimination (RFE) [5, 7, 18], will be discussed as one example that is used extensively in biology with various machine learning algorithms, such as LDA and SVMs [3, 7, 9, 22, 25, 30]. RFE is readily available in most statistical programming languages and is simple to implement. RFE is an algorithm that sequentially removes features that do not contribute, or contribute the least, in discriminating between different classes. The discrimination between classes is usually quantified as classification accuracy or the area under a receiver operating characteristic (ROC) curve (AUC) derived via cross-validation (CV) steps. RFE is a good option when evaluating all possible feature combinations is computationally restrictive. It can be easily modified to terminate at specific criteria or to evaluate all feature set sizes from P to one. An important component of feature selection for data integration, however, is that the accuracy is measured in the context of the combination of the disparate features in relation to the integrated score (Eq. 2).

5 Case Study: Model Evaluation and Visualization

The Diabetes Antibody Standardization Program (DASP) is a collective effort to evaluate and improve islet autoantibody assays of type 1 diabetes (T1D) [2, 15, 20]. T1D is diagnosed in a patient according to World Health Organization criteria and samples were collected within 14 days of starting insulin. The initial dataset contained 50 and 100 sera samples from T1D and control (CTRL) patients, respectively, but was reduced to 40 and 79 to remove potential gender and age effects. Three MS-based analyses were performed, lipidomics, metabolomics, and targeted proteomics. Gas chromatography-MS was used to quantify hundreds of lipids and metabolites via peak area estimation. For the purpose of this case study, 171 lipids and 149 metabolites that were observed with complete reproducibility in our study were evaluated (i.e., there were no missing values in either dataset). Targeted Liquid Chromatography Multiple Reaction Monitoring MS data of the relative abundance of 52 peptides across the patient cohort was collected. Peptides were also quantified by peak area. The peptide data was previously published [31].

The data is first analyzed via standard NB -P3 on the concatenated data consisting of the full 372 feature set. The NB model was fit using the "fitcnb" function in MatLab, with default parameters, which models the predictor distribution within each class as a Gaussian. The priors are derived empirically as frequencies of the classes in the training data. The posterior probabilities were obtained from the "predict" function with default parameters. Accuracy was measured as the AUC returned from CV. Since separate iterations of CV will yield different results, the CV was repeated 100 times to get a more robust metric of accuracy and the variability associated with the AUC. This yielded an AUC of ∼0.88 with a standard deviation of 0.02. An AUC for a perfect classifier is 1.0 and 0.5 for a random classifier, thus the full model based on all 372 features performs significantly better than a random model.

Feature selection was then performed via RFE using the metric of optimization also as average AUC from CV. In particular, at each iteration of RFE all remaining m features in the model are removed iteratively, i.e., $P\left(x_i \middle| c_j\right)$ is removed and m AUC values are obtained. The feature for which the maximum AUC is observed is removed and the process is repeated for the remaining m-1 features. Although rules can be utilized with RFE to determine an end point, such as terminate the feature select at the point that the AUC decreases by δ, in this case the feature space was small enough that RFE terminated with one feature. The optimal model was selected to be the one with m features at which the AUC was maximum. Figure 1 gives a heat map showing the distribution of the features selected at each iteration of the RFE as split between the three datasets. The single most discriminative feature is a peptide and the best two-feature model includes one peptide and one metabolite. In addition, this shows the clear lower performance of the lipid data.

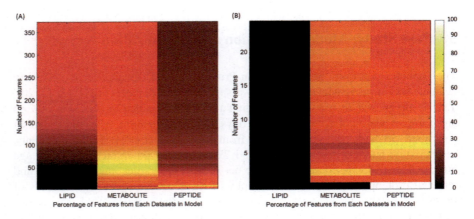

Fig. 1 Heatmap displaying the percentage of features from each of the three datasets. (**a**) All features decreasing from 372 to one feature using RFE, thus the *top row* represents ∼46, ∼40, and 14 % since the individual datasets have 171, 149, and 52 features, respectively. (**b**) The features decreasing from 20 to one feature after RFE has removed 352 features; the *pure white on the bottom right corner* indicates that the last feature remaining (100 % representation) is a peptide

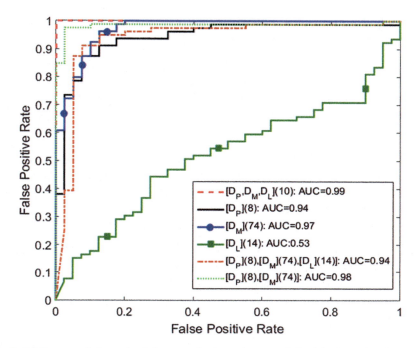

Fig. 2 ROC curve of the optimal feature selection subsets as defined by RFE, requiring only ten features for the integrated model, but requiring 14, 74, and 8 lipids (D_L), metabolites (D_M), and peptide (D_P), respectively, for the individual datasets to identify the optimal AUC via RFE. Integrating features after individual RFE improves the AUC, but it is not statistically larger than the full integrated model with ten features

Figure 2 displays an ROC curve of each individual dataset at the point they reach the maximum AUC, which is 9.5 % of the lipids, 42.7 % of the metabolites, and 15.1 % of the peptides. As seen in Fig. 2, the AUCs of both the metabolomic and proteomic datasets are quite high, greater than 0.94, however; the integrated model yields an AUC greater than 0.99 while only requiring a combination of ten peptide and metabolite features. If features are selected based on individual RFE analyses, the subset would have 96 features; 8, 74, and 14 peptides, metabolites, and lipids, respectively. In this scenario, integrating the features into a single NB classifier yields an AUC of 0.94. If only the subset of 82 peptides and metabolites is used, the AUC improves to 0.98. However, based on a Wilcoxon signed rank test the integrated model with only ten features is still significantly more accurate ($p < 1e{-}10$).

Lastly, the datasets were integrated at the model level (Eq. 2). To perform this first each full dataset was evaluated independently to identify the learning algorithms (RT, LDA, NB) that performed best as measured by average AUC. The LDA was fit using the "fitcdiscr" function in MatLab, with default parameters, which models the posterior as the product of the multivariate normal distribution and the prior probability. The priors are derived empirically in the same manner as the NB model. The RT was fit using the "TreeBagger" function in MatLab based on Brieman's method. In both cases, the posterior probabilities were obtained from the "predict" function with default parameters. The RT posterior is computed as the fraction of observations of a particular class in a tree leaf averaged across trees.

From this analysis an RT was found to work well for both the lipidomic and metabolomic datasets, while a standard LDA was selected for the proteomics data. To simplify the case study feature selection is not incorporated, but could be by simply recursive feature elimination across the dataset. Figure 3 plots the classification accuracy plots, which show the percentage of samples predicted in each class [1]. The rows represent the total number in each group, 40 and 79 for the CTRL and T1D, respectively, for this case and the colors indicate the fraction classified in either of the two classes. A perfect classifier in this visualization would have all white on the diagonal from top left to bottom right. For this case study it is observed that the total classification accuracy (i.e., the percentage of both groups correctly classified) is 69.8 %, 89.9 %, and 95.8 % for the lipid, metabolite, and protein data, respectively. Integration based on P3 attained from each dataset results in a total classification accuracy of 97.5 % (right). This type of analysis and visualization allows the researcher to evaluate which samples are being misclassified into which groups, which can be particularly powerful when more than two classes are being classified. One key interesting difference between this approach and the standard method of concatenating the features is that the AUC of the integrated model is at 0.99 prior to any feature selection. Errors in one dataset are compensated for by another.

6   Discussion

Data integration is becoming a necessity to address the complexity of the experiments being performed via MS. MS is enabling HTP experiments that can simultaneously measure various types of biomolecules and integrating these data into an interpretable form requires the development of statistical methods that can manage the differences in formats, resolutions, and data sizes from the different instruments. Data fusion has the potential in many cases to give a more complete or accurate model of the biological question of interest. For example, the diabetes case could be accurately classified by a combination of just ten metabolites and peptides while individual datasets required many more features. In general, the posterior probabilities acquired from machine learning algorithms allow the uncertainty with the classification of an individual sample to be these analyzed, as well as offering a straightforward approach to directly integrate disparate data. However, like many machine learning and integration algorithms there are many approaches that work well for one dataset, but may perform differently in new circumstances, so all analyses should be undertaken with care.

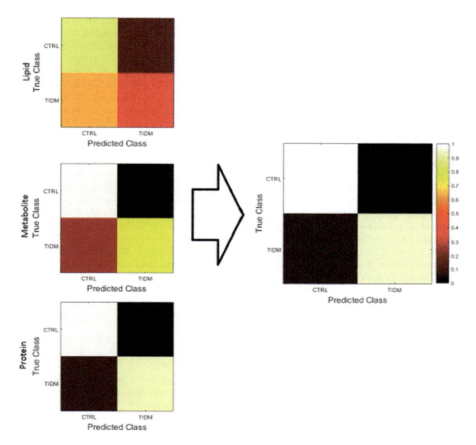

Fig. 3 Heat maps display the classification accuracy for individual datasets (*left*) and the integrated model (*right*). *White diagonal from top left to bottom right* indicates perfect accuracy

Acknowledgments This work was funded by NIH NIDDK grant R33 DK070146. Significant portions of the work were performed at the Environmental Sciences Laboratory, a national scientific user facility sponsored by the Department of Energy's (DOE) Office of Biological and Environmental Research and located at Pacific Northwest National Laboratory (PNNL) in Richland, Washington. PNNL is a multi-program national laboratory operated by Battelle for the DOE under contract DE-AC05-765RL0 1830.

References

1. Beagley, N., Stratton, K. G., & Webb-Robertson, B. J. (2010). VIBE 2.0: Visual integration for Bayesian evaluation. *Bioinformatics, 26*(2), 280–282. doi:10.1093/bioinformatics/btp639.
2. Bingley, P. J., Bonifacio, E., & Mueller, P. W. (2003). Diabetes Antibody Standardization Program: First assay proficiency evaluation. *Diabetes, 52*(5), 1128–1136.

3. Chen, X., Liang, Y. Z., Yuan, D. L., & Xu, Q. S. (2009). A modified uncorrelated linear discriminant analysis model coupled with recursive feature elimination for the prediction of bioactivity. *SAR and QSAR in Environmental Research, 20*(1–2), 1–26. doi:10.1080/10629360902724127.
4. Dai, Q., Cheng, J. H., Sun, D. W., & Zeng, X. A. (2015). Advances in feature selection methods for hyperspectral image processing in food industry applications: A review. *Critical Reviews in Food Science and Nutrition, 55*(10), 1368–1382. doi:10.1080/10408398.2013.871692.
5. De Martino, F., Valente, G., Staeren, N., Ashburner, J., Goebel, R., & Formisano, E. (2008). Combining multivariate voxel selection and support vector machines for mapping and classification of fMRI spatial patterns. *NeuroImage, 43*(1), 44–58. doi:10.1016/j.neuroimage.2008.06.037.
6. Eriksson, C., Masaki, N., Yao, I., Hayasaka, T., & Setou, M. (2013). MALDI imaging mass spectrometry-A mini review of methods and recent developments. *Mass Spectrom (Tokyo), 2*(Spec Iss), S0022. doi:10.5702/massspectrometry.S0022.
7. Gholami, B., Norton, I., Tannenbaum, A. R., & Agar, N. Y. (2012). Recursive feature elimination for brain tumor classification using desorption electrospray ionization mass spectrometry imaging. *Conference Proceedings: Annual International Conference of the IEEE Engineering in Medicine and Biology Society, 2012*, 5258–5261. doi:10.1109/EMBC.2012.6347180.
8. Hand, D. J. (1997). *Construction and assessment of classification rules.* New York: Wiley.
9. Hu, C., Wang, J., Zheng, C., Xu, S., Zhang, H., Liang, Y., et al. (2013). Raman spectra exploring breast tissues: Comparison of principal component analysis and support vector machine-recursive feature elimination. *Medical Physics, 40*(6), 063501. doi:10.1118/1.4804054.
10. Ibanez, C., Simo, C., Garcia-Canas, V., Cifuentes, A., & Castro-Puyana, M. (2013). Metabolomics, peptidomics and proteomics applications of capillary electrophoresis-mass spectrometry in foodomics: A review. *Analytica Chimica Acta, 802*, 1–13. doi:10.1016/j.aca.2013.07.042.
11. Jarman, K. H., Kreuzer-Martin, H. W., Wunschel, D. S., Valentine, N. B., Cliff, J. B., Petersen, C. E., et al. (2008). Bayesian-integrated microbial forensics. *Applied and Environmental Microbiology, 74*(11), 3573–3582. doi:10.1128/AEM.02526-07.
12. Jia, P., He, H., & Lin, W. (2005). Decision by maximum of posterior probability average with weights: A method of multiple classifiers combination. In *Proceedings of Fourth International Conference on Machine Learning and Cybernetics, Guangzhou, 2005* (pp. 1949–1954). IEEE.
13. Kruve, A., Rebane, R., Kipper, K., Oldekop, M. L., Evard, H., Herodes, K., et al. (2015). Tutorial review on validation of liquid chromatography-mass spectrometry methods: Part I. *Analytica Chimica Acta, 870*, 29–44. doi:10.1016/j.aca.2015.02.017.
14. Kruve, A., Rebane, R., Kipper, K., Oldekop, M. L., Evard, H., Herodes, K., et al. (2015). Tutorial review on validation of liquid chromatography-mass spectrometry methods: Part II. *Analytica Chimica Acta, 870*, 8–28. doi:10.1016/j.aca.2015.02.016.
15. Lampasona, V., Schlosser, M., Mueller, P. W., Williams, A. J., Wenzlau, J. M., Hutton, J. C., et al. (2011). Diabetes antibody standardization program: First proficiency evaluation of assays for autoantibodies to zinc transporter 8. *Clinical Chemistry, 57*(12), 1693–1702. doi:10.1373/clinchem.2011.170662.
16. Lanckriet, G. R., De Bie, T., Cristianini, N., Jordan, M. I., & Noble, W. S. (2004). A statistical framework for genomic data fusion. *Bioinformatics, 20*(16), 2626–2635. doi:10.1093/bioinformatics/bth294.
17. Liesenfeld, D. B., Habermann, N., Owen, R. W., Scalbert, A., & Ulrich, C. M. (2013). Review of mass spectrometry-based metabolomics in cancer research. *Cancer Epidemiology, Biomarkers and Prevention, 22*(12), 2182–2201. doi:10.1158/1055-9965.EPI-13-0584.
18. Lin, X., Yang, F., Zhou, L., Yin, P., Kong, H., Xing, W., et al. (2012). A support vector machine-recursive feature elimination feature selection method based on artificial contrast variables and mutual information. *Journal of Chromatography B, Analytical Technologies in the Biomedical and Life Sciences, 910*, 149–155. doi:10.1016/j.jchromb.2012.05.020.
19. Piao, Y., Piao, M., Park, K., & Ryu, K. H. (2012). An ensemble correlation-based gene selection algorithm for cancer classification with gene expression data. *Bioinformatics, 28*(24), 3306–3315. doi:10.1093/bioinformatics/bts602.

20. Rolandsson, O., Hagg, E., Nilsson, M., Hallmans, G., Mincheva-Nilsson, L., & Lernmark, A. (2001). Prediction of diabetes with body mass index, oral glucose tolerance test and islet cell autoantibodies in a regional population. *Journal of Internal Medicine, 249*(4), 279–288.

21. Saeys, Y., Inza, I., & Larranaga, P. (2007). A review of feature selection techniques in bioinformatics. *Bioinformatics, 23*(19), 2507–2517. doi:10.1093/bioinformatics/btm344.

22. Saligan, L. N., Fernandez-Martinez, J. L., deAndres-Galiana, E. J., & Sonis, S. (2014). Supervised classification by filter methods and recursive feature elimination predicts risk of radiotherapy-related fatigue in patients with prostate cancer. *Cancer Information, 13*, 141–152. doi:10.4137/CIN.S19745.

23. Semmar, N., Canlet, C., Delplanque, B., Ruyet, P. L., Paris, A., & Martin, J. C. (2014). Review and research on feature selection methods from NMR data in biological fluids. Presentation of an original ensemble method applied to atherosclerosis field. *Current Drug Metabolism, 15*(5), 544–556.

24. Shapiro, C. P. (1977). Classification by maximum posterior probability. *The Annals of Statistics, 5*(1), 185–190.

25. Tao, P., Liu, T., Li, X., & Chen, L. (2015). Prediction of protein structural class using trigram probabilities of position-specific scoring matrix and recursive feature elimination. *Amino Acids, 47*(3), 461–468. doi:10.1007/s00726-014-1878-9.

26. Van Oudenhove, L., & Devreese, B. (2013). A review on recent developments in mass spectrometry instrumentation and quantitative tools advancing bacterial proteomics. *Applied Microbiology and Biotechnology, 97*(11), 4749–4762. doi:10.1007/s00253-013-4897-7.

27. Webb-Robertson, B. J., Kreuzer, H., Hart, G., Ehleringer, J., West, J., Gill, G., et al. (2012). Bayesian integration of isotope ratio for geographic sourcing of castor beans. *Journal of Biomedicine and Biotechnology, 2012*, 450967. doi:10.1155/2012/450967.

28. Webb-Robertson, B. J., McCue, L. A., Beagley, N., McDermott, J. E., Wunschel, D. S., Varnum, S. M., et al. (2009). A Bayesian integration model of high-throughput proteomics and metabolomics data for improved early detection of microbial infections. *Pac Symp Biocomput* (pp. 451–463).

29. Webb-Robertson, B. J., Wiberg, H. K., Matzke, M. M., Brown, J. N., Wang, J., McDermott, J. E., et al. (2015). Review, evaluation, and discussion of the challenges of missing value imputation for mass spectrometry-based label-free global proteomics. *Journal of Proteome Research, 14*(5), 1993–2001. doi:10.1021/pr501138h.

30. Yousef, M., Jung, S., Showe, L. C., & Showe, M. K. (2007). Recursive cluster elimination (RCE) for classification and feature selection from gene expression data. *BMC Bioinformatics, 8*, 144. doi:10.1186/1471-2105-8-144.

31. Zhang, Q., Fillmore, T. L., Schepmoes, A. A., Clauss, T. R., Gritsenko, M. A., Mueller, P. W., et al. (2013). Serum proteomics reveals systemic dysregulation of innate immunity in type 1 diabetes. *Journal of Experimental Medicine, 210*(1), 191–203. doi:10.1084/jem.20111843.

Logistic Regression Modeling on Mass Spectrometry Data in Proteomics Case-Control Discriminant Studies

Bart J.A. Mertens

1 Introduction

Mass spectrometry is currently attracting much interest for its ability to simultaneously profile hundreds of proteins in tissue, urine, or serum samples [17] and across a large range of protein masses. The human proteome is a potentially rich source of diagnostic information and hence, this technology holds great promise for the construction of diagnostic rules for the detection of pathological states, by identifying differential peptide/protein expression through evaluation of mass spectra. Such research is particularly relevant for the construction of new diagnostic or screening tools in diseases where either no methods are available or when the reliability of existing approaches is poor.

Good examples where advances in medical diagnosis are urgently needed are in the early detection of cancer, such as breast and especially colon cancer. For colon cancer, this applies as current diagnostic methods tend to detect pathology at an advanced stage, which limits the impact of treatment regimes. Reliable diagnosis of muscular dystrophy with simple noninvasive procedures is another such problem, since current methods require muscle biopsies to be taken, which is a cumbersome and unpleasant procedure for the patient. Furthermore, diagnosis is difficult and prone to error, as it is based on extensive histology of the collected tissue samples.

In oncology, proteomic analysis of blood samples is of interest since it is hypothesized that malignant or pre-malignant cells will interact with their cellular environment, causing changes in enzymatic and protease activity. Such protein fragments may be diffused into the blood circulation and as a consequence,

B.J.A. Mertens (✉)
Department of Medical Statistics, Leiden University Medical Center, PO Box 9600,
2300 RC Leiden, The Netherlands
e-mail: b.mertens@lumc.nl; http://www.lumc.nl

© Springer International Publishing Switzerland 2017
S. Datta, B.J.A. Mertens (eds.), *Statistical Analysis of Proteomics, Metabolomics, and Lipidomics Data Using Mass Spectrometry*, Frontiers in Probability and the Statistical Sciences, DOI 10.1007/978-3-319-45809-0_12

213

proteomic mass spectrometry is particularly attractive in diagnostic research, since it may be based on the analysis of sample materials which are easily available within routine clinical practice, such as blood or urine samples. Similarly, muscular dystrophy has a genetic basis which causes changes in the proteome of affected muscle tissues throughout the body. By applying mass spectrometry on blood, serum, or plasma samples, researchers seek to construct diagnostic approaches which can be quickly, put to use in routine hospital practice, to either replace or augment the presently applied diagnostic methods. In all these examples, a related research objective is the identification of proteomic biomarkers.

We discuss data from two case-control experiments set up at the Leiden University Medical Center to evaluate potential of mass spectrometry-based diagnosis in the above described examples: a colon cancer study and a mouse study on muscular dystrophy. Statistical methodology is introduced for the calibration of discriminant rules on the proteomics spectra in such case-control experiments. We will use the colon cancer data to introduce the methodology. We propose and describe a post-hoc data exploratory analysis of the results from application of such methodology. Subsequently, results are contrasted with application on the experimental data in the muscular dystrophy case-control study.

1.1 Mass Spectrometry Experimental Data

Mass Spectrometry and Data

Mass spectrometric measurement is based on the principle of ionizing prepared samples (such as blood serum or plasma) through application of a high-energy laser beam. The laser-induced ionization vaporizes the sample material to a cloud of charged particles, part of which will consist of the protein fragments of interest. This charged particle mixture is then accelerated down a so-called flight tube through application of an electric field. For each particle within the vaporized mixture, the time-of-flight may be recorded by means of a detector which is placed at the end of the flight tube. The flight times themselves depend on the mass-to-charge (m/z) ratios of particles. The spectrometer will convert the observed times-of-flight to mass-to-charge values, which is achieved through a quadratic transformation, which is calibrated prior to measurement by using test samples. The unit of measurement at the mass-to-charge scale is Dalton (D).

In this manner, the spectrometer records, for each sample, a series of intensity values on a pre-defined, fixed, and ordered set of contiguous bins which spans the mass-to-charge interval of interest, which so discretizes the observed signal. For each sample, the recorded intensity for any bin corresponds to the number of particles detected within that bin's mass-to-charge range and hence, we may think of mass spectra as extremely high-dimensional histograms, as the number of bins on the full mass-to-charge range will typically be in the thousands. From a substantive scientific point of view, it is the mass-to-charge values which are of interest, as these may allow scientists to interpret spectra by identifying known combinations ("fingerprints") of intensity spikes at specified mass-to-charge values within the spectra corresponding to known molecules. In practical clinical proteomics work, however, we cannot always rely on known molecules being so readily identified based on the spectrum alone. In such a case, subsequent sequencing must identify the relevant peptides and this result must then in turn be compared with the observed mass-to-charge ratio of relevant peaks within the spectrum.

To denote that the spectra are themselves generated from observed times-of-flight, the acronym "TOF" is often used in connection to description of mass spectrometry equipment. In this paper, MALDI-TOF spectra will be investigated. The name "MALDI" is an explicit reference to the mixing of sample material with an energy-absorbing material, typically referred to as "matrix," which helps the ionization process. Figure 1 gives an example of a mass spectrum obtained from a patient blood serum sample. Clearly visible is an increase in intensities at the lower mass-to-charge range (roughly 1000 up to 4000 Dalton), which is sometimes referred to as "baseline" and gradually decays towards the higher mass-to-charge values. Superposed on this baseline, a number of "peaks" can be identified of

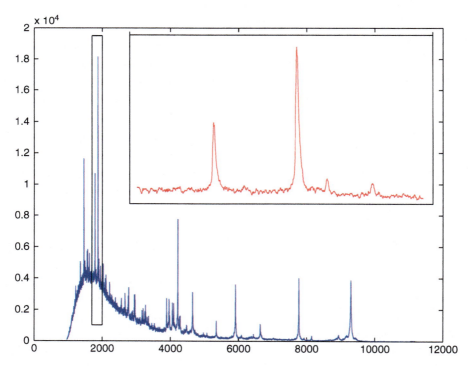

Fig. 1 Example of a MALDI spectrum. The *rectangle to the left* selects a spectral region between 1700 and 2000 Dalton, which is repeated as the enlarged rectangle in the upper right corner of the plot

varying intensities. The smaller rectangle at the left-side of the spectrum selects a region ranging from 1700 up to 2000 Dalton, which is repeated enlarged in the larger rectangle in the upper right corner of the same plot. At this scale, the so-called baseline appears almost flat and the two peaks appear as narrow spikes of approximately 20 Daltons wide, within the range selected . The main contribution to the baseline effect is thought to be due to the ionization of matrix particles and thus may be regarded as a nuisance effect which is not of direct interest to the analyst.

Proteomic Case-Control Experiments

We consider data from two case-control experiments recently carried out at the Leiden University Medical Center. The first contrasts serum samples of 50 healthy controls with those from 63 colon cancer patients in a randomized block design. Chapter "Transformation, Normalization and Batch Effect in the Analysis of Mass Spectrometry Data for Omics Studies" in this volume discusses the same data and shows the experimental design (see Table 1, [15]) of this study. For each sample, the corresponding spectrum was generated using MALDI-TOF mass spectrometry (specifically an Ultraflex TOF/TOF instrument, Bruker Daltonics, equipped with a SCOUT ion source which was operated in linear mode), after a series of serum pre-treatment steps (bead-based fractionation, the primary purpose of which is to isolate a specific group of peptides from the serum mixtures for further analysis). The entire signal was discretized on a pre-defined and fixed grid of contiguous small bins, which are about 1 Dalton wide at the lower end of the mass/charge axis, increasing towards a bin width of about 2 Dalton at the upper end of the mass scale ranging from 1000 up to 5000 Dalton. This gives a grid consisting of 2346 bins. Some mild pre-processing steps were applied to the data prior to subsequent analysis. The first step consisted of baseline removal through application of an asymmetric least squares algorithm [7], in order to remove the matrix effect at lower mass-to-charge values. Subsequently, spectra were standardized by subtracting the within-spectrum medians and re-scaling using the within-spectrum interquartile ranges. Finally, a log-transform was applied to these transformed spectral intensities. The latter steps were taken to stabilize the variances which range several orders of magnitude across the spectral range and to remove effects from differences in ionization levels between samples. We refer to de Noo et al. [4] or the supplementary materials for the precise details on design, measurement protocols, and sample processing.

Figure 2 shows a plot contrasting the sample mean spectra obtained from the colon cancer case-control experiment, subsequent to the above pre-processing steps. The plot reveals a large number of localized intensity peaks and graphical inspection suggests several potential areas of differential expression. The selected area between 1300 and 1900 Dalton with corresponding subplot represents a spectral region for which the more formal analyses presented subsequently in this paper confirms differences in intensity levels between cases and controls.

The muscular dystrophy study has similar design and measurement procedures and we will introduce details further on in the paper.

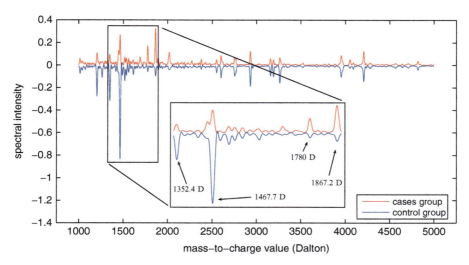

Fig. 2 Mean spectra for each group separately within the colon cancer case-control experiment, after pre-processing. We plot negative intensity value for the control group (bottom mean spectrum). A number of spectral locations are identified which correspond to peaks with different intensity levels between case and control samples

1.2 Statistics and Mass Spectrometry Proteomics

A key problem in proteomic spectra data analysis is the absence of good physical models for the data generation process of mass spectral data, though some authors have investigated model-based simulation exercises (see [3]). Should such models be available, then they might be used as the basis for the data-analytic approach. Good models are not available, due to the complicated nature of both the sample preparation steps and the measurement generation processes on which mass spectrometry is based. For example, a precise understanding of the origins of the baseline is lacking, although it is hypothesized that the matrix solution added to the sample material is the main contributor to this effect. Likewise, the ionization process is highly complex and subject to many effects, among which ionization suppression for low abundant peptides, due to saturation effects caused by highly abundant proteins within the sample mixtures (e.g., albumin in blood plasma).

This paper proposes an approach to calibration of logistic regression discriminant models, based on an explicit parametrization of the vector of regression coefficients. This allows us to incorporate knowledge on the nature of the spectral signal and leads to more parsimonious models that tend to be more easily interpretable, while maintaining predictive performance in comparison to standard logistic ridge regression implementations. The remainder of the paper, first introduces the model description as well as application in analysis of the colon cancer case-control study discussed above [4]. After presentation of the first results, we consider a sensitivity analysis. From the sensitivity study, a key feature of the methodology may be demonstrated which is that the model can identify distinct competing predictors with similar predictive performance. We then contrast the methodology with results obtained from more traditional forms of analysis referred to above, specifically linear discriminant analysis with component reduction as well as double cross-validatory penalized logistic ridge regression. Subsequently, we introduce the muscular dystrophy case-control studies and evaluate the method on the mass spectral data from these trials. The final discussion provides additional information on computational issues and the connection with the wider functional data analysis and chemometrics literature.

2   Conditional Modeling of Class Outcome Using Mass Spectra

Let $y_i \in \{0, 1\}$, $i = 1, \ldots, n$ be the binary class indicators denoting presence or absence of the clinical condition of interest and n the sample size. We will write $x_i = (x_{i1}, \ldots, x_{iq})$ for the associated intensities of the features in the mass spectrum from each ith sample (after pre-processing), which is an ordered sequence of intensity values on the grid selected to record the spectrum. The grid is a contiguous set of small bins which span the range of the spectrum and is defined in advance of the experiment and kept fixed across all samples. Similarly, we write $\mathbf{m} = \{m_1, \ldots, m_q\}$ for the ordered set of ordinates of the bins along the mass axis corresponding to the sequence of q measured intensity values from each sample. We consider the binary regression model

$$y_i \sim \text{Bernoulli}(p_i),$$

with

$$g(p_i) = \eta_i,$$

where $\eta_i = x_i\boldsymbol{\beta}$ is a linear predictor, $p_i = P(y_i = 1)$ is the probability of being a case, and g denotes a link function with $\boldsymbol{\beta}$ a vector of regression parameters. Specifying the link function as the logit $g(p) = \log(p/(1 - p))$ delivers a logistic regression model on mass spectra. Other link functions might also be considered, such as $g(p) = \Phi^{-1}(p)$, with Φ the cumulative standard normal distribution function, which renders a probit model.

We parameterize the vector of regression parameters $\boldsymbol{\beta}$ in terms of a set of Gaussian basis functions along the mass/charge axis

$$\psi_j(x, \mu_j, \sigma_j) = e^{-\frac{1}{2}\frac{(x-\mu_j)^2}{\sigma_j^2}}, \text{ for } j = 1, \ldots, k,$$

of finite dimension k, where the parameters μ_j and σ_j denote the locations and widths (expressed as standard deviation) for $j = 1, \ldots, k$. Note that k is unknown and thus itself a parameter of the model which is estimated. We now write the regression parameter vector $\boldsymbol{\beta} = (\beta_1, \ldots, \beta_q)$ as a linear combination

$$\boldsymbol{\beta} = \sum_{j=1}^{k} \alpha_j \boldsymbol{\Psi}_j,$$

with

$$\boldsymbol{\Psi}_j = (\psi_j(m_1), \ldots, \psi_j(m_q))^T, \text{ for } j = 1, \ldots, k.$$

The latter functions may be interpreted as operators on x and hence, we will write $\boldsymbol{\Psi}_j(x) = x\boldsymbol{\Psi}_j$ for $j = 1, \ldots, k$. With this formulation, the linear predictor equation may be rewritten as

$$\eta_i = x_i \boldsymbol{\beta} = x_i \sum_{j=1}^{k} \alpha_j \boldsymbol{\Psi}_j = \sum_{j=1}^{k} \alpha_j \boldsymbol{\Psi}_j(x_i),$$

which reveals the linear predictor as a linear combination of basis functions in x. For full generality, we will also include an intercept term into the model by introducing the constant term $\boldsymbol{\Psi}_0(x_i) = 1$, such that the linear predictor becomes

$$\eta_i = \alpha_0 + \sum_{j=1}^{k} \alpha_j \boldsymbol{\Psi}_j(x_i) = \sum_{j=0}^{k} \alpha_j \boldsymbol{\Psi}_j(x_i),$$

where α_0 is the intercept term which augments the vector of regression parameters $\boldsymbol{\alpha} = (\alpha_0, \alpha_1, \ldots, \alpha_k)^T$. In the remainder of this paper and in order to simplify terminology, we will also refer to the basis functions as "basis components," though it must be clearly understood that this is shorthand terminology at all times for "basis function components within the regression coefficients" and that these constitute an assumption at the level of the regression parameters.

2.1 Hierarchical Specification of Logistic Models

For reasons of computational efficiency, we employ a "linearizing" representation of the previously described models based on auxiliary variables, instead of the explicit formulation in terms of link function, as described in [10] and [1]. This is achieved by rewriting the logistic model as

$$y_i = 1 \text{ if } z_i > 0 \text{ and } 0 \text{ otherwise,}$$

for an auxiliary variable $z_i = \mathbf{x}_i\boldsymbol{\beta} + \epsilon_i$ and $\epsilon_i \sim N(0, (2\omega_i)^2)$, where ω_i is a random variable with the Kolmogorov–Smirnov distribution [6]. (Note the assumption $\epsilon_i \sim N(0, 1)$ would be simpler and give the probit model instead.) Within a fully Bayesian paradigm, we will consider the quantities k, $\boldsymbol{\theta} = \{\mu_j, \sigma_j; j = 1, \dots, k\}$ and $\boldsymbol{\alpha}$ as random variables. For logistic regression, we may then factorize the joint distribution as

$$P(\mathbf{y} \mid z, \boldsymbol{\alpha}, \boldsymbol{\theta}, \boldsymbol{\omega}, k)P(z \mid \boldsymbol{\alpha}, \boldsymbol{\theta}, \boldsymbol{\omega}, k)P(\boldsymbol{\alpha} \mid \boldsymbol{\theta}, \boldsymbol{\omega}, k)$$

$$P(\boldsymbol{\theta} \mid \boldsymbol{\omega}, k)P(\boldsymbol{\omega} \mid k)P(k),$$

where $\mathbf{y} = (y_1, \dots, y_n)$, $z = (z_1, \dots, z_n)$ and $\boldsymbol{\omega} = (\omega_1, \dots, \omega_n)$. We have that

$$P(\mathbf{y} \mid z, \boldsymbol{\alpha}, \boldsymbol{\theta}, \boldsymbol{\omega}, k) = P(\mathbf{y} \mid z),$$

$$P(z \mid \boldsymbol{\alpha}, \boldsymbol{\theta}, \boldsymbol{\omega}, k) = P(z \mid \boldsymbol{\alpha}, \boldsymbol{\theta}, \boldsymbol{\omega}),$$

$$P(\boldsymbol{\alpha} \mid \boldsymbol{\theta}, \boldsymbol{\omega}, k) = P(\boldsymbol{\alpha} \mid \boldsymbol{\theta}, k),$$

$$P(\boldsymbol{\theta} \mid \boldsymbol{\omega}, k) = P(\boldsymbol{\theta} \mid k)$$

$$P(\boldsymbol{\omega} \mid k) = P(\boldsymbol{\omega}).$$

Furthermore, we will assume that $P(\boldsymbol{\alpha} \mid \boldsymbol{\theta}, k) = P(\boldsymbol{\alpha} \mid k)$, which yields the hierarchical specification of the joint distribution of the logistic model

$$P(\mathbf{y} \mid z)P(z \mid \boldsymbol{\alpha}, \boldsymbol{\theta}, \boldsymbol{\omega})P(\boldsymbol{\alpha} \mid k)P(\boldsymbol{\theta} \mid k)P(\boldsymbol{\omega})P(k).$$

It is important to note that, except for the first term $P(\mathbf{y} \mid z)$ and conditional on k, the basis functions $\boldsymbol{\theta}$ and $\boldsymbol{\omega}$, the remainder of the hierarchical structure specifies an ordinary linear model with known variance. Figure 3 shows a representation of the above hierarchical specification as a directed acyclic graph.

2.2 Prior Structure

To complete the hierarchical model, we specify a normal prior

$$(\alpha_1, \dots, \alpha_k) \sim N_k(0, \tau \upsilon^2 \mathbf{I}_k),$$

on the regression parameters with υ a known scale factor and $\tau = 1/\varsigma$ a randomly distributed re-scaling factor, such that ς has a Gamma(a, b) distribution, with a and b two positive real numbers. We will restrict choice of these hyperparameters to choices $a = b$, such that the prior mean on ς will equal $a/b = 1$ with variance a/b^2. The intercept is given a weakly informative normal prior $\alpha_0 \sim N(0, 10^2)$. Note that the dimensionality parameter k must also be given a prior belief, since it will itself be estimated within the variable-dimension MCMC sampler. We specify a discrete uniform hyperprior distribution for k on the set of integers $\{0, 1, 2, .., k_{\max}\}$ with

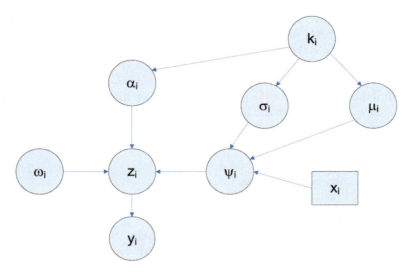

Fig. 3 Graphical representation of the conditional dependence structure within the joint distribution for the logistic regression model

$k_{\max}$ a large positive integer. (An alternative specification here might be to give k a Poisson distribution $Poisson(\lambda)$ with possibly fixed small hyperparameter value λ.) For the prior distribution of location parameters $\mu_j, j = 1, \ldots, k$, we assume a uniform discrete prior on the grid of location values $\boldsymbol{m}$. For the prior distribution on the width parameters $\sigma_j, j = 1, \ldots, k$, we postulate a truncated normal distribution

$$p(\sigma_j) \sim \phi(\sigma_j - \gamma)I(\sigma_j),$$

for all $j = 1, \ldots, k$ with $I()$ the heaviside function and $\phi()$ the standard normal density. γ is a positive parameter which denotes the prior mode of the width of basis components within the regression parameter vector. We will return to some aspects of prior choice in the data analysis section.

2.3 Variable Dimension MCMC Sampling for Logistic Regression

A number of publications have discussed hybrid MCMC sampling methods for variable dimension models over the past few years. We will therefore only describe the basic ingredients and refer to [9, 19], and [5] among others, for more details. Our approach essentially follows the methodology outlined by Holmes and Held [10], and Denison et al. [5] for linear and probit models.

Our hybrid MCMC sampler is based on a random choice between three simple steps which either add (birth), remove (death), or move a basis component to the model. Let $0 < b_k < 1$ and $0 < d_k < 1$ be probability threshold values for birth and death, respectively, such that $b_k + d_k \leq 1$ for each $k = 0, \ldots, k_{\max}$. We choose $b_0 = 1$, $b_{k_{\max}} = 0$, $d_0 = 0$, $d_{k_{\max}} = 0.5$, and $b_k = d_k = 1/3$ for all other k. We will write $p_k = p(k)$ for the prior probabilities on model dimensionality k for $k = 0, \ldots, k_{\max}$. Then the basic step within the algorithm is to carry out a random choice between possible move types conditional on the current state of the sampler in the following manner:

Sample u from a $U(0, 1)$ distribution
    If $u \leq b_k$ carry out a *birth step*
    Else if $b_k < u \leq b_k + d_k$ carry out a *death step*
    Else carry out a *move step*

For the birth step, we first propose a new component location μ_{k+1} from a discrete uniform density on the set $\mathbf{m} \setminus \{\mu_1, \ldots, \mu_k\}$, where $\{\mu_1, \ldots, \mu_k\}$ is the set of locations which are currently selected to the model and subject to the additional restriction that the new proposal must at least be 9 Dalton (D) removed from any of the k locations $\{\mu_1, \ldots, \mu_k\}$ which are already present. If, subject to this restriction, q^* is the number of locations where we can still place a new component, retaining all others, then the acceptance probability of a Metropolis–Hastings step for accepting the proposal is

$$\min \left(1, \frac{p_{k+1}}{p_k} \frac{q^*}{q} \frac{d_k}{b_{k+1}} BF \right),$$

where BF is the ratio of marginal likelihoods of the new proposed model to that of the old, conditional on ς, z and the parameters σ_j, $j = 1, \ldots, k$ and ω_i, $i = 1, \ldots, n$. For the death step, we analogously propose one of the existing k components at random for deletion. The acceptance probability is identical to that for the birth step, except that we invert the ratio inside the minimum. Likewise, for the move step we pick one of the existing components at random and move it to an arbitrary choice within the set $\mathbf{m} \setminus \{\mu_1, \ldots, \mu_k\}$ and subject to the previously described restriction. The acceptance ratio is now simply $\min(1, BF)$, with the BF the ratio of marginal likelihoods for the model with the newly proposed location to that of the old model.

Subsequent to the above described Metropolis step we update the remaining random variables σ_j, $j = 1, \ldots, k$, ω_i, $i = 1, \ldots, n$, z and the (re-)scaling factor ς, through Gibbs steps. It is of interest to note that the regression coefficients $\boldsymbol{\alpha}$ are only required when updating z and the scale variables ω_i, $i = 1, \ldots, n$ and ς, as they can be integrated out of the conditional Bayes factors. Hence, this allows for some computational efficiency by only recomputing these terms every tenth iteration (for example).

3   Data Analysis and Application

3.1   The Colon Cancer Data

We applied the above model to the analysis of the colon cancer mass spectral data within the mass/charge range between 1000 and 5000 Dalton, which contained $q = 2346$ bins in our example and using $k_{max} = 150$. Prior choice of γ is based on the observation that peaks within the spectral data itself will have a typical width of about 3 Dalton (as SD of a Gaussian). More precisely, there will tend to be a gradual increase in peak width from lower to higher masses, but this is not a large effect. It is important to acknowledge that γ itself represents a prior assumption on the regression coefficients instead. However, it is reasonable to assume that within the confines of our model and if there is any differential expression of peaks within the data and between groups this should give rise to components of differential expression within the regression coefficient vector which are of similar width. Hence we use $\gamma = 3$ as prior choice, irrespective of the location of any regression coefficient component proposed. Our prior choice of υ originates from the knowledge that *approximately* in the logistic model and conditional on the true model we have that the variance–covariance structure of the regression parameter vector $(\alpha_1, \dots, \alpha_k)$ is

$$n * (\mathbf{\Psi}^T \mathbf{\Psi})^{-1},$$

with $\mathbf{\Psi} = [\mathbf{\Psi}_1, \dots, \mathbf{\Psi}_k]$. The inverse cross-product for any basis function component with width equal to 3 placed anywhere along the mass-charge axis is of order 10^{-3}. Hence, taking into account that the sample size $n = 113$ suggests a prior guess of $\upsilon^2 = 0.1$. For the re-scaling parameter, we used $a = b = 1$ for the analyses presented in the paper. However, rerunning the algorithm for other choices showed that results were insensitive to choices in the range $a = b = 0.001$ up to $a = b = 3$. The prior on ς is likely to become extremely influential as a or b becomes either much smaller than 0.0001 or larger than 3.

Identification of Differentially Expressed Mass Spectral Regions

Applying the above model to the colon cancer dataset, Fig. 4 shows for each potential basis function location within the set $\mathbf{m} = \{m_1, \dots, m_q\}$ the number of times that location was selected into the model, across all models simulated and expressed as a percentage of the total number of models considered. The plot identifies two major discriminating sources of variation in the data, the first centered at 1352.4 Dalton (corresponding basis function selected to nearly 25 % of all models) and the second at 1867.2 Dalton (almost 15 % of all models). The corresponding Fig. 5 shows the marginal mean of logistic regression coefficients

224   B.J.A. Mertens

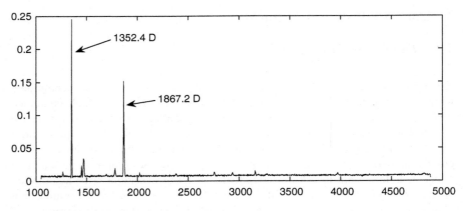

Fig. 4 Marginal posterior density of peak component locations

$$\sum_{m \in M} \beta_m / M = \sum_{m \in M} \sum_{j=1}^{k} \alpha_{mj} \Psi_{mj} / M,$$

across all simulated models, where m is an indicator identifying the model within the set M of all simulated models. M is the total number of models simulated. The plot reveals that the discriminating information corresponds to a contrast between bin intensities centered at 1352.4 and 1867.2 Dalton. This interpretation can be easily confirmed through a scatter plot of bin intensities at these locations (Fig. 6, left plot). As can be seen, large intensities at 1352.4 Dalton for one group correspond to small intensities at 1867.3 Dalton for the other group, and vice versa. We note Figs. 4 and 5 also identify a number of other interesting locations in the spectra, most notably a small contrasting in coefficients which corresponds to differential spectral contrasting at locations 1450 and 1467.7 Dalton. There is some indication of other separating regions in the spectra at 1265.7 Dalton and at 1780 Dalton, though the evidence is not as strong. The corresponding bin locations are selected to less than 5 % of all simulated models. For ease of comparison, the "inset" graph in the lower right-hand corner of Fig. 5 displays and enlarges the fitted mean discriminant coefficients within the region of 1300 up to 1900 Dalton, together with the within-group mean spectra (below) (as in Fig. 2 also). As may be seen, the identified discriminant locations correspond to spectral locations which may— in hindsight—be confirmed as regions with large between-group differences in mean spectral intensity. It is, however, of interest that the peak at 1352.4 Dalton has smaller between-group separation in comparison to its immediate neighbor at 1467.7 Dalton, while the calibrated discriminant weights are, however, typically larger at this location (1352.4 D). Also noteworthy is that the identified locations at 1450 and 1467.7 Dalton seem to correspond to two distinct peak intensities close together in the spectra, which separate in the mean spectra of the cases group, while they are not easily distinguished for the control group mean spectrum.

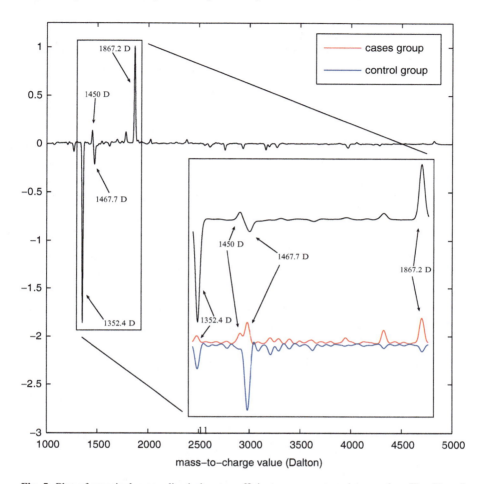

Fig. 5 Plot of marginal mean discriminant coefficients versus mass-charge value. The *"inset"* graph in the lower right-hand corner enlarges the fitted mean discriminant coefficients within the region of 1300 up to 1900 Dalton, together with the within-group mean spectra (*below*) (as in Fig. 2)

Evaluation of Classification Performance

To assign observations, we calculate for each observation the mean a-posteriori class probabilities of group-membership. That is, we compute

$$P(y_i = 1 \mid x_i) = \sum_{m \in M} P_m(y_i = 1 \mid x_i)/M,$$

for all $i = 1, \ldots, n$, where P_m denotes the a-posteriori class probability calculated from the mth model simulated within the MCMC chain and the sum is across all

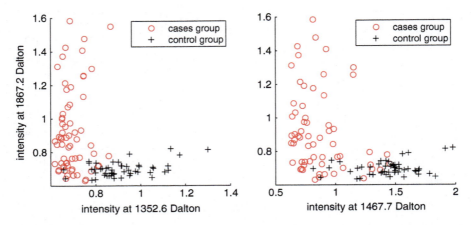

Fig. 6 Scatter plots of spectral intensity values at 1352.6 and 1467.7 Dalton versus those at 1867.2 Dalton, using distinct plotting symbols to distinguish cases (*closed circle*) from controls (*plus symbol*)

Table 1 Confusion table summarizing classification performance of the logistic regression discriminant model, based on the mean a-posteriori class probabilities across all simulated models

Predicted class cases			Controls	
True class	Cases	58	5	63
	Controls	4	46	50
		62	51	113

models simulated. Using a cut-off value of 0.5 (e.g.) to assign observations based on these mean a-posteriori class probabilities, we may now summarize classification performance as shown in Table 1. Based on this assessment, we find a sensitivity and specificity of both 0.92, such that the global misclassification error rate equals 0.08. The Brier distance, defined as

$$B = \frac{1}{n} \sum_i [1 - P(c(i) \mid \mathbf{x}_i)]^2,$$

equals 0.066, where $c(i)$ denotes the true class label of the ith observation with $P(c(i) \mid \mathbf{x}_i)$ the mean a-posteriori class probability for the ith observation for the true class. Figure 7 shows a Receiver Operating Curve (ROC) for the marginal mean a-posteriori class probability. The Area Under the Curve (AUC) equals 0.978.

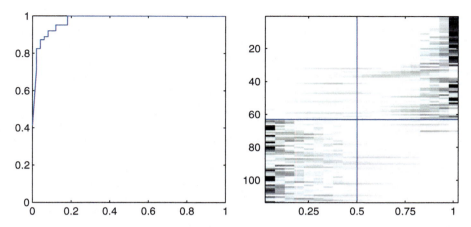

Fig. 7 Empirical Receiver Operating Curve (ROC) for the marginal mean a-posteriori class probability (*left plot*) and marginal density plots of calibrated a-posteriori class probabilities for each observation separately displayed as gray-scale image. The first 63 observations are colon cancer cases. The added *horizontal* and *vertical line* represents the cross-over from case to control samples, and the 0.5 cut-off value, respectively

Sensitivity Analysis and Model Comparisons

The prior variance $\zeta = \tau v^2$ on the regression parameters $(\alpha_1, \ldots, \alpha_k)$ is highly influential on the complexity of models simulated. This will happen, due to the ridge-type prior $N(0, \tau v^2 I_k)$, which will eventually force all regression parameters towards zero as $\zeta \to 0$. Similarly, however, we will also tend to reject more complex models (in the sense of "more basis functions" and thus larger dimensionality k) as $\zeta \to \infty$. This is because the ratio of marginal likelihoods (*BF*) of any $(k + 1)$-dimensional model versus any k-dimensional model is inversely proportional to the square root of the prior variance $\sqrt{\zeta}$. This effect (sometimes referred to as the Lindley effect [2, 11, 13]) will become stronger the greater the difference in dimensionality between models compared and the larger ζ becomes.

There are several reasons why it is of interest to implement a sensitivity study for the effect of this parameter. In this analysis, we repeat the above modeling procedure, but this time keeping the variance ζ fixed and this for a number of values across a finite grid. Table 2 shows some results across a grid ranging from 0.0001 up to 100,000, each time increasing the magnitude by one order. Note that the extremes of this table correspond to variances which would never be contemplated in practice as reasonable prior choices, given our argument explained above in Sect. 3. The first six columns give information on the typical model dimensionality of the simulated models across 500,000 simulations. We give the minimum model dimensionality observed, as well as the first, 2.5th, 5th, and 25th percentile and the median model dimensionality k. The last four columns give the sensitivities, specificities, total recognition rates, and Brier scores of the models. The distribution of k is right

Table 2 Sensitivity analysis of logistic modeling results, using $\tau = 1$ and recelebrating the model for different fixed values of υ (and thus also $\zeta \equiv \upsilon$)

ζ	min	1	2.5	5	25	50	Se	Sp	T	B
0.0001	0	0	0	1	4	9	100	0	50	0.357
0.001	0	0	0	1	4	9	100	0	50	0.359
0.01	0	0	0	1	4	9	100	0	50	0.358
0.1	1	3	4	6	13	20	87.3	82.0	84.6	0.115
1	2	5	6	8	15	21	92.1	92.0	92.0	0.0647
10	2	6	7	9	15	20	98.4	94.0	96.2	0.0204
100	4	7	8	9	16	20	100	100	100	0.0204
1000	3	6	8	9	13	16	100	100	100	0.000375
10,000	2	2	2	2	2	5	100	94.0	97.0	0.0179
100,000	1	1	1	1	1	1	93.7	88.0	90.1	0.0731
Full model	1	4	6	7	14	20	92.0	92.0	92.0	0.066

The last line corresponds to results on the full model (ζ variable)

skewed and tends to favor larger values, which is a not unexpected feature of the approach we will return to in more detail in the discussion.

As ζ increases from 0.0001 upwards, model dimensionality first increases, and then tends to decrease again, from a variance of about 100 onwards. A-posteriori plot summaries of the marginal posterior density of basis function component locations selected (as in Fig. 4) for the first two models ($\zeta = 0.0001$ up to 0.01) have the appearance of a band of white noise (all locations equally unlikely a-posteriori) (not shown—see supplementary materials). These models are clearly underfitting the data, which can also be seen from the classification performance measures selected. All observations are assigned to the largest group only for these three models, which is due to the fact that no discriminant information is allowed into the model due to the extreme penalization inherent in the choice of the variances. As a consequence, we are effectively classifying with an "intercept-only" model, which explains assignment to the largest group.

On the other hand, models corresponding to variances in the range 0.1 up to 10 do give interesting calibrations and are able to detect discriminant information present in the data, as may also be seen from sensitivity and specificity measures. Specifically, the output from the model corresponding to $\zeta = 1$ is effectively the same as shown before for our "full" model. We thus identify the same separating functional components (centered at 1352.4 and 1867.2 Daltons) and marginal mean regression vector β as from the full model and shown in Figs. 4 and 5. This is not surprising, because the typical a-posteriori variance simulated for the full model is close to 1. Figure 8 displays a histogram of the a-posteriori simulated values on ζ for the full model, which shows a clear shift upwards from our initial prior guess 0.1, with a mode at about 0.9.

The model corresponding to $\zeta = 0.1$ is of interest as well, as the solution contained within this calibration is remarkably similar to that from a "standard" maximum likelihood logistic regression fit on a principal components reduction of

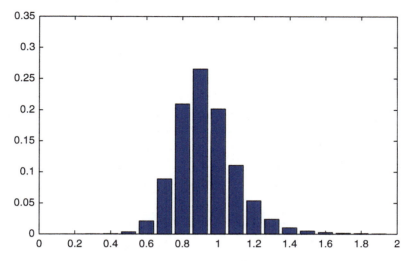

Fig. 8 Histogram of a-posteriori simulations of $\zeta = \tau \upsilon^2$

the full spectral data. Figure 9 shows a plot of the a-posteriori marginal means of regression coefficients obtained from this model. Below, we plot the first two principal components computed from the complete matrix of spectral intensity measurements (Fig. 10). As we can see, the first principal component summarizes a strong peak at about 1467.7 Dalton, together with lesser contributions at 1352.4, 1207.1, 1867.2, and 1778.9 Dalton. The second component identifies a particularly strong peak at 1867.2 Dalton, together with smaller contributions at 1778.9 and 1467.7 Dalton (among others). [Note chapter "Transformation, Normalization and Batch Effect in the Analysis of Mass Spectrometry Data for Omics Studies" of this volume [15] uses the same data in Figs. 2 and 3 and the discussion in the text that refers to these.] The logistic discriminant fit on the full component decomposition shows a similar behavior as observed in our model with ζ fixed at 0.1. This solution contrasts locations at 1467.7 and 1867.2 Daltons, instead of the contrast between 1867.2 and the "smaller" contribution at 1352.4 Dalton as in our "full" solution (ζ variable). This somewhat different contrasting within the maximum likelihood logistic model comes from the fact that the fit is mainly driven by the first two principal components. Note that the contrasting between the peaks found is already inherent in the first two components. From this perspective, our initial guess on the regression coefficients variance $\zeta = 0.1$ would thus appear to correspond (roughly) to a principal components-based solution. Calibration of our full logistic model represents replacing the component at 1467.7 Daltons in favor of the smaller contribution at 1352.4 Daltons, both of which may be identified within the first principal component, however.

As we increase the variance ζ even further from a value of 10 upwards, we first get increasingly complex model fits, which are likely to represent over-fitting of

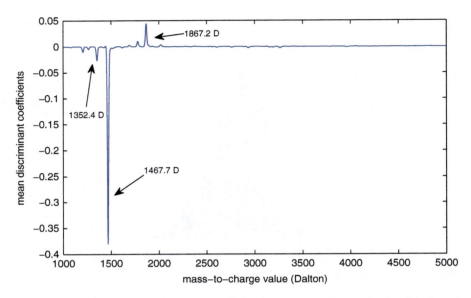

Fig. 9 Plot of marginal mean discriminant coefficients versus mass-charge value, conditioning on $\zeta=0.1$

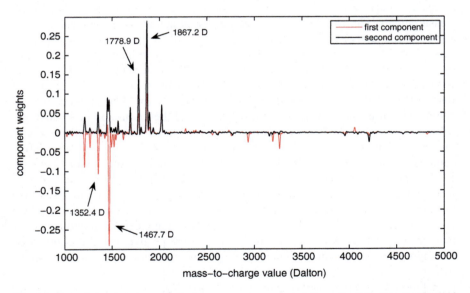

Fig. 10 Plot of the first two principal components computed from the full data matrix (113 samples by 4483 bins) of mass spectral intensity measurements

the data. Eventually, the effect of the prior variance becomes so strong (Lindley effect) that the model dimensionality must collapse again and we are simulating single-location models only (ζ=100,000). As Fig. 8 shows, however, these choices are effectively excluded from our full model fit.

Comparison with Classical Linear Discrimination

It is of interest to compare our results with those obtained from a more traditional analysis derived from classical Fisher linear discrimination. Such an analysis can be implemented in the mass spectrometry context by formulating the discriminant problem as assignment of an observation $\mathbf{x}$ to the group for which the smallest distance measure

$$D_g(\mathbf{x}) = (\mathbf{x} - \boldsymbol{\mu}_g)\boldsymbol{\Sigma}^{-1}(\mathbf{x} - \boldsymbol{\mu}_g)^T,$$

is observed, where g denotes the group indicator, $\boldsymbol{\mu}_g$ the population means, and $\boldsymbol{\Sigma}$ the population within-group dispersion matrix which we assume equal across groups. A formulation which is known to work well in spectrometry applications is to estimate $\boldsymbol{\Sigma}$ through the observed pooled dispersion matrix $\mathbf{S}$ and then first compute the component decomposition $\mathbf{S} = \mathbf{Q}\boldsymbol{\Lambda}\mathbf{Q}^T$ where $\mathbf{Q}$ and $\boldsymbol{\Lambda} = \text{diag}(\lambda_1, \ldots, \lambda_r)$ are the matrices of principal component weights (or loadings) and variances, respectively, with $\lambda_1 > \ldots > \lambda_r > 0$ and r the rank of the pooled covariance matrix. The covariance matrix is then re-estimated by retaining only the first $1 \leq l \leq r$ components from the decomposition, which gives us the sample-based linear discriminant distance

$$\widehat{D}_g(\mathbf{x}) = (\mathbf{x} - \bar{\mathbf{x}}_g)\mathbf{S}_{(l)}^{-1}(\mathbf{x} - \bar{\mathbf{x}}_g)^T,$$

where $\mathbf{S}_{(l)} = \mathbf{Q}_{(l)}\boldsymbol{\Lambda}_{(l)}\mathbf{Q}_{(l)}^T$ is the covariance matrix estimate based on the first l components only. To simplify the discriminant rule even further, we will restrict $\boldsymbol{\Lambda}_{(l)} = \mathbf{I}_{(l)}$ which reduces the allocation to linear discrimination using Euclidean distance metric along the first l components. Dimension reduction methods based on (principal) components have a long history of successful application in spectroscopy applications generally and we refer the reader to texts by Stone and Jonathan [20, 21], Martens and Naes [14], Krzanowski et al. [12] and Naes et al. [16] among others, for an introduction to the field and further references.

We applied this method to the data using a combined double-cross-validatory (leave-one-out) approach, which removes each observation in turn from the data, after which the number of components to be retained (l) is estimated through a secondary cross-validatory layer within the left-over set. The resulting discriminant rule is then applied to the left-out datum in the "outer layer" within the double-cross-validatory analysis. This procedure is then repeated across all observations in order to calculate a fully validated error rate. Results from this analysis are similar to those shown above for the full model and give a sensitivity of 0.87, a specificity of

0.9, and total error rate of 0.89 (and roughly halfway between classification results for our two "reduced models" corresponding to $\zeta = 0.1$ and $\zeta = 1$, respectively).

Double-cross-validation is joint calibration and validation of a (discriminant) model and thus also gives information on the manner in which observations are allocated. In this case, we can investigate the numbers of components l that are chosen for the allocation of each observation. We find that for all observations, double-cross-validation selects the first 2 components, except for 2 out of 113 observations for which only the first component is retained. In this way, double-cross-validation identifies the first two principal components as summarizing the main discriminatory information in the data. See Fig. 10 plotted above for a graphical representation of these components in the previous section on our logistic regression analysis on dimension reduction.

Investigation of the fitted discriminant weights based on a re-fitting of the above described discriminant model on the first two principal components and the full data gives near identical results as mentioned previously for the maximum likelihood logistic discriminant model (and thus mainly identifying and contrasting peak locations at 1467.7 and 1867.2 Dalton). As can be seen, these analyses are much influenced by the first few principal components here. It must be noted, however, that this property will likely apply to many of the classical discriminant methods that could be applied here, besides the ones discussed. Similar behavior will tend to be observed for quadratic discriminant methods or logistic regression methods as pointed out by Goldstein and Smith [8], for example. In contrast, our method is more "local" in the sense that it can investigate sets of contiguous bins without any explicit or implicit reliance on the way in which peak variation and correlation within the spectra is summarized within-components. On the other hand, cross-validatory analysis of our approach would be a much more challenging task, although it is not too difficult to calculate validated estimates of error based on a separate test set, when available.

Model Dimensionality

An aspect of our model which is of interest is the a-posteriori distribution of k. The skew of this distribution is large, but not unexpected. It is partly due to the high number of "peak" contributions which may be identified within the first two (separating) principal components (see discussion above) which indicates existence of a larger number of distinct peak locations within the spectral data which correlate. Another reason is the limited a-priori information on likely component locations within the model. The phenomenon is similar to that in Bayesian calibration of model dimensionality in mixture problems, and has also been discussed by Richardson and Green [19], for example. It must be noted that this problem will occur more generally than just within the full Bayes context alone, as remarked by Cox, D.R. in discussion of the Richardson and Green paper [19]. This is due to the intrinsic ill-conditioning of the dimension calibration within mixture models and related-type problems. It implies we may only hope to interpret the left-hand side of this distribution with any reliability. The right-hand side is likely to become ever more sensitive to prior assumption. For the present example and analysis, perhaps the strongest conclusion which can be drawn from the posterior distribution on k is that any reasonable model tends to rule out an intercept-only model.

On the other hand, there is no need to explicitly choose k as shown in the above analysis in order to interpret the model, as such interpretation may be based on evaluation of the marginal posterior distributions of regression coefficients. In addition, analysis may always be augmented by a-posteriori evaluation which departs from fully Bayesian paradigm, by restricting to specific values of k (or ζ). For example, we may pick a minimal value of k and explore the changes within the a-posteriori deviances

$$-2\log(P(y \mid z)),$$

for small perturbations to k around that chosen value. Doing this for the full model, we find that the a-posteriori distributions of deviances are near-completely separated between the intercept-only models ($k = 0$) and those for $k = 1$. In contrast, a-posterior deviance distributions are overlapping for all models with $k > 0$. We refer to [19] for further ideas along these lines and discussion by other authors on this and other approaches.

3.2 The Dystrophic Mouse Data

For comparison, we investigated application of the same Bayesian logistic regression model to MALDI-TOF proteomic spectra from 18 dystrophin-deficient mdx mice and 74 healthy control mice. Spectra were recorded from m/z 962 up to 11,223 Dalton, on a grid of 11,239 points. Similar pre-processing was applied as previously explained for the colon cancer data prior to analysis. Figure 11 shows the mean spectra as shown before for the colon cancer data. A number of peaks are highlighted which are identified as differentially expressed between both groups based on subsequent analysis.

We ran the algorithm using same prior settings as before (particularly $\upsilon^2 = 0.1$ and $a = b = 1$ and a flat prior on model dimensionality k). Figures 12 and 13 show the marginal densities of peak location selection and marginal regression coefficient vector, respectively.

The results suggest only the two peaks at locations 3908 and 6830 Daltons can suffice to summarize the discrimination between both groups. For comparison, we show the regression coefficients (Fig. 14) from calibration of a standard logistic ridge regression model, with the penalization parameter chosen from leave-one-out cross-validation. The logistic ridge solution is—as one would expect—much less sparse, though it does identify the two above found peaks as major contributors to the regression coefficient vector. Several other locations, however, also have large coefficients, especially in the high-mass region above 8000 Dalton. Repetition of same calibration using the lasso penalty selects a "6-peaks model" model at

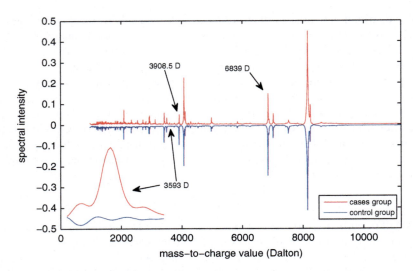

Fig. 11 Mean spectra for mdx versus control mice, after pre-processing. We plot negative intensity value for the control group (bottom mean spectrum). A number of spectral locations are identified which correspond to peaks with different intensity levels between case and control samples

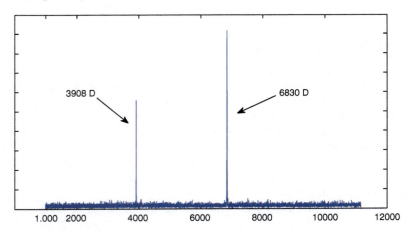

Fig. 12 Marginal posterior density of peak component locations

locations 3113, 3909, 6839, 6840, 7510, and 8147. Both peaks at 3908 and 6830 Dalton turned out to be strong discriminators as seen in a scatterplot (Fig. 15). We investigated the other identified peaks from the logistic ridge and lasso analysis also, but did not find them to be of further interest as we could not confirm these to add significantly to the discrimination.

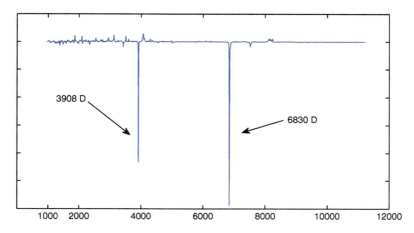

Fig. 13 Plot of marginal mean discriminant coefficients versus mass-charge value

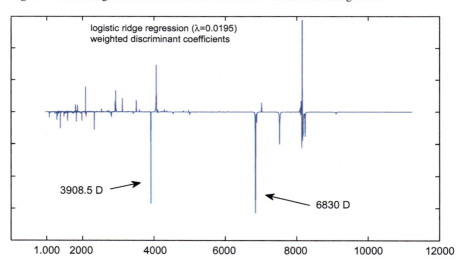

Fig. 14 Regression coefficients from logistic ridge regression optimized using leave-one-out cross-validation

4   Discussion

4.1   *Computation*

We spent little effort trying to improve computational efficiency of our code. Even so, calculations could easily be performed on a standard desktop personal computer within one to 2 h at most. We were surprised at the speed and consistency of convergence, even when starting from distinct initial solutions. Even smaller details

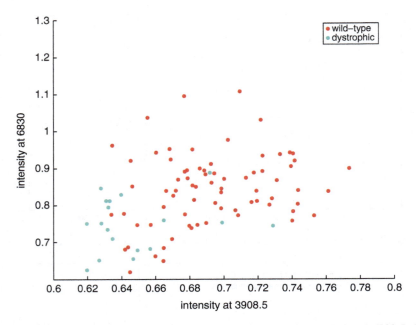

Fig. 15 Scatterplot contrasting proteomic expression between peaks at 3908 and 6830 Dalton between mdx and control mice

within the marginal posteriors, such as the local contrasting between 1448.6 and 1471.5 Dalton, for example, visible in Fig. 5, are established in short runs of 50,000 simulations already. Results in this paper were based on much larger runs of up to one million simulations.

4.2 Functional Data Analysis

There is an extensive literature on functional data analysis, which is of potential interest to researchers in mass spectrometry proteomics. Our approach could perhaps be termed a "pseudo-functional data analysis" [18] because we pose the assumption on the behavior of the vector of regression coefficients.

Our experience has been that within the broad class of functional data analysis approaches, the classical methods from chemometrics and analytical chemistry for analyzing functional spectral data—typically based on some form of component reduction and summarization—can be very powerful within mass spectroscopy proteomics as well. Much of that work has been inspired by near infrared and infrared spectroscopy, but not exclusively. As shown and for this experiment, our results are close but not quite identical to those derived with such methods. In general we would expect the Bayesian method presented here to lead to more sparse solutions, which may be a significant advantage when interpreting results.

Acknowledgements This work was supported by funding from the European Community's Seventh Framework Programme FP7/2011: Marie Curie Initial Training Network MEDIASRES ("Novel Statistical Methodology for Diagnostic/Prognostic and Therapeutic Studies and Systematic Reviews," www.mediasres-itn.eu) with the Grant Agreement Number 290025 and by funding from the European Union's Seventh Framework Programme FP7/ Health/F5/2012: MIMOmics ("Methods for Integrated Analysis of Multiple Omics Datasets," http://www.mimomics.eu) under the Grant Agreement Number 305280.

References

1. Albert, J. H., & Chib, S. (1993). Bayesian analysis of binary and polychotomous response data. *Journal of the American Statistical Association, 88*(422), 669–679.
2. Bartlett, M. S. (1957). A comment on D. V. Lindley's statistical paradox. *Biometrika, 44*, 533–534.
3. Coombes, K. R., Kooman, J. M., Baggerly, K. A., & Kobayashi, R. (2005). Improved peak detection and quantification of mass spectrometry data acquired from surface-enhanced laser desorption and ionisation by denoising spectra with the undecimated discrete wavelet transform. *Proteomics, 5*, 4107–4117.
4. de Noo, M. E., Mertens, B. J., Ozalp, A., Bladergroen, M. R., van der Werff, M. P. J., van de Velde, C. J., et al. (2006). Detection of colorectal cancer using MALDI-TOF serum protein profiling. *European Journal of Cancer, 42*(8), 1068–1076.
5. Denison, D. G. T., Holmes, C. C., Mallick, B. K., & Smith, A. F. M. (2002). *Bayesian methods for nonlinear classification and regression.* New York: Wiley.
6. Devroye, L. (1986). *Non-uniform random variate generation.* New York: Springer.
7. Eilers, P. H. (2004). Parametric time warping. *Analytical Chemistry, 76*, 404–11.
8. Goldstein, M., & Smith, A. F. M. (1974). Ridge-type estimators for regression analysis. *Journal of the Royal Statistical Society, B, 36*(2), 284–291.
9. Green, P. (1995). Reversible jump Markov chain Monte Carlo computation and Bayesian model determination. *Biometrika, 82*, 711–732.
10. Holmes, C. C., & Held, L. (2006). Bayesian auxiliary variable models for binary and multinomial regression. *Bayesian Analysis, 1*(1), 145–168.
11. Jeffreys, H. (1967). *Theory of probability.* Oxford: Oxford University Press.
12. Krzanowski, W. J., Jonathan, P., McCarthy, W. V., & Thomas, M. R. (1995). Discriminant analysis with singular covariance matrices: methods and applications to spectroscopic data. *Applied Statistics, 44*(1), 101–115.
13. Lindley, D. V. (1957). A statistical paradox. *Biometrika, 44*, 187–192.
14. Martens, H., & Naes, T. (1989). *Multivariate calibration.* Chichester: Wiley.
15. Mertens, B. J. A. (2016). Transformation, normalization and batch effect in the analysis of mass spectrometry data for omics studies. In *Statistical analysis of proteomics, metabolomics, and lipidomics data using mass spectrometry.* New York: Springer.
16. Naes, T., Isaksson, T., Fearn, T., & Davies, T. (2002). *A user-friendly guide to multivariate calibration and classification.* Chichester: NIR Publications.
17. Petricoin, E. F. III, Ardekani, A. M., Hitt, B. A., Levine, P. J., Fusaro, V. A., Steinberg, S. M., et al. (2002). Use of proteomic patterns in serum to identify ovarian cancer. *Lancet, 359*, 572–577.
18. Ramsay, J. O., & Silverman, B. W. (1997). *Functional data analysis.* New York: Springer.

19. Richardson, S., & Green, P. (1997). On Bayesian analysis of mixtures with an unknown number of components (with discussion). *Journal of the Royal Statistical Society, B, 59*, 731–792.
20. Stone, M., & Jonathan, P. (1993). Statistical thinking and technique for QSAR and related studies. Part 1: General theory. *Journal of Chemometrics, 7*, 455–475.
21. Stone, M., & Jonathan, P. (1994). Statistical thinking and technique for QSAR and related studies. Part 2: Specific methods. *Journal of Chemometrics, 8*, 1–20.

Robust and Confident Predictor Selection in Metabolomics

J.A. Hageman, B. Engel, Ric C. H. de Vos, Roland Mumm, Robert D. Hall,
H. Jwanro, D. Crouzillat, J. C. Spadone, and F.A. van Eeuwijk

1 Introduction

Metabolomics is the simultaneous analysis of hundreds of metabolites present in crude extracts, such as those from plants [1]. The so-called untargeted metabolomics approach takes signals (e.g., Mass Spectrometry, NMR) from both known and (many yet) unknown metabolites into consideration for data processing and

J.A. Hageman (✉) • B. Engel
Biometris-Applied Statistics, Wageningen University, P.O. Box 16, 6700 AA Wageningen,
The Netherlands

Centre for BioSystems Genomics, P.O. Box 98, 6700 AB Wageningen, The Netherlands

Netherlands Metabolomics Centre, Einsteinweg 55, 2333 CC Leiden, The Netherlands
e-mail: jos.hageman@wur.nl

R.C.H. de Vos
Centre for BioSystems Genomics, P.O. Box 98, 6700 AB Wageningen, The Netherlands

Netherlands Metabolomics Centre, Einsteinweg 55, 2333 CC Leiden, The Netherlands

Plant Research International, P.O. Box 619, 6700 AP Wageningen, The Netherlands

R. Mumm
Centre for BioSystems Genomics, P.O. Box 98, 6700 AB Wageningen, The Netherlands

Plant Research International, P.O. Box 619, 6700 AP Wageningen, The Netherlands

R.D. Hall
Centre for BioSystems Genomics, P.O. Box 98, 6700 AB Wageningen, The Netherlands

Netherlands Metabolomics Centre, Einsteinweg 55, 2333 CC Leiden, The Netherlands

Plant Research International, P.O. Box 619, 6700 AP Wageningen, The Netherlands

Netherlands Consortium for Systems Biology, P.O. Box 94215, 1090 GE Amsterdam,
The Netherlands

© Springer International Publishing Switzerland 2017
S. Datta, B.J.A. Mertens (eds.), *Statistical Analysis of Proteomics, Metabolomics,
and Lipidomics Data Using Mass Spectrometry*, Frontiers in Probability and the
Statistical Sciences, DOI 10.1007/978-3-319-45809-0_13

interpretation. This provides detailed insight into metabolite differences and similarities between crude extracts with a different sample background (genotype, treatment). Variation in metabolite composition has been linked to genetic variation [2], physiology of fruit ripening [3], quality of fresh food [4], key compounds in hydrolysates [5], and processed food products [6, 7].

One application of untargeted metabolomics is to select metabolites (strongly) correlating to sensory or phenotypical traits (response traits) that can be used as predictive (bio)markers [8]. In subsequent research, metabolites selected in this way can be monitored in a targeted way. They can, for example, be used to predict quality of a crop or indicate whether a process has reached optimal product quality [8]. Typically, the goal is to select one or a few metabolites that can accurately predict a sensory or other phenotypic trait [8].

Metabolic studies typically comprise small numbers of experimental units and large numbers of measured metabolites. This increases the risk of selecting spurious key metabolites that apparently have predictive value for the relatively small data set at hand, but show poor predictive ability for future experimental units [9, 10].

A metabolite with the "best" predictive properties may not be the optimal candidate for quantitative screening of new samples. The compound may be unknown or there may be a lack of commercially available authentic standard for quantification. Therefore, it is worthwhile to find alternative (identified) metabolites with comparable predictive value. Because most metabolites are components in biological pathways, e.g., a common biosynthesis pathway, it is expected that there will be several metabolites showing similar predictive value [11].

As food chemists are generally interested in a small set of predictive metabolites, some form of variable selection is needed. In the approach proposed in this chapter we put an upper bound on the number of metabolites used for prediction in concert with forward selection and linear regression, as a fast and straightforward method [12, 13].

Typically, predictive power is inflated when the same observations that have been used for selection of metabolites and model building are also used for model validation [14]. In cross-validation, predictive quality is assessed by using part of the observations for selection and model building and the remaining (small) part for model validation. This offers a more reliable impression of the predictive value of

H. Jwanro • D. Crouzillat
Centre R&D Nestlé, Plant Science and Technology, 37097 Tours CEDEX 2, France

J.C. Spadone
Nestle Research Center, Vers-chez-les-Blanc, CH-1000 Lausanne 26, Switzerland

F.A. van Eeuwijk
Biometris-Applied Statistics, Wageningen University, P.O. Box 16, 6700 AA Wageningen, The Netherlands

Centre for BioSystems Genomics, P.O. Box 98, 6700 AB Wageningen, The Netherlands

Netherlands Metabolomics Centre, Einsteinweg 55, 2333 CC Leiden, The Netherlands

Netherlands Consortium for Systems Biology, P.O. Box 94215, 1090 GE Amsterdam, The Netherlands a set of selected metabolites. Here we use an approach which comprises a double cross-validation procedure [15] combining an inner cross-validation for control over model complexity (the maximum number of selected metabolites), which is an integral part of the variable selection procedure, with an outer cross-validation for evaluation of predictive value.

To show the benefits of the proposed method, it has been compared to two alternative methods. The first comparison is to a related standard variable selection technique: stepwise regression. The second comparison is with a procedure in which stepwise regression is combined with a single leave-one-out cross-validation. In the next section, stepwise regression, leave-one-out cross-validation, and the proposed method (LOO2CV-regression) are explained.

Although the proposed method yields a single final prediction model, in the outer cross-validation procedure a multitude of regression models is generated (as will be discussed in detail in the next section), each model comprising its own set of selected metabolites. Metabolites that are selected more often, i.e., feature in more of the generated regression models, are potentially more stable predictors for a trait [10, 16, 17]. In the spirit of stability selection, a high frequency of inclusion can be considered as an indication for the importance of these metabolites.

2 Theory

To decide which metabolites are most important for prediction of a trait, a regression model is assumed that relates concentrations of the metabolites (or any subset of metabolites) to the specific trait:

$$y = X\beta + \varepsilon \tag{1}$$

Here, y is a vector with the response trait values, X is a matrix containing the metabolite concentrations, β is a vector of regression coefficients, and ε is a vector of error terms. Inspection of the regression coefficients suggests which metabolites (potentially) have good predictive properties [10, 18].

When the number of units is relatively low compared to the number of metabolites, regression coefficients β in Eq. 1 cannot be estimated by ordinary least squares and techniques such as ridge regression, principal component regression (PCR), or partial least squares (PLS) may be considered instead [19]. Methods such as PCR or PLS complicate the selection of highly predictive metabolites because of the use of latent variables. Both the number of latent variables and the contribution of metabolites to the latent variables need to be assessed. This adds an extra layer of complexity to the prediction problem. However, when the number of metabolites in the model is restricted to a relatively small subset (well below the number of units), regression coefficients can be estimated again by ordinary least squares. Since we aim for a computationally fast procedure, and cross-validation is computationally demanding, we did not consider the use of elaborate variable selection techniques such as those based on global optimizers [20], but restricted attention to the more traditional methods for selection in regression. The present implementation of the proposed approach uses forward selection. Alternatives in the same vein are stepwise regression, or all subsets regression, possibly in the form of a fast approach like the branch and bound method of Furnival and Wilson [21].

2.1 Variable Selection by Forward and Stepwise Regression

Stepwise regression starts with an "empty" model. In subsequent steps metabolites are added to the model or removed later on when their predictive power diminishes in the presence of other selected metabolites. Addition or removal of metabolites follows from F-tests with predefined significance levels. This iterative procedure of adding and removing variables is repeated until the model cannot be improved anymore (R^2 does not change significantly). Stepwise regression has been shown to be appropriate for selecting the smallest number of predictors that can fit the data as well as if the whole data set was used, also in comparison to more "modern" methods [18, 22]. For a more detailed explanation of stepwise regression see [12, 19, 23]. Stepwise regression without the possibility of removing metabolites is forward selection.

Our proposed procedure, in the form of forward selection with additional control of model complexity (details will follow later), has been compared to stepwise regression with the commonly chosen significance levels of 0.05 and 0.10 of F-tests for inclusion of a metabolite or exclusion of an already selected metabolite. The latter approach offers no explicit control on model complexity. Our procedure will also be compared to stepwise regression coupled to a single cross-validation for assessing predictive power, again without explicit control on model complexity.

2.2 Model Validation by Leave-One-Out Cross-Validation

An often used criterion for the performance of a regression model is (adjusted) R^2, which shows how much of the total variation in the observations of the response trait can be explained by the metabolites. Part of the observations will be set aside as a validation set. After fitting the model, observations in the validation set are predicted and prediction errors are inspected. In metabolomics data sets, with only a modest number of units, the reduction of units available for model building, i.e., variable selection, has little appeal. To mitigate this problem, leave-one-out cross-validation (LOOCV) can be used [19].

In LOOCV each observation is left out once as a validation set, while the remaining $(N-1)$ observations are used for selection of metabolites and model building (where N is the total number of observations). Using this selected model, the left out observation (y) is predicted ($\hat{y}$) and the prediction error, the deletion residual ($y - \hat{y}$), is calculated. This procedure (also called Jackknife [24]) is repeated for all N observations, meaning that N regression models with different selected sets of metabolites and N associated deletion residuals are obtained which are stored for further use.

Assessment of the Predictive Quality

Using the deletion residuals, the predictive quality is expressed in terms of the Q^2 value [14]:

$$Q^2 = 1 - \frac{\Sigma (y_i - \hat{y}_i)^2}{\Sigma (y_i - \bar{y})^2} \qquad (2)$$

This quantity is somewhat similar to R^2 but is expressed in terms of the mean square error of prediction, rather than the residual sum of squares. A high value for Q^2 (with 1 as maximum) means good predictive capacity for the procedure. Since Q^2 is derived from predictions, it can be negative, indicating that no proper model for prediction can be identified by the selection procedure.

Meta-Analysis of Generated Models

A leave-one-out cross-validation scheme yields as many regression models as there are observations. In general, this allows for comparisons between the different models and potentially gives insight into the quality of selected metabolites. One way to perform this is to use the concept of stability-based selection [10, 17]. Frequencies by which metabolites were selected in the LOOCV models are determined. A metabolite with a relatively high frequency is deemed to be more important as a predictor.

The spread in the regression coefficients of metabolites that are selected in multiple models will be summarized with 5 and 95 % percentiles, indicating the sign and range for these coefficients.

2.3 Construction of the Prediction Formula and Model Validation

In the proposed method, there is explicit control of the model complexity (the number of selected metabolites in the model). This is most easily achieved for forward selection, which is presently implemented. Firstly, we will describe the construction of the prediction model. This will involve an LOOCV: an inner LOOCV to determine the optimal model complexity. Secondly, we will describe how the predictive value of the prediction model is evaluated. This will involve an extra LOOCV: the outer LOOCV. It has to be stressed that the inner LOOCV is an integral part of the construction of the prediction formula, while the outer LOOCV is the "usual" cross-validation to evaluate the predictive value. So, two separate calculations are made (1) the construction of the prediction formula, involving the inner LOOCV only, and (2) the LOO2CV, involving both the inner and outer LOOCV, to evaluate the predictive value. A flowchart of the proposed LOO2CV procedure is given in Fig. 1 and schematically depicted in Fig. 2 and is explained in more detail below. Pseudo code is also provided for the LOO2CV.

Fig. 1 Flowchart of the
double cross-validation
model. The inner loop is
shown in the *grey box*

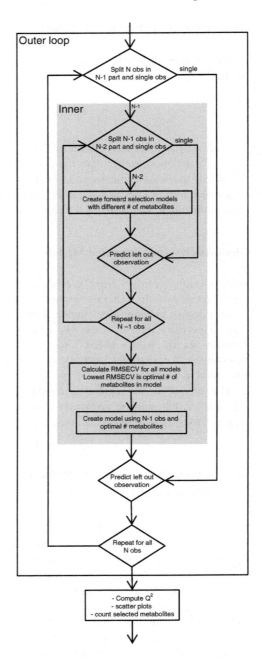

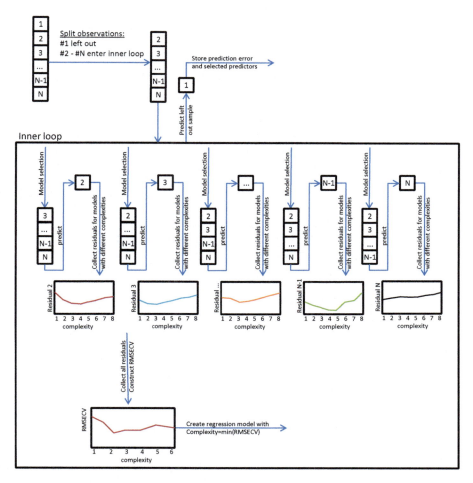

Fig. 2 Schematic representation of the first iteration of the outer loop including all corresponding inner loops. See text for details

```
Pseudo code: description of the double cross-validation
//start outer loop
for i=1 to N
    set aside i^th observation for prediction
    use other N-1 observations to enter inner loop

    //start inner loop
    for j=1 to N-1
      set aside j^th observation for selection of model
          complexity
```

```
      use other N-2 observations for model building

      for h=1 to maximum of model complexity
           using forward selection, create model with
                complexity h using N-2 observations
           get prediction and calculate residual for jth
                observation
           store residual and its corresponding model
                complexity h
      end
   end


      calculate RMSECV with stored residuals for the
   different model complexities
      determine model complexity that corresponds to the
   lowest RMSECV value
      using forward selection, create model based on N-1
   observations with the chosen model complexity

      //end inner loop

      use final inner loop model to predict the ith
   observation
      store the prediction for the ith observation
      store the selected metabolites in this model

   end
   //end outer loop

   calculate Q2 with the stored predictions from the outer
        loops
   for all metabolites count how often they were selected
        in outer loop models
   create scatter plots from observations and predictions
        from outer loops
```

Construction of the Prediction Model

An LOOCV is used where each of the N units is left out in turn. A maximum of $\sqrt{N}$ (rounded to the nearest integer below) selected metabolites are set for the model complexity [13]. For the remaining $(N-1)$ units, $\sqrt{N}$ regression models with complexities running from 1 to $\sqrt{N}$ (8 in our case) are constructed with forward selection. The observation that is left out is predicted with each of these eight models. Deletion residuals are calculated as the differences between the observation that was left out and each of the eight predictions. For each complexity, the root mean square error of cross-validation (RMSECV) is calculated as the square root of the average of the associated N squared deletion residuals. The final complexity chosen is the one with the smallest RMSECV. The prediction model is constructed by forward selection with the final complexity applied to all N observations.

Evaluation of Prediction: Outer and Inner Cross-Validation

The outer LOOCV is for determining the predictive value of the prediction formula for individual observations. Each observation is left out once, and the remaining $(N-1)$ observations are used for model construction.

Figure 2 shows the first iteration of the outer loop, where observation 1 is set aside and the observations $2 \ldots N$ enter the inner loop. The inner loop results in a regression model with a chosen model complexity as explained in the previous section (but this time with one observation less). In the boxed area of Fig. 2, deletion residuals (connected by line segments) are visualized in the residual vs complexity plots. The resulting model is now used to predict the one observation that was left out in the outer loop, e.g., observation 1 (top part of Fig. 2). The selected metabolites and the deletion residual for the left out observation are stored and used later on in the assessment of prediction and predictor quality. So, in the outer loop, each time an observation is left out, a (slightly) different set of units enters the inner loop. For each of these N sets, the optimal model complexity is determined afresh, and the chosen complexity may differ from one set to another (because evaluation of prediction error should include dimension selection).

2.4   Materials

The data used in this chapter originate from a study on liquors of 76 different cocoa genotypes, provided by Nestle Research Centre. For all these genotypes, metabolite profiles of the liquors, using both LCMS- and GCMS-based untargeted metabolomics platforms, have been determined. For the liquors, sensory traits were also determined by a trained sensory panel. We refer to [6, 25, 26] for the untargeted data generation and pre-processing workflow. This consisted of (1) the LCMS and GCMS analyses of the cocoa samples, (2) the Metalign-based untargeted peak picking, and (3) alignment followed by grouping of individual mass features into the so-called reconstructed metabolites. For the present study, the sensory traits of liquors such as fruity, spicy, and acidic are predicted using LCMS data (obtained in negative ionization mode). Each metabolite from the LCMS data set is numbered and preceded by an "LCMS neg" label, indicating the metabolite has been measured using LCMS in negative ionization mode. This LCMS dataset consisted of 229 reconstructed metabolites.

A simulated trait was added to the three sensory traits. This simulated trait was the sum of three arbitrarily selected metabolites labelled by LCMS neg 8, 17, and 199. Normally distributed noise was added such that the three metabolites explained 80 % of the total sum of squares (around the mean) for each sensory trait (and the noise the remaining 20 %).

The proposed method and the stepwise regressions used for comparison were programmed in MATLAB version 7.14 [27]. All calculations were performed on an Intel Xeon CPU at 2.67 GHz.

3   Results and Discussion

All traits have been analyzed with "plain" stepwise regression, with stepwise regression in combination with LOOCV, and with the proposed method. Table 1 shows the results from the "plain" stepwise regression models, so without cross-validation, applied to the liquor data obtained with LCMS for the traits fruity, spicy, acidic, and the simulated trait. All selected metabolites are significant at the 0.05 significance level.

To predict the trait fruity, ten metabolites have been selected explaining up to 82 % of the variance. The biggest contribution comes from metabolites 994 and 1480. Trait spicy is predicted by just one metabolite (1036) with 31 % explained variance, which is rather low. Trait acidic has 14 metabolites in the regression model and 90 % explained variance. Metabolites 2126 and 44 explain most of the variance. The model for the simulated trait has eight metabolites in it, including the three metabolites (199, 17, and 8) that were used in the construction of the trait, which shows that stepwise regression is indeed able to pick up relevant metabolites. Metabolites 199, 17, and 8 account for 80 % explained variance which is exactly what was put into this simulated trait. In the models for fruity, acidic, and the simulated trait a rather large number of metabolites were selected. Typically the first few metabolites have a large R^2, while the metabolites added subsequently only explain up to a few percent per metabolite. This suggests overfitting; metabolites that explain only a few percent are unlikely to improve prediction of future observations.

Stepwise regression was also used in combination with LOOCV. For the resulting N models it is counted how often metabolites are selected. Corresponding regression coefficients are collected and the 5th and 95th percentiles are calculated. For the four traits, Table 2 shows an overview of selected metabolites together with

Table 1 Selected metabolites from stepwise regression for the four selected traits

Trait	Metabolite	β	$SE(\beta)$	R^2
Fruity	LCMS neg 994	0.89	0.14	48 %
	LCMS neg 1480	−1.11	0.22	62 %
	LCMS neg 2073	0.73	0.18	66 %
	LCMS neg 1073	−0.37	0.15	69 %
	LCMS neg 1644	−0.44	0.14	71 %
	LCMS neg 2436	0.25	0.14	75 %
	LCMS neg 2174	−0.9	0.19	78 %
	LCMS neg 2213	0.29	0.13	80 %
	LCMS neg 872	−1.04	0.49	81 %
	LCMS neg 925	0.48	0.23	82 %
Spicy	LCMS neg 1036	0.76	0.13	31 %
Acidic	LCMS neg 2126	−0.76	0.16	63 %
	LCMS neg 44	−0.75	0.27	74 %
	LCMS neg 994	1.08	0.25	78 %
	LCMS neg 2238	1.3	0.15	81 %
	LCMS neg 2117	−1.34	0.3	82 %
	LCMS neg 685	2.24	0.28	83 %
	LCMS neg 801	−3.35	0.48	85 %
	LCMS neg 1068	−2.72	0.52	85 %
	LCMS neg 1781	2.43	0.36	87 %
	LCMS neg 1592	−1.04	0.29	88 %
	LCMS neg 1108	0.72	0.23	87 %
	LCMS neg 2019	−0.68	0.25	89 %
	LCMS neg 1009	1.22	0.48	90 %
Simulated	LCMS neg 199	0.8	0.07	25 %
	LCMS neg 17	0.84	0.07	44 %
	LCMS neg 8	0.83	0.07	80 %
	LCMS neg 1889	0.11	0.03	82 %
	LCMS neg 1938	−0.11	0.03	83 %
	LCMS neg 1285	0.11	0.04	85 %
	LCMS neg 1892	0.04	0.02	86 %

Metabolites are listed in the order as included in the model by stepwise regression. β indicates the corresponding regression coefficient, $SE(\beta)$ is the associated standard error, and R^2 is the percentage variance explained for a given metabolite, including all preceding ones

the selection frequencies and percentiles. Besides the metabolites selected by the "plain" stepwise model (indicated with an asterisk), other metabolites are now also selected. Typically most have low occurrences (less than 25 %), but some have rather high occurrences (e.g., 1518 and 865 with 46 % and 45 %, respectively). The metabolites already selected by "plain" stepwise regression do not always have high occurrences in the LOOCV. Metabolites that are not often selected may have poor prediction ability for future observations. Metabolites that are chosen relatively often are deemed more important predictors.

Table 3 shows the Q^2 values obtained with stepwise regression and LOOCV. For fruity Q^2 is 0.20, for spicy Q^2 is -0.058, indicating that fruity is hard to predict, and that for spicy no suitable model could be found. For acidic and the simulated trait, Q^2 values are higher: 0.51 and 0.68, respectively. New observations can be predicted reasonably well for the simulated trait. This probably stems from the way it was created: the simulated trait is just the sum of three metabolites with added normally distributed noise. The Q^2 value of 0.68 approaches the explained variance R^2 of 0.80 of the three metabolites that were used to create the trait. The other sensory traits cannot be predicted up to this level, indicating that these traits are probably more complex and possibly not well explained by an additive linear model. Based on the R^2 values from the stepwise models (see Table 1) one could expect sizeable predictive ability for the four traits, since most R^2 values are close to one. However, based on the Q^2 values, predictive ability for acidic and the simulated trait may be considered as acceptable, but for spicy it is poor. The somewhat low Q^2 value for fruity indicates that this trait can only be predicted with poor accuracy for new experimental units.

Significance levels for inclusion and exclusion were set at commonly chosen values. Looking at the many metabolites that were selected with low occurrences, there is the suggestion that this could be improved. The many metabolites with low occurrences suggest some interchangeability between these metabolites for explaining a part of the variance. These metabolites are probably not good stable predictors for future experimental units and their selection is likely caused by the chosen significance levels in the stepwise regression being too "lenient."

The proposed approach (LOO2CV) attempts to remedy that situation by controlling the model complexity (the number of included metabolites in the model). Stepwise regression is now replaced by forward selection, and model complexity is determined with a separate LOOCV.

Table 4 shows the chosen metabolites and regression coefficients for the prediction models for the four traits. These models are created using forward selection for complexities optimized using a single LOOCV procedure. From these LOOCV procedures, the optimal complexities were determined to be two metabolites for fruity, one for spicy, four for acidic, and five for the simulated trait. The prediction models can be used in cases where it is desirable to yield a single model for, e.g., future predictions of (truly) new observations.

Table 5 shows the metabolites from the LOO2CV procedure together with occurrences and 5 and 95 % percentiles for the regression coefficients. The selected metabolites were also selected by the LOOCV procedure and nearly all by the "plain" stepwise model. Occurrences still vary somewhat for selected metabolites, but many of them have occurrences larger than 50 %. Looking at the Q^2 values in Table 3 we can see that all Q^2 values have gone up compared to results from the stepwise regression. In particular, for spicy the difference is rather striking, from a negative Q^2 value to (an admittedly still modest) Q^2 of 0.20. Controlling the model complexity with LOO2CV reduces the number of selected metabolites and increases prediction quality, suggesting that the more relevant metabolites are selected.

Table 2 Metabolites selected by stepwise regression in concert with LOOCV

Trait	Metabolite	Occurrence	5th percentile	95th percentile
Fruity	LCMS neg 994*	99 %	0.726	1.245
	LCMS neg 2073*	89 %	0.573	0.988
	LCMS neg 1073*	71 %	−0.803	−0.356
	LCMS neg 1644*	68 %	−0.621	−0.333
	LCMS neg 1480*	66 %	−1.397	−0.851
	LCMS neg 2436*	61 %	0.246	0.487
	LCMS neg 2174*	55 %	−0.961	−0.377
	LCMS neg 2213*	49 %	0.208	0.385
	LCMS neg 872*	36 %	−2.258	−1.009
	LCMS neg 1376	24 %	0.474	1.106
	LCMS neg 985	22 %	−1.167	−0.486
	LCMS neg 1592	21 %	−1.435	−0.808
	LCMS neg 1009	20 %	−3.431	−1.255
	LCMS neg 440	14 %	1.092	2.442
	LCMS neg 925*	14 %	0.464	0.696
	LCMS neg 1068	13 %	−3.393	−1.321
	LCMS neg 1534	12 %	−1.717	−0.967
	LCMS neg 2227	11 %	0.371	0.617
Spicy	LCMS neg 1036*	76 %	0.734	0.781
	LCMS neg 1781	16 %	0.899	1.613
	LCMS neg 2126	14 %	−0.558	−0.193
	LCMS neg 2324	11 %	−0.763	−0.259
Acidic	LCMS neg 2126*	83 %	−1.082	−0.51
	LCMS neg 2238*	61 %	0.939	1.779
	LCMS neg 2117*	55 %	−2.325	−1.111
	LCMS neg 685*	54 %	1.5	2.485
	LCMS neg 801*	53 %	−3.472	−1.797
	LCMS neg 994*	53 %	0.762	1.341
	LCMS neg 1068*	51 %	−5.161	−1.648
	LCMS neg 44*	46 %	−4.332	−0.705
	LCMS neg 1518	46 %	0.752	2.752
	LCMS neg 865	45 %	0.971	2.298
	LCMS neg 2209*	45 %	−1.276	0.905
	LCMS neg 1108	43 %	0.702	1.204
	LCMS neg 2019*	42 %	−1.276	−0.641
	LCMS neg 1781*	39 %	1.849	3.836
	LCMS neg 1592*	33 %	−1.582	−0.833
	LCMS neg 1009*	30 %	1.009	1.716
	LCMS neg 1690	24 %	−1.367	−0.493
	LCMS neg 423	16 %	−2.845	−0.809
	LCMS neg 872	14 %	−2.054	−0.827
	LCMS neg 2397	14 %	−0.503	0.377

(continued)

Table 2 (continued)

Simulated	LCMS neg 8*	100 %	0.825	0.934
	LCMS neg 17*	100 %	0.829	1.001
	LCMS neg 199*	100 %	0.739	0.842
	LCMS neg 1938*	80 %	−0.185	−0.066
	LCMS neg 1285*	75 %	0.089	0.118
	LCMS neg 1889*	70 %	0.1	0.127
	LCMS neg 1892*	61 %	0.032	0.044
	LCMS neg 995	34 %	0.229	0.385
	LCMS neg 1184	23 %	−0.413	−0.205
	LCMS neg 2183	21 %	−0.154	−0.053
	LCMS neg 2024	20 %	0.075	0.152
	LCMS neg 1614	16 %	0.144	0.268
	LCMS neg 1690	15 %	−0.166	−0.117

Occurrence indicates how often a metabolite has been selected. 5th and 95th percentiles of associated regression coefficients are presented as well. Metabolites marked with an asterisk were also chosen in the "plain" stepwise approach

Table 3 Summary of Q^2 values for stepwise regression with LOOCV and the proposed method consisting of forward selection and LOO2CV

Trait	R^2 stepwise	Q^2 LOOCV	Q^2 LOO2CV
Fruity	0.82	0.20	0.24
Spicy	0.31	−0.058	0.15
Acidic	0.90	0.51	0.58
Simulated trait	0.86	0.68	0.71

R^2 values of stepwise regression are shown for reference

Every inner loop ends with a single model, for which the model complexity is optimized. Figure 3 shows how often a certain model complexity has been selected for any of these models. For the traits fruity and acidic, the complexity most often chosen is two. For spicy, the most chosen model complexity is one. For the simulated trait the most often chosen model complexity is five.

The construction and modelling of the simulated trait has been repeated a number of times, each time with a new simulated noise. For these repeated simulated traits the most often chosen model complexities were four and five (results not shown). This indicates that the LOO2CV model is overfitting, since only three metabolites were used in construction of it. Although the proposed approach mitigates the problem of selection of spurious metabolites, it cannot completely remove the problem. The latter is practically impossible: with many metabolites, there remains ample opportunity to "model" part of the noise in the response trait.

For the simulated trait, the three metabolites used in construction of the trait were selected in 100 % of all outer loop models. The 5th and 95th percentiles of the coefficients of these metabolites are close to 1.0, which is in close agreement with how this trait was established, as the sum of three metabolites, i.e., each with coefficient 1.0. For the simulated trait, some extra metabolites were selected that

Table 4 Selected
metabolites in the prediction
model, together with
regression coefficients (β)
and their standard errors
($SE(\beta)$)

Trait	Metabolite	β	$SE(\beta)$	R^2
Fruity	LCMS neg 994	0.79	0.19	48 %
	LCMS neg 1480	−2.07	0.48	62 %
Spicy	LCMS neg 1036	−0.62	0.21	31 %
Acidic	LCMS neg 865	−0.50	0.31	63 %
	LCMS neg 2209	1.87	0.22	74 %
	LCMS neg 1518	1.20	0.53	78 %
	LCMS neg 2126	−1.07	0.12	81 %
Simulated	LCMS neg 199	0.50	0.09	25 %
	LCMS neg 17	0.54	0.07	44 %
	LCMS neg 8	0.66	0.06	80 %
	LCMS neg 1938	−0.23	0.03	82 %
	LCMS neg 1285	0.13	0.04	83 %

R^2 is the percentage variance explained for each
metabolite including preceding ones

Table 5 Selected metabolites by the LOO2CV procedure

Trait	Metabolite	Occurrence	5th percentile	95th percentile
Fruity	LCMS neg 994	100 %	0.973	1.418
	LCMS neg 1480	76 %	−1.461	−0.91
	LCMS neg 1073	20 %	−0.552	−0.438
	LCMS neg 2073	20 %	0.635	0.896
	LCMS neg 2436	11 %	0.238	0.549
Spicy	LCMS neg 1036	88 %	0.732	0.785
	LCMS neg 1781	12 %	0.883	1.085
Acidic	LCMS neg 865	95 %	1.758	2.21
	LCMS neg 2209	91 %	−1.382	−0.568
	LCMS neg 1518	50 %	0.862	2.133
	LCMS neg 2126	41 %	−0.893	−0.699
Simulated	LCMS neg 8	100 %	0.847	0.97
	LCMS neg 17	100 %	0.852	0.985
	LCMS neg 199	100 %	0.753	0.897
	LCMS neg 1938	79 %	−0.111	−0.039
	LCMS neg 1889	70 %	0.101	0.115
	LCMS neg 995	25 %	0.182	0.335
	LCMS neg 1285	15 %	0.093	0.116
	LCMS neg 2183	14 %	−0.141	−0.057

Occurrence indicates how often a metabolite has been selected. 5th and 95th
percentiles of the regression coefficients of metabolites are shown as well

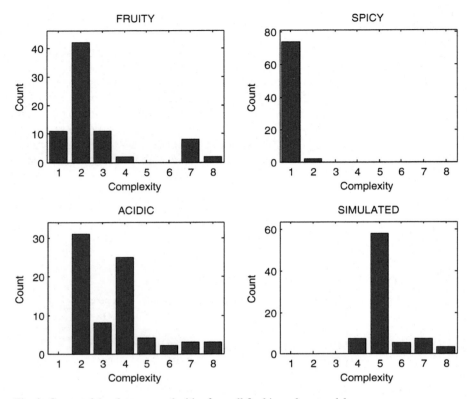

Fig. 3 Counts of the chosen complexities from all final inner loop models

were not involved in the construction of the trait. They were chosen presumably because they seem to model part of the 20 % added noise.

When we compare the metabolites chosen in the prediction models with the results from the LOO2CV procedure, we can see that all metabolites from the prediction models are also present in the metabolites selected by the LOO2CV method. The difference is that the list of metabolites obtained with the LOO2CV method is somewhat longer but the extra metabolites all have low occurrences. When we combine the final model with the frequencies of inclusion from Table 5, we can create a threshold rule for determining which metabolites are (most) valuable. We suggest setting this threshold to 75 %, to include interesting metabolites with some confidence while excluding metabolites that are not very stable. Using this threshold, for fruity, the interest would be for metabolites 994 and 1480. For predicting spicy, only metabolite 1036 is potentially interesting. The trait acidic would be best predicted using metabolites 865 and 2209 (dropping metabolites 1518 and 2126 from the final model). For the simulated trait metabolites 8, 17, 199, and 1938 would be selected (dropping 1285 from the final model). Within each set of metabolites, if limited resources are available for, e.g., follow-up research on selected metabolites, the interest would be in the metabolite with the highest occurrence. When considering all dropped metabolites from the final models, we can see that these metabolites have an added R^2 value of 7 % or less, indicating that these metabolites do not contribute that much in the final model.

We have discussed the list of chosen metabolites based on the outer loop because it may provide extra information. For the data that we analyzed the list supports the prediction models resulting from our proposed method. However, variation between selected sets of metabolites induced by leave-one-out may not be large enough to draw useful conclusions. Maybe leave-one-out should be reserved for model validation and, e.g., leave 10 % out should be introduced in a third calculation to obtain a more useful list of selected metabolites.

4   Conclusions

A novel approach has been proposed to select metabolites for their predictive value for a trait. This method is based on forward selection in a multiple linear regression setting and controls the number of selected metabolites in order to reduce the risk of overfitting. The approach comprises two nested leave-one-out cross-validation (LOOCV) steps; one inner LOOCV for control of the number of selected metabolites (the complexity of the model) and the other outer LOOCV for evaluation of predictive value of the selected prediction model (in terms of root mean squared error).

By counting how often metabolites are selected in any of the models created in the outer LOOCV, a list of metabolites with their frequency of inclusion can be obtained. Metabolites in this list with a high frequency are considered more reliable predictors. For follow-up research, e.g., targeted monitoring, those metabolites with high frequencies are likely candidates to use. Further research will have to show whether such a list, perhaps based on separate calculations with leave 10 % out, provides information that might change the choice of variables in the final prediction formula.

These techniques have been demonstrated on three real sensory traits in combination with LCMS data from cocoa liquor and one simulated trait. The simulated trait confirms that relevant metabolites are picked up. The proposed approach yields a more robust analysis, allowing for more confident predictor selection: compared to "plain" stepwise regression and stepwise regression with LOOCV, fewer metabolites are selected resulting in higher Q^2 values.

Funding This project was financed by the Netherlands Metabolomics Centre (project NMC-BS2a "Power analysis of metabolomics studies") which is part of the Netherlands Genomics Initiative/Netherlands Organization for Scientific Research.

References

1. Hall, R. D. (2011). Biology of plant metabolomics. In R. D. Hall (Ed.), *Annual plant reviews* (Vol. 43). Oxford: Wiley.
2. Keurentjes, J. J. B., et al. (2006). The genetics of plant metabolism. *Nature Genetics, 38*(7), 842–849.
3. Moing, A., et al. (2011). Extensive metabolic cross-talk in melon fruit revealed by spatial and developmental combinatorial metabolomics. *New Phytologist, 190*(3), 683–696.
4. Tikunov, Y. M., et al. (2010). A role for differential glycoconjugation in the emission of phenylpropanoid volatiles from tomato fruit discovered using a metabolic data fusion approach. *Plant Physiology, 152*(1), 55–70.
5. Gupta, A. J., et al. (2014). Chemometric analysis of soy protein hydrolysates used in animal cell culture for IgG production - An untargeted metabolomics approach. *Process Biochemistry, 49*(2), 309–317.
6. Lindinger, C., et al. (2009). Identification of ethyl formate as a quality marker of the fermented off-note in coffee by a nontargeted chemometric approach. *Journal of Agricultural and Food Chemistry, 57*(21), 9972–9978.
7. Capanoglu, E., et al. (2008). Changes in antioxidant and metabolite profiles during production of tomato paste. *Journal of Agricultural and Food Chemistry, 56*(3), 964–973.
8. Hendriks, M., et al. (2011). Data-processing strategies for metabolomics studies. *Trac-Trends in Analytical Chemistry, 30*(10), 1685–1698.
9. Jelizarow, M., et al. (2010). Over-optimism in bioinformatics: An illustration. *Bioinformatics, 26*(16), 1990–1998.
10. Wehrens, R., et al. (2011). Stability-based biomarker selection. *Analytica Chimica Acta, 705*(1–2), 15–23.
11. Hageman, J. A., et al. (2008). Simplivariate models: Ideas and first examples. *PLoS One, 3*(9).
12. Montgomery, D., & Peck, E. (1982). *Introduction to linear regression analysis.* Wiley.
13. Efron, B., & Tibshirani, R. (1993). *An introduction to the bootstrap.* Monographs on statistics and applied probability (Vol. 57). Chapman & Hall.
14. Westerhuis, J. A., et al. (2008). Assessment of PLSDA cross validation. *Metabolomics, 4*(1), 81–89.
15. Smit, S., et al. (2007). Assessing the statistical validity of proteomics based biomarkers. *Analytica Chimica Acta, 592*(2), 210–217.
16. Abeel, T., et al. (2010). Robust biomarker identification for cancer diagnosis with ensemble feature selection methods. *Bioinformatics, 26*(3), 392–398.
17. Meinshausen, N., & Buhlmann, P. (2010). Stability selection. *Journal of the Royal Statistical Society Series B-Statistical Methodology, 72*, 417–473.
18. Menendez, P., et al. (2012). Penalized regression techniques for modeling relationships between metabolites and tomato taste attributes. *Euphytica, 183*(3), 379–387.
19. Vandeginste, B. G. M., et al. *Handbook of chemometrics.* Data handling in science and technology (Vol. 20B). Amsterdam: Elsevier.
20. Hageman, J. A., et al. (2003). Wavelength selection with tabu search. *Journal of Chemometrics, 17*(8–9), 427–437.
21. Furnival, G. M., & Wilson, R. W. (1974). Regressions by leaps and bounds. *Technometrics, 16*(4), 499–511.
22. Hammami, D., et al. (2012). Predictor selection for downscaling GCM data with LASSO. *Journal of Geophysical Research-Atmospheres, 117*, 1–11.
23. Neter, et al. (1996). *Applied linear statistical models.* Irwin.
24. Tukey, J. W. (1958). Bias and confidence in not-quite large samples. *Annals of Mathematical Statistics, 29*(2), 614.

25. De Vos, R. C. H., et al. (2007). Untargeted large-scale plant metabolomics using liquid chromatography coupled to mass spectrometry. *Nature Protocols, 2*(4), 778–791.
26. Tikunov, Y. M., et al. (2012). MSClust: A tool for unsupervised mass spectra extraction of chromatography-mass spectrometry ion-wise aligned data. *Metabolomics, 8*(4), 714–718.
27. Mathworks, I. (2008). *Matlab 7.1.*

On the Combination of Omics Data
for Prediction of Binary Outcomes

Mar Rodríguez-Girondo, Alexia Kakourou, Perttu Salo, Markus Perola,
Wilma E. Mesker, Rob A.E.M. Tollenaar, Jeanine Houwing-Duistermaat,
and Bart J.A. Mertens

1 Introduction

Proteomics, metabolomics, and related omics research fields are revolutionizing biomolecular research by the ability to simultaneously profile many compounds within either patient blood, urine, tissue, or other. Increasingly, clinical studies include several sets of such omics measures available for each patient, measuring different levels of biology. In the last decade, much work has been done on accommodating one single high-dimensional source of (omic) predictors for prognosis. Nowadays, one of the main challenges in predictive research is the integration of different sources of omic biomarkers for the prediction of health traits.

Prediction with several omic sources involves a number of difficulties. First of all, omic sets of predictors are typically high-dimensional ($n < p$, n sample size

---

The authors "Mar Rodríguez-Girondo" and "Alexia Kakourou" contributed equally to this work.

M. Rodríguez-Girondo (✉) • A. Kakourou • B.J.A. Mertens
Department of Medical Statistics and Bioinformatics, Leiden University Medical Center,
Leiden, The Netherlands
e-mail: M.Rodriguez_Girondo@lumc.nl

P. Salo • M. Perola
National Institute For Health and Welfare, Helsinki, Finland

W.E. Mesker • R.A.E.M. Tollenaar
Department of Surgery, Leiden University Medical Center, Leiden, The Netherlands

J. Houwing-Duistermaat
Department of Medical Statistics and Bioinformatics, Leiden University Medical Center,
Leiden, The Netherlands

Department of Statistics, Leeds University, Leeds, UK

© Springer International Publishing Switzerland 2017
S. Datta, B.J.A. Mertens (eds.), *Statistical Analysis of Proteomics, Metabolomics,
and Lipidomics Data Using Mass Spectrometry*, Frontiers in Probability and the
Statistical Sciences, DOI 10.1007/978-3-319-45809-0_14

and p the number of predictors), and correlation between features is typically high which requires the use of model building techniques beyond classical regression-based methods. Furthermore, several potential predictor sets measured on the same subjects may share (part of) the underlying biological information, which thus introduces correlation between the distinct omic sources. Moreover, differences on scales, normalization procedures, and other technical issues inherent to omic research may play a role when trying to integrate information provided by several omic sources. These issues may dramatically affect the gain in predictive ability of a hypothetical "naive" combination, based on stacking all the available features, and ignoring their different origin, specially in high-dimensional settings. For example, one of the sources of predictors may be obscured by another one due to their different dimensionality, noise level, and correlation between them.

In this work, we propose to replace the original (high-dimensional) sources of predictors by their corresponding predicted values of the outcome based on single-source prediction models and to combine those, with the intention of outperforming single-omic prediction models. This approach has become relatively popular nowadays for combining predicted values obtained from different methods but based on a common source of predictors [12, 14, 29] but it has been less applied in the context of combination of predictions of a common outcome based on different sources of predictors [20]. An important issue when fitting a model for combining predictions which are themselves fitted is that it requires the calibration of each of the single-omic-based predictions as well as the combined model using the same set of observations. In this situation, the calibration of the resulting combination will depend on the prior calibration using the single sources. This issue can be handled by appropriate use of cross-validation techniques.

The former setting refers to the parallel combination of omic sources, in the sense that both single-omic predictions are jointly considered, without imposing any "a priori" hierarchy or different importance among them. Alternatively, combination can be regarded as the sequential augmentation with new biomarkers of previously calibrated prediction models based on a given omic dataset. In this way, the first source of predictors is prioritized. Several examples may motivate such an asymmetric approach to combination, such as the addition of more expensive predictors to potentially improve the performance of a prediction rule based on more economic sources or the addition of less reliable sources to models based on stable and well-established markers. It seems clear that when a new molecular marker set emerges due to (technical) advances in the field, it must then prove its worth in the face of existing knowledge on the predictive capacity of an established biomarker set. However, if a sequential augmentation outperforms the parallel combination in terms of predictive performance is unclear.

Next, we present and discuss different strategies for the combination and augmentation of single-omic prediction models in the context of classification problems (binary outcome). We discuss their relations and differences and we compare the considered methods by means of the analysis of two real datasets. On the one hand, we revisit the problem of calibrating an early diagnosis tool for breast cancer based on MS-based proteomics profiling. Specifically, we consider the combination of two different sets of predictors, each obtained by different techniques of fractionation of the same spectrometry data (see for details [4, 20]). On the other hand, we consider data from a population-based cohort, the Dietary, Lifestyle, and Genetic determinants of Obesity and Metabolic syndrome (DILGOM) study, sampled from the Helsinki area, in Finland [9]. In this case, we consider the combination of transcriptomics and metabolomics sources in the prediction of obesity.

2 Methods

2.1 Double Cross-Validation Prediction

Let the observed data be given by $(\mathbf{y}, \mathbf{X}_1, \mathbf{X}_2)$, where $\mathbf{y} = (y_1, \ldots, y_n)^\mathsf{T}$ is a binary outcome, $y_i \in \{0, 1\}$ for $i = 1, \ldots, n$ independent individuals and $\mathbf{X}_1$ and $\mathbf{X}_2$ are two matrices of dimension $n \times p$ and $n \times q$, respectively, representing two omic predictor sources with p and q features. We assume that we are in a high-dimensional setting $(p, q > n)$ and that our objective is to enrich single-omic predictions $p_{ik} = \widehat{P}(y_i = 1|\mathbf{X}_{ki})$, $k = 1, 2$ by the combined use of the two distinct sources.

Two crucial difficulties in high-dimensional prediction problems are the control of the optimal level of shrinkage (or in general, any tuning parameter λ associated with the statistical model f used to obtain estimates of $\mathbf{y}$ based on a single high-dimensional source of predictors) and the quantification of the error of the resulting predictions in new data. Double cross-validation algorithms [1, 11, 18–20, 25], consisting of two (or more) nested loops, allow to handle both issues. In the inner loop a cross-validated grid-selection is used to determine the optimal prediction rule, i.e., for model selection, while the outer loop is used to estimate the prediction performance by application of models developed in the inner loop part of the data (training sets) to the remaining unused data (validation sets). In this manner, double cross-validation is capable of jointly calibrating and assessing models in a predictive sense, while also avoiding the bias in estimates of predictive ability which would result from use of a single-cross-validatory approach only.

Next, we present two approaches for combination of omic-based predictions of binary health traits, both based on the replacement of the original (high-dimensional) sources of predictors by their corresponding estimated values of the outcome based on single-source prediction models and the double cross-validation principle. On the one hand, we consider *parallel* combination methods, in which the outer cross-validation loop of the "double" cross-validation procedure is used to calculate, in a first step, predictions of the outcome $\mathbf{y}$ of interest based on each of the omic sources of predictors, $\mathbf{X}_1$ and $\mathbf{X}_2$, which are combined in a second step. On the other hand, we consider a *sequential* combination of $\mathbf{X}_1$ and $\mathbf{X}_2$ in which the double cross-validated predictions of $\mathbf{y}$ based on $\mathbf{X}_1$ (considered as primary source) enter as an offset term (not refitted) in a second step in which $\mathbf{X}_2$ is evaluated as an additional set of variables to predict $\mathbf{y}$.

2.2   Parallel Combination of Predictions

Firstly, we consider a parallel combination approach based on replacing the original sets of predictors X_1 and X_2 with the sets of their corresponding estimated class probabilities $\mathbf{p}_1 = (p_{11}, \ldots, p_{1n})^\mathsf{T}$ and $\mathbf{p}_2 = (p_{21}, \ldots, p_{2n})^\mathsf{T}$. In a first stage, the double cross-validated probabilities $\mathbf{p}_1$ and $\mathbf{p}_2$ are estimated by calibrating the prediction model using each single predictor source X_1 and X_2 as input variables. Subsequently, we combine the double cross-validated class probabilities either by considering convex combinations of the estimated class probabilities or by using them as new input variables for the construction of a final combined model.

Convex Combination via Linear Mixtures

A simple way to combine the cross-validated estimates $\mathbf{p}_1$ and $\mathbf{p}_2$ is to consider linear mixtures of the separately calibrated class probabilities

$$\mathbf{p}_C = w\mathbf{p}_1 + (1 - w)\mathbf{p}_2 \tag{1}$$

with $\mathbf{p}_C = (p_{C1}, \ldots, p_{Cn})^\mathsf{T}$ the newly combined class probabilities vector and w some number in the interval $[0, 1]$. Since different choices of w result in different prediction rules, we should choose w so that it optimizes the final prediction rule. The parameter w can be considered as the optimal balance between the predictions based on the two distinct sources X_1 and X_2. Note that the two extreme choices, for which w is either 0 or 1, result in excluding X_1 or X_2 completely from estimating the combined class probabilities vector $\mathbf{p}_C$.

In many applications, it has been observed that the predictive performance of linear combinations of predictors are often insensitive to the exact values of their weights w, i.e., quite large deviations from the optimal set of weights w often lead to predictive performance not substantially worse than those obtained using optimal weights. This phenomenon has been termed as the flat maximum effect [5, 6]. Specifically, if the correlations between $\mathbf{p}_1$ and $\mathbf{p}_2$ are high, the simple average ($w = 0.5$) will be highly correlated with any other weighted sum of $\mathbf{p}_1$ and $\mathbf{p}_2$, and hence the choice of weights will make little difference. Thus, averaging across the estimated class probabilities is expected to result in improved estimates in many omic settings. The main advantage of this approach lies in its simplicity, due to the fact that no further optimization or cross-validatory scheme is required in order to obtain an unbiased estimate of the predictive performance, since the double cross-validatory nature of the estimated class probabilities is preserved entirely.

Model-Based Combination

A somewhat more sophisticated way to combine the cross-validated estimates, in a parallel fashion, is to fit a (semi)parametric model, such as an ordinary logistic regression model, to the set of double cross-validated class probabilities $(\mathbf{p}_1, \mathbf{p}_2)$.

In this case, the original set of predictors is replaced with the set of estimated class probabilities, reducing the dimensionality of the original data to a low-dimensional space. The final combined class probabilities can then be derived by fitting the logistic model

$$\text{logit}(p_{Ci}) = \alpha + \beta_1 \text{logit}(p_{1i}) + \beta_2 \text{logit}(p_{2i}) \tag{2}$$

where $\text{logit}(p_{ki}) = \log(\frac{p_{ki}}{1-p_{ki}})$. To maintain the double cross-validatory nature of the predictive performance evaluation we embed the logistic model calibrations within an additional single cross-validatory loop, leaving out each cross-validated pair (p_{1i}, p_{2i}) in turn and fitting the logistic model using the remaining pairs to obtain the final (cross-validated) class probabilities $\mathbf{p}_C = (p_{C1}, \ldots, p_{Cn})^{\mathsf{T}}$.

Attention must be brought at this point to the fact that, in the case of model-based combination, the class probabilities $(\mathbf{p}_1, \mathbf{p}_2)$ are not only combined, as in the case of convex combination, but also re-calibrated as suggested by Cox [3]. This re-calibration aspect should be taken into account when interpreting the results from this combination approach and when comparing them directly to the performance measures of the calibrated, yet not re-calibrated model, based on $\mathbf{X}_1$ or $\mathbf{X}_2$ only. This is of particular importance since fitting a model based on the set of the calibrated estimates from a single predictor source, instead of the single predictor source itself, could alter the final estimates. A more fair comparison thus would be the comparison between the cross-validated predictions of the logistic model combination and the cross-validated predictions of the re-calibrated logistic models using the estimates from the first source only $(\mathbf{p}_1)$ or the estimates from the second source only $(\mathbf{p}_2)$ such that

$$\text{logit}(p_{ki}^R) = \alpha + \beta_1 \text{logit}(p_{ki}) \tag{3}$$

for $k = 1, 2$, with p_{ki}^R the re-calibrated probabilities for the kth source. This type of comparison would assess the extent of the re-calibration effect and would give us insight in whether the improvement in prediction performance is due to combining the cross-validated estimates using each different source or due to re-calibration.

2.3 Sequential Combination of Prediction

An alternative approach to the aforementioned parallel combination approach is to consider the problem of combination of omic predictors in an "asymmetric" manner by sequentially fitting prediction models based on different omic datasets. If we focus on the study of two omic datasets, we can proceed as follows. Firstly, the double cross-validated predictions of the outcome $\mathbf{y}$ based on the primary source, $\mathbf{X}_1$, and a given model specification, $\mathbf{p}_1 = (p_{11}, \ldots, p_{1n})^{\mathsf{T}}$ are estimated. Then, in a second step we construct a second model based on $\mathbf{X}_2$ as predictor and devoted to predict the variation of $\mathbf{y}$ which remains unexplained by $\mathbf{X}_1$. In the logistic regression context, this can be implemented by considering $\text{logit}(\mathbf{p}_1)$ as an offset term in a regularized regression based on $\mathbf{X}_2$ for predicting $\mathbf{y}$ as follows:

$$\text{logit}(p_{Ci}) = \text{logit}(p_{1i}) + f(\mathbf{X}_{2i}) \tag{4}$$

The first source of omic predictors is hence prioritized by its inclusion as offset, and hence not refitted in the second stage. This means that the second source of predictors $\mathbf{X}_2$ is "added" to the previous fit based on $\mathbf{X}_1$, since the single-omic prediction $\mathbf{p}_1$ is fixed in the second step.

As was the case in the model-based parallel combination presented in section "Model-Based Combination," the model in expression (2) contains $\mathbf{p}_1$ which is fitted itself. Hence, following the lines of Sect. 2.1, we embed the estimation of $\mathbf{p}_1$ in the double cross-validation loop to guarantee a realistic estimation of the predictive performance of the sequential combination. Namely, we leave out each cross-validated p_{1i} and corresponding y_i and $\mathbf{X}_{2i}$ in turn and fit the regularized regression model based on the remaining observations in $\mathbf{X}_2$ to get the final $\mathbf{p}_C = (p_{C1},\ldots,p_{Cn})^{\mathsf{T}}$.

3 Performance Evaluation

The performance of prediction models can be summarized in several ways. Traditionally, two aspects have been considered as crucial when evaluating prediction models for binary outcomes: calibration and discrimination (see [24], for a review). Calibration refers to the quantification of how close predictions are to the actual outcome, while discrimination focuses on determining to what extent individuals with positive outcome have higher risk predictions than those with negative outcome. The relation and differences among them has been object of extensive research in the past years (see [22], and the references therein). Beyond calibration and discrimination, other measures such as reclassification [21] and clinical usefulness [30] have been proposed, but an exhaustive comparison of performance measures falls beyond the scope of this work.

3.1 Calibration Measures

To evaluate the predictive performance of the different combination strategies described in Sect. 2, and to compare them with single-omic predictive models, we considered several calibration measures.

Denote by $\text{PRESS}(\mathbf{y}, \mathbf{p}) = \sum_{i=1}^{n}(y_i - p_i)^2$ the prediction sum of squares based on some arbitrary vector of predictions $\mathbf{p} = (p_1,\ldots,p_n)^{\mathsf{T}}$. The prediction sum of squares is also denoted as Brier score, and it is usually used for reporting model performance.

Consider $\mathbf{p}_0$, the simplest cross-validated predictor of $\mathbf{y}$ based on an intercept-only logistic model, corresponding to the classification rule based on assigning all the observations to the majority class. Denote by $\mathrm{CVSS}(\mathbf{p}_1, \mathbf{p}_2) = \sum_{i=1}^{n}(p_{1i} - p_{2i})^2$ the cross-validated sum of squared differences of two vectors of predictions $\mathbf{p}_1$ and $\mathbf{p}_2$.

For any vector $\mathbf{p}$ of predicted values of the outcome of interest, we define

$$Q_{\mathbf{p}}^2 = \frac{\mathrm{CVSS}(\mathbf{p}, \mathbf{p}_0)}{\mathrm{PRESS}(\mathbf{y}, \mathbf{p}_0)} = \frac{\sum_{j=1}^{J}\sum_{i \in S^{(j)}} (p_i - p_{0i})^2}{\sum_{j=1}^{J}\sum_{i \in S^{(j)}} (y_i - p_{0i})^2}. \tag{5}$$

where the computations of p_{0i}, p_{1i}, and p_{2i}, $i \in S^{(j)}$ for each of the $j = 1, \ldots, J$ random splits $S^{(j)}$ of the sample S are based on the observations not belonging to $S^{(j)}$. Intuitively, $Q_{\mathbf{p}_1}^2$ and $Q_{\mathbf{p}_2}^2$ can be regarded as cross-validation equivalents of the R^2, in which the performance of the model-based predictions are compared to the naive predictions based on the prevalence of the outcome variable y. Analogously, $Q_{\mathbf{p}C}^2$ represents the predictive ability of the combination of $\mathbf{X}_1$ and $\mathbf{X}_2$, obtained with any of the methods presented in Sect. 2.

Additionally, we evaluate the calibration of each of the combinations strategies and for each individual source of predictors in terms of the deviance given by

$$\mathrm{Deviance}_{\mathbf{p}} = -2\sum_{i=1}^{n} y_i \log p_i + (1-y_i)(\log(1-p_i)) = -2\sum_{i=1}^{n}\log(1-|y_i-p_i|) \tag{6}$$

and which is evaluated in the cross-validated predicted probabilities.

3.2 Discrimination Measures

To quantify the discrimination ability of the different methods introduced in Sect. 2, we use the c-statistic, the most commonly used summary of discrimination. The c-statistic accounts for the proportion of individuals with $y = 1$ with higher predicted probability than individuals with $y = 0$, among all possible pairs. It also can be defined as the area under the receiver operating characteristic (ROC). Specifically:

$$c-\mathrm{index}_{\mathbf{p}} = \frac{1}{n_1 n_2}\sum_{i \in G_1}\sum_{j \in G_2}\left[I\left(p_i > p_j\right) + 0.5 \times I\left(p_i = p_j\right)\right] \tag{7}$$

where G_1 and G_2 are the index sets for $y = 1$ and $y = 0$, respectively, and n_1 and n_2 their respective sizes. The c-statistic is a measure of discrimination, that is, of the extent the double cross-validated predictions are higher for individuals in the groups defined by the outcome variable y and it varies from 0 and 1. A value of 0.5

indicates that probabilities are distributed randomly among the two groups given by $y = 1$ and $y = 0$, while a value equal to 1 indicates that all predicted probabilities for individuals with $y = 1$ are higher than the probabilities assigned to individuals with $y = 0$. A c-index below 0.5 indicates reverse ordering, i.e., probabilities for individuals with $y = 1$ are lower than estimated probabilities for individuals with $y = 0$. Because the c-index is invariant under monotone transformations, it provides no information about calibration, however, separation among classes defined by the outcome variable y is of great interest in diagnosis applications, for example.

4 Application

4.1 Data Presentation

Breast Cancer Data

The first study we consider is a clinical proteomics study conducted in the Leiden University Medical Center, The Netherlands, which comprises 307 women, from which 105 are breast cancer patients and 202 are healthy controls. In order to classify participants as cancer cases or healthy controls, we use two different subsets of proteins. Both were processed by MALDI-TOF mass spectrometry but they differ in the techniques used for extraction, which yields different subsets of proteins suitable for detection. On the one hand, we consider 48 protein measures resulting from the use of weak-cation exchange magnetic beads for protein extraction (WCX). On the other hand, the use of reversed-phased C18 magnetic beads (C18) resulted in 42 different measures (see [4, 20], for details).

We carried out two distinct analyses using the combination approaches described in Sect. 2. As a first analysis, we adopted a parallel combination approach, and we derived combinations of WCX and C18 bead processing measures based on a linear mixture and model-based combinations, presented in Sect. 2.2. Additionally, we also constructed a naive combination by stacking all the 90 available features without distinguishing if they come from the same pre-processing method. Secondly, we analyzed the data using the sequential approach revisited in Sect. 2.3. We considered the WCX method as state-of-the-art technique which is treated as primary omic source and we evaluated the added value of the features obtained with the C18 processing method. Alternatively, we turned around the roles of the available sources by firstly fitting a model based on the C18 source of proteomic predictors.

DILGOM Data

In a second case study, we consider data from a population-based cohort, the Dietary, Lifestyle, and Genetic determinants of Obesity and Metabolic syndrome (DILGOM) study, sampled from the Helsinki area, in Finland [9, 10]. We are interested in getting insight in the role of serum NMR metabolites measures and gene expression levels in the prediction of obesity (defined in terms of the Body Mass Index (BMI), specifically, an individual was considered to be obese if BMI$\geq$ 30). The metabolomic predictor data consists of quantitative information on 139 metabolic measures, mainly composed of measures on different lipid subclasses, but also amino acids, and creatine (see [9], for details). The gene expression profiles were derived from Illumina 610-Quad SNParrays (Illumina Inc., San Diego, CA, USA). Initially, 35,419 expression probes were available after quality filtering (see [10], for pre-processing details). In addition to the pre-processing steps described by Inouye et al. [10], we conducted a prior filtering approach and removed from our analyses those probes with extremely low variation (see [16], for details on the conducted pre-processing). As a result, we retained measures from 7380 beads for our analyses. The analyzed sample contained $n = 406$ individuals for which both types of omic measurements and the outcome of interest were available. From them 78 (19 %) were obese.

As for the breast cancer data, we performed both parallel and sequential combinations of metabolome and gene expression for the prediction of obesity. Specifically, we firstly obtained a naive, linear mixture, and model-based parallel combinations of transcriptomics and metabolomics. Secondly, we analyzed the data using the sequential approach revisited in Sect. 2.3. We considered the metabolic profile as primary omic source for the prediction of obesity and evaluated the added value of the blood gene expression profiles. Alternatively, we turned around the roles of the omic sources, first fitting a model based on gene expression and then adding the metabolome.

4.2 Model Choice: Logistic Regularized Regression

Several statistical methods are available to derive prediction models of binary outcomes in high-dimensional settings. In this work, we focus on regularized logistic regression models [7, 8, 15, 26]. For a given omic source of predictors $\mathbf{X}, f(\mathbf{X}) = \mathbf{X}\boldsymbol{\beta}$ and logit$(p_i) = \mathbf{X}_i\boldsymbol{\beta}$, with $p_i = \widehat{P}(y_i = 1|\mathbf{X}_i)$. The estimation of $\boldsymbol{\beta}$ is conducted by maximizing the penalized log-likelihood $\sum_{i=1}^{n} [y_i \log(p_i) + (1 - y_i) \log(1 - p_i)] - \lambda\text{pen}(\boldsymbol{\beta})$. The penalty parameter λ regularizes the $\boldsymbol{\beta}$ coefficients, by shrinking large coefficients in order to control the bias-variance trade-off. We consider two different (and widely used) penalization types. On the one hand, we use the ridge penalty [8, 15], with pen$(\boldsymbol{\beta}) = ||\boldsymbol{\beta}||_2^2 = \sum_{j=1}^{p} \beta_j^2$, i.e., a ℓ_2 type penalty. On the other hand, we consider lasso regression [26], with pen$(\boldsymbol{\beta}) = ||\boldsymbol{\beta}||_1 = \sum_{j=1}^{p} |\beta_j|$, i.e., lasso uses a ℓ_1 type penalty.

Note that other model building strategies for prediction of binary outcomes could have been used in this framework, such as the elastic net penalization [31] by setting $\alpha = 0.5$, boosting methods [2, 13, 27], or random forests [1] among others.

4.3   Results

The main findings from the analysis of the breast cancer data are summarized in Tables 1 and 2, while Tables 3 and 4 show the results corresponding to the DILGOM data. The top parts of the tables refer to the results obtained by using ridge regression as model to derive the double cross-validated predictions and the bottom parts of the tables refer to lasso regression. For both datasets, we provide results of the individual performance (in terms of calibration and discrimination) of each of the two considered omic sources, jointly with the evaluation of the previously introduced strategies for (parallel and sequential) combination. In Tables 1 and 3, the single-omic sources are evaluated in terms of the (non-re-calibrated) predictions p_1 and p_2. Tables 2 and 4 contain the results for each single-omic source based on re-calibrated probabilities, along the lines of expression (3).

Table 1 Breast cancer data

Single source				Combination methods				
					Parallel		Sequential	
		C18	WCX	Naive	Average	Model-based	WCX|C18	C18|WCX
Ridge	Brier score	0.128	0.111	0.113	0.109	0.093	0.088	0.092
	Deviance	253.36	227.25	231.07	226.92	201.00	188.50	199.55
	c-index	0.879	0.911	0.900	0.925	0.922	0.933	0.922
	Q^2	0.304	0.360	0.354	0.286	0.559	0.519	0.537
Lasso	Brier score	0.120	0.091	0.092	0.087	0.079	0.086	0.083
	Deviance	253.46	198.50	204.41	190.30	179.85	209.63	192.16
	c-index	0.877	0.929	0.919	0.939	0.939	0.920	0.939
	Q^2	0.522	0.567	0.582	0.463	0.609	0.673	0.703

Table 2 Breast cancer data

Re-calibrated single source				Combination methods				
					Parallel		Sequential	
		C18	WCX	Naive	Average	Model-based	WCX|C18	C18|WCX
Ridge	Brier score	0.126	0.107	0.113	0.109	0.093	0.088	0.092
	Deviance	249.69	224.47	231.07	226.92	201.00	188.50	199.55
	c-index	0.875	0.906	0.900	0.925	0.922	0.933	0.922
	Q^2	0.441	0.500	0.354	0.286	0.559	0.519	0.537
Lasso	Brier score	0.123	0.093	0.092	0.087	0.079	0.086	0.083
	Deviance	254.12	205.96	204.41	190.30	179.85	209.63	192.16
	c-index	0.872	0.926	0.919	0.939	0.939	0.920	0.939
	Q^2	0.430	0.549	0.582	0.463	0.609	0.673	0.703

Table 3 DILGOM data

Single source				Combination methods				
				Parallel			Sequential	
		Transc	Metab	Naive	Average	Model-based	*Metab|Transc*	*Transc|Metab*
Ridge	Brier score	0.142	0.116	0.134	0.120	0.114	0.113	0.114
	Deviance	363.66	306.95	340.64	314.47	300.04	298.32	304.20
	c-index	0.716	0.811	0.790	0.837	0.827	0.829	0.815
	Q^2	0.057	0.256	0.090	0.102	0.285	0.284	0.286
Lasso	Brier score	0.146	0.121	0.132	0.123	0.121	0.123	0.124
	Deviance	371.95	311.09	337.77	318.37	311.09	319.62	315.15
	c-index	0.682	0.806	0.768	0.808	0.806	0.808	0.803
	Q^2	0.108	0.285	0.201	0.128	0.285	0.359	0.295

Table 4 DILGOM data

Re-calibrated single source				Combination methods				
				Parallel			Sequential	
		Transc	Metab	Naive	Average	Model-based	*Metab|Transc*	*Transc|Metab*
Ridge	Brier score	0.142	0.117	0.134	0.120	0.114	0.113	0.114
	Deviance	361.63	313.08	340.64	314.47	300.04	298.32	304.20
	c-index	0.711	0.805	0.790	0.837	0.827	0.829	0.815
	Q^2	0.111	0.247	0.090	0.102	0.285	0.284	0.286
Lasso	Brier score	0.147	0.122	0.132	0.123	0.121	0.123	0.124
	Deviance	374.38	314.38	337.77	318.37	311.09	319.62	315.15
	c-index	0.676	0.801	0.768	0.808	0.806	0.808	0.803
	Q^2	0.084	0.245	0.201	0.128	0.285	0.359	0.295

Breast Cancer Data

In the breast cancer setting, we observe a slightly better performance of the protein fractionation WCX than the alternative C18, according to the two studied model specifications (ridge and lasso regression) and regarding both re-calibrated and non-re-calibrated results. Both sets of markers show similar and very good performance in terms of discrimination (c − index around 0.91 for WCX and slightly inferior for C18, with c − index around 0.88). WCX also outperforms C18 in terms of Brier score, deviance, and Q^2. In terms of model specification, lasso regression seems to provide better results in the individual evaluation of each of the omic sources (Table 1), specially in terms of calibration, but the two model specifications behave similar when compared in terms of re-calibrated probabilities (Table 2). Interestingly, re-calibration of the single-omic predictions is beneficial when using ridge regression in terms of calibration, while discrimination becomes slightly worse. On the other hand, re-calibration of single-source lasso-based predictions is not beneficial (all the summary measures worsen with re-calibration).

With regard to the combination strategies, we observe that the naïve combination (stacking the two sources of predictors and conduct a new regularized regression with common penalty) provides, in general, worse results compared to the alternative strategies for parallel combination (averaging and model-based), in terms of both, calibration and discrimination measurements. In fact, the naïve combination provides worse results than using WCX only.

The model-based parallel combination outperforms the (simpler) parallel combination based on averaging in terms of calibration (smaller values of Brier score and deviance and larger values of Q^2), for both ridge and lasso specifications. The difference between average and model-based is almost twofold for ridge regression, while the difference is smaller for lasso regression ($Q^2 = 0.286$ for averaging, $Q^2 = 0.559$ for the logistic regression combination.) Alternatively, if we focus on the differences in discrimination ability, averaging seems to show a slightly better performance than the model-based combination. The right part of Table 1 provides the results from a sequential approach to the combination of omic predictors. Firstly, we observe that, especially for the lasso specification, the order of introduction of the predictors influences the resulting performance of the sequential combination. We denote by C18|WCX the sequential combinations of WCX and C18, introduced in Sect. 2.3, which treats WCX as primary source. Alternatively, WCX|C18 refers to the sequential combinations of WCX and C18 which considers C18 as the primary source. We observe that for the lasso specification, the sequential combination C18|WCX provides better results than WCX|C18, both in terms of calibration and discrimination. These results agree with the intuition that the preferable sequential procedure is the one which starts using the omic source which optimizes the performance when considered individually. Results are less conclusive when using ridge regression as model to sequentially generate the predictions. The results are less influenced by the order in the sequential procedure, and the summary measures disagree with regard to the preferable strategy. WCX|C18 presents better results in terms of Brier score, deviance, and c − index, while the Q^2 is larger for C18|WCX.

Interestingly, the sequential combinations may outperform the parallel model-based alternative. For ridge regression, we observe that the lowest Brier score and deviance are reached by the sequential combination WCX|C18 (BS $= 0.088$, Deviance $= 188.50$ for WCX|C18 versus BS $= 0.093$, Deviance $= 201.00$ for the model-based parallel combination), and also the largest c-index (c − index $= 0.933$ for WCX|C18 versus c − index $= 0.922$ for the parallel combination). The model-based parallel combination presents the best performance in terms of Q^2 ($Q^2 = 0.559$ for the parallel combination versus $Q^2 = 0.537$ for the sequential combination C18|WCX). When using lasso regression, the parallel combination outperforms the sequential approach in terms of Brier score, and deviance, while the sequential combination C18|WCX presents the best results in terms of Q^2 ($Q^2 = 0.703$ for C18|WCX versus $Q^2 = 0.609$ for the model-based parallel combination). The results in terms of discrimination measured through the c-index are the same for both model-based parallel and sequential C18|WCX.

Finally, note that the model-based combinations, both parallel and sequential outperform the single-omic models, even after accounting for re-calibration of those. The average-based parallel combination seems also beneficial when focusing on discrimination, but its appropriateness is questionable when focusing on calibration, specially in terms of Q^2, its performance is worse than the observed for single-omic models.

DILGOM Data

In the DILGOM data, each of the considered sources of predictors (transcriptomics and metabolomics) presents a considerably different performance. In terms of discrimination, the metabolome itself presents a c-index around 0.81 while the discriminatory ability of the transcriptomics is notably lower with c-index around 0.70. In the same line, we observed that the metabolomics predictors also outperform transcriptomics in terms of calibration. Note that the Q^2 of metabolomics is more than twice larger than the Q^2 for transcriptomics. These differences are observed with both studied methods, ridge and lasso regression.

The differences between metabolomics and transcriptomics remain approximately the same when focusing on re-calibrated single-omic predictions (Table 4). In the DILGOM setting, re-calibration only improves the results obtained for transcriptomics based on ridge regression. The performance of the re-calibrated ridge-based model for metabolomics is slightly worse than the crude ones. As we observed in the breast cancer settings, the re-calibration of single-source lasso-based predictions is not beneficial.

The performance of the naïve method is clearly unsatisfactory, specially in terms of calibration for the ridge specification. Focusing on Q^2 as summary measure, the behavior of the naïve combination ($Q^2 = 0.090$) is far from the performance of the best single-source model, based on metabolome, with $Q^2 = 0.256$. Similar results were found with regard to the Brier score and the deviance. Regarding calibration, the performance of the naïve combination approach, even if sub-optimal, is better when it relies on a lasso specification. For example, the naïve combination based on lasso regression presents $Q^2 = 0.201$, while for the metabolome-based model, we found $Q^2 = 0.285$. In terms of c-index, both lasso and ridge naïve combinations behave similarly.

As in the breast cancer setting, we observe a better performance of the averaging parallel combination in terms of discrimination, while the model-based parallel combination outperforms averaging in terms of calibration. The sequential approaches behave similar to the model-based parallel combination, with a slight outperformance of the sequential approaches based on using transcriptomics first.

Specifically, even if the differences are slight, it seems that the strategy of first fitting a model based on transcriptomics and adding metabolome in a second step is the preferable one, for both ridge and lasso specifications. Moreover, lasso seems to provide better estimates of calibration than ridge ($Q^2 = 0.359$ for lasso versus

$Q^2 = 0.284$ for ridge), while the ridge specification provides larger discrimination ability (c-index=0.829 for ridge versus c-index=0.808 for lasso). Nevertheless, for the ridge specification the maximum c-index is reached by the average parallel combination (c-index=0.837). For lasso regression, both average parallel and the sequential combination *Metab|Transc* provide the same maximum c-index=0.808.

However, the benefit of using a combination of transcriptomics and metabolomics instead of using metabolomics alone in order to classify individuals as obese or not is arguable from the discrimination point of view, since the differences in c-index are small. The improvement in calibration is also not clear, only in terms of Q^2 the combination of both sources seems to outperform the prediction based on metabolites only.

5 Summary and Discussion

In this work, we have addressed the problem of integrating several omic sets in the context of prediction of binary outcomes. Several methods for the combination of single-source predictions have been presented, all relying on regularized regression.

First of all, we have considered two "parallel" combination approaches, in which a new vector of predictions is obtained as a weighted sum of single-source predictions of the outcome of interest. The specific weight may be fixed beforehand (for example, averaging the single-source predictions) or estimated (model-based parallel combination). For the latter, we considered a logistic regression model using the individual predictions as covariates. Given that the single-source predictions are fitted themselves, the model-based combination requires to be embedded in the cross-validatory setting in order to obtain unbiased final combined predictions.

As an alternative to parallel approaches, we have considered a "sequential" combination method consisting of choosing beforehand one of the omic sources as "primary." Namely, we propose to introduce the vector of individual predictions based on the "primary" source as an extra covariate with fixed weight when fitting a prediction model based on the "secondary" source of omic predictors. In the context of logistic regularized regression models, this is implemented by considering the vector of predictions based on the "primary" source as an offset term. As in the model-based parallel combination, the use as covariate of a vector of predictions (which are fitted themselves) requires an extra layer of cross-validation to embed the uncertainty of calibrating the first source of predictors in the procedure.

We have applied the studied combination methods to two omic applications. Firstly, we have revisited the problem of the combination of different fractionations in proteomic spectrometry for breast cancer diagnosis. Secondly, we have evaluated the possible combination of transcriptomics and metabolomics for the prediction of obesity.

Our results show that better predictions can be obtained by combining predictions based on different omic sources, outperforming single-omic predictions. This seems to be the case for the first of our applications, as combining two proteomic fractions benefits breast cancer classification, from both discrimination and calibration points of view. With respect to the DILGOM study, our results are less conclusive. The sequential approach suggests that transcriptomics are of little use for improving the predictive performance of a ridge or lasso model based on metabolomics only. The reverse is not true, as enriching a transcriptomic-based predictor with metabolomics measures leads to more accurate predictions.

The preferable method to conduct such combination seems to depend on the aspect we focused on for the evaluation of the resulting models. In this work, we have evaluated the resulting predictions in terms of discrimination and calibration. For improving discrimination, measured through c-index, averaging single predictions seems to be enough, and in fact, this simple method provides slightly better results than the more sophisticated model-based parallel approach. In terms of calibration, both parallel and sequential model-based approaches present better and comparable (between them) performance than more simple approaches, as averaging of individual predictions. As expected, the naïve approach, consisting of stacking the omic sources ignoring their different origin, is highly misleading.

We have focused on combination approaches for the integration of different sets of omic predictors. An alternative route, still based on regularized regression, may be to consider different penalizations for each different omic dataset [17, 28]. A systematic comparison of these methods with combination approaches is left as future research. Also, the analysis in this context of alternative model building techniques which (to some extent) rely on the idea of combination of simple classifiers such as random forests [1] and boosting [2, 27] is left as interesting line of future research. Also, it would be interesting to consider modifications of the currently used model-based parallel combination. For example, positively restricted regression coefficients [12] or non-linear combinations of single-source predictions could be considered.

Finally, we would like to highlight that we have focused on evaluating the predictive ability of combinations of different omic-based predictions. Formally testing the added predictive value of a given omic dataset on top of an established one falls beyond the scope of this work. Recently, a test for added value based on the sequential approach presented on Sect. 2.3. has been proposed for continuous outcomes [23]. Its extension to binary outcome is left as a promising line of future research.

Acknowledgements The work is supported by funding from the European Community's Seventh Framework Programme FP7/2011: Marie Curie Initial Training Network MEDIASRES with the Grant Agreement Number 290025 and by funding from the European Union's Seventh Framework Programme FP7/Health/F5/2012: MIMOmics under the Grant Agreement Number 305280. We thank Sigrid Jusélius Foundation and Yrjö Jahnsson Foundation for providing the DILGOM expression data and Yrjö Jahnsson Foundation for providing DILGOM NMR metabolomics.

References

1. Breiman, L. (2001). Random forests. *Machine Learning, 45*, 5–32.
2. Bühlmann, P., & Hothorn, T. (2007). Boosting algorithms: Regularization, prediction and model fitting. *Statistical Science, 22*, 477–505.
3. Cox, D. R. (1958). Two further applications of a model for binary regression. *Biometrika, 45*, 562–565.
4. de Noo, M. E., Deelder, A. M., Mertens, B. J. A., Ozalp, A., Bladergroen, M. R., van der Werff, M. P. J., & Tollenaar, R. A. E. M. (2005). Detection of colorectal cancer using MALDI-TOF serum protein profiling. *European Journal of Cancer, 42*, 1068–1076.
5. Hand, D. J. (1997). *Construction and assessment of classification rules*. Chichester: Wiley.
6. Hand, D. J. (2006). Classifier technology and the illusion of progress. *Statistical Science, 21*, 1–18.
7. Hastie, T., Tibshirani, R., & Friedman, J. (2001). *Elements of statistical learning: Data mining, inference, and prediction*. Springer series in statistic. New York: Springer
8. Hoerl, A. E., & Kennard, R. (1970). Ridge regression: Biased estimation for nonorthogonal problems. *Technometrics, 12*, 55–67.
9. Inouye, M., Kettunen, J., Soininen, P., Silander, K., Ripatti, S., et al. (2010). Metabonomic, transcriptomic, and genomic variation of a population cohort. *Molecular Systems Biology, 6*, 441.
10. Inouye, M., Silander, K., Hamalainen, E., Salomaa, V., Harald, K., Jousilahti, P., et al. (2010). An immune response network associated with blood lipid levels. *Plos Genetics, 6*, e1001113. doi:10.1371/journal.pgen.1001113.
11. Jonathan, P., Krzanowski, W. J., & McCarthy, M. V. (2000). On the use of cross-validation to assess performance in multivariate prediction. *Statistics and Computing, 10*, 209–229.
12. Kakourou, A., Vach, W., & Mertens B. (2014). Combination approaches improve predictive performance of diagnostic rules for mass-spectrometry proteomic data. *Journal of Computational Biology, 21*, 898–914.
13. Kneib, T., Hothorn, T., & Tutz, G. (2009). Variable selection and model choice in geoadditive regression models. *Biometrics, 65*, 626–634.
14. Leblanc, M., & Tibshirani, R. (1996). Combining estimates in regression and classification. *Journal of the American Statistical Association, 91*, 1641–1650.
15. Le Cessie, S., & van Houwelingen, J. C. (1992). Ridge estimators in logistic regression. *Applied Statistics, 41*, 191–201.
16. Liu, H., DÁndrade , P., Fulmer-Smentek, S., Lorenzi, P., Kohn, K. W., Weinstein, J. N., Pommier, Y., & Reinhold, W. C. (2010). mRNA and microRNA expression profiles of the NCI-60 integrated with drug activities. *Molecular Cancer Therapeutics, 9*, 1080–1091.
17. Meier, L., van de Geer, S., & Bühlmann, P. (2008). The group lasso for logistic regression. *Journal of the Royal Statistical Society B, 70*, 53–71.
18. Mertens, B. J. A. (2003). Microarrays, pattern recognition and exploratory data analysis. *Statistics in Medicine, 22*, 1879–1899
19. Mertens, B. J. A., de Noo, M. E., Tollenaar, R. A. E. M., & Deelder, A. M. (2006). Mass spectrometry proteomic diagnosis: Enacting the double cross validatory paradigm. *Journal of Computational Biology, 13*, 1591–1605.
20. Mertens, B. J. A., van der Burgt, Y. E. M., Velstra, B., Mesker, W. E., Tollenaar, R. A. E. M., & Deelder, A. M. (2011). On the use of double cross-validation for the combination of proteomic mass spectral data for enhanced diagnosis and prediction. *Statistics and Probability Letters, 81*, 759–766.
21. Pencina, M. J., D'Agostino, R. B. Sr., D'Agostino, R. B. Jr., & Vasan, R. S. (2008). Evaluating the added predictive ability of a new marker: From area under the ROC curve to reclassification and beyond. *Statistics in Medicine, 27*, 157–172.
22. Pepe, M. S., Kerr, K. F., Longton, G., & Wang, Z. (2013). Testing for improvement in prediction model performance. *Statistics in Medicine, 32*, 1467–1482.

23. Rodríguez-Girondo, M., Salo, P., Burzykowski, T., Perola, M., Houwing-Duistermaat, J. J., & Mertens, B. (2016) Sequential double cross-validation for augmented prediction assessment in high-dimensional omic applications. *Working Paper in ArXiv*. arXiv:1601.08197v1.
24. Steyerberg, E. W., Vickers, A. J., Cook, N. R., Gerds, T., Gonen, M., Obuchowski, N., Pencina, M. J., & Kattan, M. W. (2010). Assessing the performance of prediction models: A framework for some traditional and novel measures. *Epidemiology, 21*, 128–138.
25. Stone, M. (1974). Cross-validatory choice and assessment of statistical predictions (with discussion). *Journal of the Royal Statistical Society. Series B, 36*, 111–147.
26. Tibshirani, R. (1996). Regression shrinkage and selection via the lasso. *Journal of the Royal Statistical Society. Series B (Methodological), 58*, 267–288.
27. Tutz, G., & Binder, H. (2006). Generalized additive modeling with implicit variable selection by likelihood-based boosting. *Biometrics, 62*, 961–971.
28. van de Wiel, M. A., Lien, T. G., Verlaat, W., van Wieringen, W. N., & Wilting, S. M. (2015). Better prediction by use of co-data: Adaptive group-regularized ridge regression *Statistics in Medicine, 35*, 368–381.
29. van der Laan, M. J., Polley, E. C., & Hubbard, A. E. (2007). Super learner. *U.C. Berkeley Division of Biostatistics Working Paper Series*, Working Paper 222.
30. Vickers, A. J., & Elkin, E. B. (2006). Decision curve analysis: A novel method for evaluating prediction models. *Medical Decision Making, 26*, 565–574.
31. Zou, H., & Hastie, T. (2005). Regularization and variable selection via the elastic net. *Journal of the Royal Statistical Society. Series B, 67*, 301–320.

Statistical Analysis of Lipidomics Data in a Case-Control Study

Bart J.A. Mertens, Susmita Datta, Thomas Hankemeier, Marian Beekman, and Hae-Won Uh

1 Introduction

Lipids are essential components for cell functions such as storing energy, signaling, and act as structural components of cell membranes. They can be found in bodily fluids such as blood and plasma. These lipids are mostly molecules including fats, waxes, sterols, fat-soluble vitamins (such as vitamins A, D, E, and K), monoglycerides, diglycerides, triglycerides, phospholipids, and others. Lipids are obtained, for example, through our diet and can be modified by gene-coded enzymes. There exist about 10,000 different lipids which are annotated in the most comprehensive lipid database, LIPID MAPS [14, 24]. The LIPID MAPS consortium introduced a unique 12-digit identifier to represent each lipid molecule [5, 6]. Tandem mass spectrometry (MS/MS) fragment information for several lipid molecular species is also available through LIPID MAPS. Lipids are not encoded by genes. In the past decade, lipidomic and metabolomic research was overshadowed by the progress of genomics. Many recent research projects, however, involve lipidomics research

B.J.A. Mertens (✉) • M. Beekman • H.-W. Uh
Department of Medical Statistics and Bioinformatics, Leiden University Medical Center,
PO Box 9600, 2300 RC Leiden, Netherlands
e-mail: b.mertens@lumc.nl

S. Datta
Professor, Department of Biostatistics, University of Florida, Gainesville, FL, USA
e-mail: susmita.datta@ufl.edu

T. Hankemeier
Faculty of Science, Leiden Academic Centre for Drug Research,
Analytical BioSciences, Leiden, Netherlands

© Springer International Publishing Switzerland 2017
S. Datta, B.J.A. Mertens (eds.), *Statistical Analysis of Proteomics, Metabolomics, and Lipidomics Data Using Mass Spectrometry*, Frontiers in Probability and the Statistical Sciences, DOI 10.1007/978-3-319-45809-0_15

277

which aims to establish their biological role with respect to the expression of proteins involved in lipid metabolism and function, which includes gene regulation [13, 28].

Lipids have been implicated in many complex diseases such as cardiovascular disease (CVD) due to atherosclerosis, the leading cause of fatality in the USA and most developed western countries [29] and many other diseases. Hence, early detection of patients through lipidomic biomarkers may improve prognosis and reduce intervention-related fatality and morbidity with complex diseases such as CVD. The overall lipidomic profile of a patient can be studied by analyzing the lipid concentrations in bodily fluid samples, such as plasma, through mass spectrometry.

Differences in whole-body lipid metabolism between men and women have long been recognized. For example, lower-body fat accumulation in women has been observed, while there is more marked accumulation of intra-abdominal visceral fat in men. Gender differences can typically be observed when comparing men with premenopausal women. Total cholesterol, LDL-cholesterol, and triacylglycerol (TG) concentrations are lower for premenopausal women, while HDL-cholesterol concentration is higher as compared to men [9]. Subsequently, sex-specific approaches for the prevention, diagnosis, and treatment of diseases have been examined for dyslipidemia, cardiovascular disease, metabolic syndrome, and other conditions [2, 3, 18]. It is well known that total TG concentrations are higher in men than in women. Given that the plasma lipidome contains dozens of different triglyceride species that constitute the parameter of total TG [19, 20], lipid differences between sexes at the molecular level are rather limited. Sugiyama and Agellon [25] discussed how differences in the acquisition, trafficking, and subcellular metabolism of fats in major organ systems can create overt sex-specific phenotypes. Gonzalez-Covarrubias et al. [7] investigated which specific lipids contribute to familial longevity and explored gender differences. Female offspring shows higher levels of ether phosphocholine (PC) and sphingomyelin (SM) species (3.5–8.7%) and lower levels of phosphoethanolamine PE (38:6) and long-chain triglycerides (TG) (9.4–12.4%). These results were obtained by rather simple statistical analysis using either univariate linear regression or logistic regression depending on the outcome variable. In this chapter we discuss some aspects of data analysis of lipidomics using mass spectrometry.

The structure of the remainder of this chapter is as follows. First we introduce the Leiden Longevity study and the lipidomics data which is a component of this study. The problem of missing values and their imputation is important for lipids data and is discussed next. We then use the Leiden Longevity lipidomics study data to illustrate key aspects of lipidomic data-based analysis. We explore the problem of lipid selection and co-selection in a case-control design generated within the Leiden Longevity Study, using a Bayesian model formulation. For our analysis, we restrict to *the females-only subset* of the full dataset, in order to avoid the systematic between-gender differences present in lipidomic expression discussed above.

2 The Leiden Longevity Study

The Leiden Longevity Study included 421 Caucasian families consisting of long-lived siblings together with their offspring and the spouses of the offspring. In total 2415 members of long-lived families have been recruited consisting of the offspring of long-lived siblings ($n = 1671$) and the partners of the offspring as controls ($n = 744$). The Medical Ethical Committee of the Leiden University Medical Center approved the study and informed consent was obtained from all subjects. The final study data consists of 2201 participants, of which 1526 are offspring of nonagenarian siblings (59 years $\pm$ 6.7), and 675 are controls (59 years $\pm$ 7.5). Citrate plasma was available and after quality control, plasma lipidome measurements were generated. Lipidome analysis was performed in citrate plasma by ultra-high pressure liquid chromatography coupled to mass spectrometry (UPLC-MS) using an optimized version of the method reported by Hu et al. [8] and has been described previously. Validation parameters were: linearity LPC (19:0), $r^2 > 0.99$; PC (34:0), $r^2 > 0.97$; PE (34:0), $r^2 > 0.98$; TG (45:0), $r^2 > 0.99$; repeatability and reproducibility, RSD < 15 %. The r^2 notation refers to r-squared statistics computed between repeated measurements for the above set of lipids. The RSD is also calculated based on these repeated measures. Lipid names and abbreviations were assigned according to LIPID MAPS nomenclature (http://www.lipidmaps.org). The following accepted abbreviations were used: phosphocholines, PC; sphingomyelins, SM; triglycerides, TG; and phosphoethanolamines, PE. The alkyl ether linkage is represented by the "O-" prefix, for example, PC (O-34:1), TG (O-50:2), the numbers following the hyphen refer to the total number of carbons of the fatty acyl chains followed by the number of double bonds of all the chains. Relative lipid levels were calculated for a total of 131 lipids from six different classes: TG, SM, PE, cholesteryl esters, lysophosphatidylcholines, and PC. Most lipid species showed a right-skewed distribution. Hence, data were logarithmically transformed prior to statistical analyses. Further, missing lipids values were imputed using the Multivariate Imputation by Chained Equations (MICE) and location (mean) batch adjustment was employed to correct for batch effects. For the current analysis, 332 female controls and 685 female offspring were analyzed, in which the status of offspring and partner was used as the binary outcome. Hence, we considered the relative levels of these 131 lipids with regard to the binary outcome (the status, offspring of nonagenarians, and controls) to identify lipid species that associate with familial longevity. For further details we refer to Gonzalez-Covarrubias et al. [7]. To take into account the high correlation among lipid variables [7, Fig. 3], we investigated use of a Bayesian model selection method instead of a "one-lipid-at-a-time" method.

3 Missing Value Imputation in Lipidomics

Another example of the use of multiple imputation (MI) in lipidomics data is found in the LUdwigshafen RIsk and Cardiovascular Health (LURIC) study [12]. The main goal of this study was to identify network structure changes and lipidomic biomarkers associated with cardiac arrest for patients with coronary artery disease (CAD). In analyzing the data, the authors identified a large number of missing values in the mass spectrometry data. One option to deal with such missing values is to consider only the complete case analysis, which means we only use patients with complete observation records. This can lead to a great reduction in sample size and hence loss of precision. In most mass spectrometry platforms, missingness is commonplace for proteomic, metabolomic, and lipidomic data.

Two sources of missingness are typically observed with mass spectral data. One could be viewed as missingness purely at random while the other is due to either the lower limit of quantification (LLOQ) of the instrument or the operator-determined denoising cutoff. Very low-level concentrations are then considered too imprecise when values are only known to be somewhere between zero and a known lower limit of quantification (LLOQ). In such situations, this gives rise to left-censored observations. Recent research by Kakourou et al. [10, 11] provides another good example and extensive discussion on censored observations in (Fourier Transform) mass spectrometry.

3.1 Imputation for Lipidomic Mass Spectrometry Data

In the presence of observations which are either completely missing or "partially observed," imputation can sometimes be performed prior to subsequent data analysis. We assume the mass spectrometry data is properly pre-processed prior to data imputation and that we have a case-control study. In general, we observe a vector of binary responses $Y = (y_1, y_2, \ldots, y_n)'$, with the values y_i, $i = 1, \ldots, n$ indicating whether a patient is either case or control. The matrix X contains the log-transformed lipid concentrations of the n patients. Let x_{ij} denote an individual (log-transformed) concentration of lipid j for patient i within the matrix X. Commonly, a substantial number of elements x_{ij} are not available due to either partial or complete missingness. Hence, we may divide the $x's$ in two parts as x^{obs} and x^{miss} denoting the observed and the missing values, respectively. In the presence of an assigned LLOQ, denoted by (say) l_j, some of these values will be below l_j and thus left-censored, also referred to as non-detects. The LLOQ thresholds are set for each platform and can vary between lipid species. The other type of missing values are due to the elimination of observations not fulfilling the quality control standards or not observed due to some random error. It is reasonable to regard the later to be missing completely at random. We take these distinct types of missing values into account in the imputation algorithm by imputing them in two different ways but under the same imputation scheme.

The above approach is often implemented in the statistical technique "Multiple Imputation" (MI). Its theoretical foundation is well developed and is fully described in several excellent references such as [23] or [4, 16, 17], among many others. It has been widely used for omics data. In essence, MI calibrates the conditional distribution of the unobserved data, and then generates random draws (simulations) from this density which are then used to impute (replace) the non-observed data. The imputation itself is repeated $M \geq 1$ times to account for the randomness (uncertainty) of the imputation—hence the term "multiple" imputation. Thus, M copies (or versions) of the imputed dataset are generated and each is then analyzed separately—in principle—as if each were the complete data set. For example, should we be interested in regression parameter estimates for some type of regression model based on the (lipid) data, then M copies of the relevant parameter estimates would be obtained with the above procedure. These need to be combined across the M analyses on the respective imputed data sets, usually by taking the mean. Likewise, an estimate of uncertainty of the combined estimate across imputations should be computed, which adjusts for the uncertainty in imputation of the unknown data. For mean estimates or regression parameter effects, standard errors can usually be generated by suitable combination of within-imputation with between-imputation variances, using Rubin's rules [23]. We provide more details on the imputation strategy below.

The construction of an appropriate (conditional) imputation distribution (or model)

$$f(x^{\text{miss}} | x^{\text{obs}}, y, \theta),$$

where θ denotes a set of unknown parameters, is key in multiple imputation. The model in essence describes what values are credible for the unobserved data, given the observed measurements. A simple approach to calibrate this model is by using a set of univariate regression models (the imputation models), by considering each incomplete variable at a time and then generate a prediction model for each such partially observed variable. For continuous incomplete variables, this can be done by fitting a linear regression of the target variable on the set of all other variables within the problem (including the case-control outcome). For binary variables one can use a logistic regression model. Note that in case-control studies (but similar for ordinary linear regression or survival analysis, etc.), the outcome variable should always be included in the predictor set for each of these imputations models, in order to respect the association between outcome and predictors (the latter being the lipids or omics data in our case). The above approach thus leads to imputation through multiple chained (prediction) equations and is implemented in the package (MICE) [21, 26, 27].

The procedure incorporates a random aspect, in that random draws are taken from the conditional densities (regression models) for the unknown values and using some suitably chosen starting values, but also iterative, in that the sequence is left running to generate a number of updates sequentially, before the final imputed values are stored. Thus, one first fills the data matrix with suitable starting values, and then simulates random draws for the unobserved entries and for each (partially observed) variable at a time from the relevant corresponding regression model, each time taking the full data matrix on all other measures (which includes the outcome variable) as the predictor set. At each iteration, this predictor set contains the current set of imputed values and the actually observed entries (which are of course never updated and kept fixed throughout). This process is left to run until some convergence criterion is met [1, 21]. For observations subject to a lower detection limit LLOQ l_j, the above procedure must be slightly modified to adjust for this known limit. This can be done, for the left-censored values, by sampling from the relevant conditional distribution (regression model), but only accepting the thus simulated value if it falls below the known limit l_j. If the condition is not met, then one simply keeps sampling until the condition is satisfied.

3.2 Stacking the Multiple Imputed Data Sets

In the case of the LURIC data analysis in Kujala et al. [12] the ultimate goal of the data analysis was to construct networks based on the lipidomics data and finally find the differences between the networks of the case and control groups. So the question was to determine how the distinct M multiple imputed data sets and their respective analyses should then be combined, as it is not clear how to generalize Rubin's rules to this scenario. The concept of averaging a network does not appear trivial. To address this issue, the authors applied the stacking-method proposed by Wood et al. [30]. Instead of performing the downstream data analysis separately for M sets of imputed values, one constructs a single data set containing Mn rows which simply stacks the M simulated datasets (n is the sample size in the original data). It can be shown that the stacking keeps the sample mean unchanged but reduces the variability [12].

4 Case-Control Modeling for Lipids Data

4.1 Model Specification

Logistic Regression Model for Case-Control Outcome

We analyze the Leiden Longevity lipids case-control data. The objective of the analysis is to assess whether the lipid data has added-value in the calibration of the case-control probabilities, after correction for age and Framingham risk score, which is known to be associated with the outcome. To implement this assessment, we use an adaptation of the variable-dimension reversible jump logistic regression model introduced in the chapter "Logistic regression modeling on mass spectrometry data in proteomics case-control discriminant studies" [15].

The variable-dimension model will allow lipids to be either selected, or completely de-selected from the model, after correction for age and Framingham score.

As before, we have $y_i \sim$ Bernoulli(p_i), $i = 1, \ldots, n$ with $n = 1017$ the sample size and y_i is the binary case-control indicator for each ith patient. In contrast to the chapter "Logistic regression modeling on mass spectrometry data in proteomics case-control discriminant studies" we now specify the regression directly on the set of 131 (pre-processed) lipid measures as

$$\text{logit}(p_i) = \alpha + \text{Age}\beta_{\text{Age}} + \text{FR}\beta_{\text{FR}} + Z_F\tilde{\beta}_F + Z_B\tilde{\beta}_B + X_L\tilde{\beta}_L$$

where p_i is the case-probability for the ith observation. The terms Age and FR denote the ages and Framingham risk scores of individuals, with β_{Age} and β_{FR} the corresponding regression effects. The term X_L represents the unknown set of lipid measures which are associated with the outcome, with regression coefficient vector

$$\tilde{\beta}_L = (\beta_1, \ldots, \beta_k)$$

which is of *unknown* dimension k. The fact that this dimensionality is considered unknown is an intrinsic component of the Bayesian approach to modeling the lipid data. This will allow us to assess the added-value of the lipids as described further on. Note also that there is no use of basis functions as described in the chapter "Logistic regression modeling on mass spectrometry data in proteomics case-control discriminant studies" as regression is directly on the processed lipid summary measures, which summarize the observed spectral data. The model consists of a permanent adjustment

$$\alpha + \text{Age}\beta_{\text{Age}} + \text{FR}\beta_{\text{FR}} + Z_F\tilde{\beta}_F + Z_B\tilde{\beta}_B$$

and a variable-dimension component

$$X_L\tilde{\beta}_L$$

with corresponding regression coefficient vector $\tilde{\beta}_L$, which is also of variable dimension. Note that the dimension k could be zero, i.e., X_L is the empty set $X_L = []$. Indeed this hypothesis is a primary objective of the analysis, which is to investigate the probability support given by the data for the presence of an additional regression effect, after correction for both age and the Framingham risk score. In the Bayesian paradigm, we thus view the set X_L, as well as the dimension parameter $0 \leq k \leq 131$ and $\tilde{\beta}_L$ as unknown parameters of the model, which must be estimated from the data. The terms

$$Z_F\tilde{\beta}_F$$

and

$$Z_B \tilde{\beta}_B$$

denote random effects for family and batch effect, respectively. The family random effect can be interpreted as correcting for systematic differences in outcome between families, while the batch random effect is included to adjust for potential between-batch variation left unaccounted for by the prior data processing. The term α denotes an intercept.

Prior Structure

We assume flat independent normal priors for the intercept, age, and Framingham risk score regression coefficients. For the random batch and family effects, we assume separate exchangeable normal priors centered at zero with separate inverse chi-squared priors on the batch and family random effect variances. A discrete uniform prior $U(0, 1, 2, .., 131)$ is used for the model dimension k. For the lipid regression parameters we use the exchangeable priors

$$\beta_{Lj} \sim N(0, \sigma_L^2), \text{ for } j = 1, \ldots, k$$

with common inverse chi-squared prior on the variance.

MCMC and Convergence

The reversible jump implementation is analogous to that described previously in the chapter "Logistic regression modeling on mass spectrometry data in proteomics case-control discriminant studies" [15]. Specifically we use three distinct moves, consisting of either adding, deleting, or interchanging (swapping) a randomly chosen lipid from the current model state. Proposal move acceptances are obtained from the conditional ratios of integrated likelihoods between the current and new proposal regression models, conditional on the current values of k, σ_L^2, σ_B^2, and σ_F^2. Note that with the above prior specification, normal inverse scaled chi-squared updating can be applied, conditional on the variance hyper-parameters. This implies that the "conditional" acceptance rates do not require actual calculation of the regression parameters on either the old or proposal models, as closed forms are available for the integrated likelihood functions conditional on the above variance terms. It leads to significant gains in computational speed. We use the auxiliary variable construction with Kolmogorov–Smirnov prior on the normal variance to obtain the logistic regression model form (see [15] for the details of the same approach).

We generated 200,000 initial burn-in samples which were discarded. Subsequently, we sampled three additional sets of simulations of 400,000 updates each and sequentially, within this single chain. We compared the last set of 400,000 simulations with the first to assess convergence of simulations. Q-Q plots between both sets were investigated, as well as kernel densities and autocorrelations on model parameters (regression parameters, model variances, and selected model dimensionality k) as well as on the Bernoulli deviance. Consistency of lipid selection between both sets was investigated as well as mean marginal regression parameters. No inconsistencies were detected. For inference, the three sets of updates were combined into a single set of simulations (1.2 million samples) for posterior model inference.

4.2 Posterior Inference for Lipids Regression Modeling

Model Dimension

The primary target of the posterior model inference is to establish evidence for the presence of a lipid regression effect, after correcting for age and Framingham risk score. Figure 1 shows to posterior density on model dimensionality k across simulations. The posterior probabilities on the NULL and non-NULL dimensions are $P(k = 0 \mid \text{Data}) = 0.027$ and $1 - P(k = 0 \mid \text{Data}) = 0.973$. Table 1 shows percentiles of the posterior density of k. The lower 5 % bound indicates a minimum dimensionality of 1.

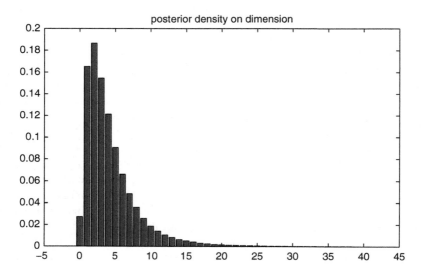

Fig. 1 Posterior probabilities of selected model dimensionalities plotted versus model dimensionality

Table 1 Posterior percentiles of model dimensionalities

Percentile	1.0	2.5	5.0	10.0	25.0	50.0	75.0	90.0	95.0
Dimensionality k	0	0	1	1	2	3	6	9	11

Table 2 Posterior probabilities of model dimensionality and ratios between subsequent model dimensionality probabilities

k	Posterior	Exploratory "Bayes factor" (BF)
0	0.03	0.00
1	0.17	6.08
2	0.19	1.13
3	0.15	0.83
4	0.12	0.78
5	0.09	0.75
6	0.07	0.73
7	0.05	0.73
8	0.04	0.74
9	0.03	0.71
10	0.02	0.72

Use of Bayes factors to compare the between-dimensionalities posterior probabilities is cumbersome in this example, as these are themselves affected by the within-model variances and assumptions thereon. Table 2 shows posterior probabilities on dimension in tabular form with ratios between subsequent posterior probabilities denoted as "exploratory Bayes factor." Note how only the transition from a NULL to a one-dimensional model is associated with a substantial increase in probability as expressed through the BF ratio. For all subsequent dimension increments, these exploratory BF ratios are approximately 1 or smaller. The mean, median, and modal posterior dimension k are 4.3, 3.7, and 2, respectively. All model dimensions $k > 4$ have prob $< \max(\text{prob}(k))$.

An alternative investigation of the impact of dimension on model fit can be obtained from an analysis of the associated model deviances. We define the model deviance as the deviance of any individual model within the sequence of MCMC model updates. This sequence of deviances can then be investigated as any other posterior summary. This is shown in Fig. 2 which plots kernel density estimates for the posterior deviances between (conditional on) distinct model dimensionality, starting from dimension $k=0$ and up to a dimensionality of ten lipids. The plots confirm the summary measures from the posterior summaries on the dimension parameter k above in that most improvement in model fit seems obtained from the transition of a zero to a one-dimensional model, with marginal gains obtained from a two-dimensional lipids model. We can thus conclude that there appears to be evidence that the lipids are associated with the case-control outcome, after correction for age and Framingham score and that a model with one or at most two lipids should suffice to account for this extra information.

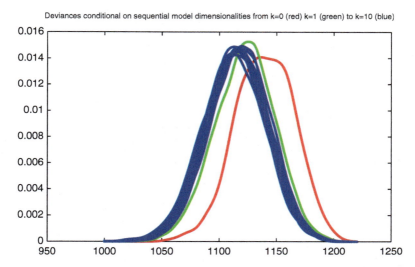

Fig. 2 Kernel density estimates of model deviances conditional on model dimensionality. The NULL (no lipids selected) model is plotted as *red. Green* is used for one-dimensional model deviances and *blue* for all other dimensionalities

Lipid Selection

Having established the presence of a lipid effect, we may now address the question "where does this extra information come from?" Figure 3 shows the marginal posterior probabilities of inclusion of any specific lipid plotted versus the lipid number (within the data matrix) (top plot) as well as a histogram of these posterior probabilities (bottom plot). As can be seen there is only moderate evidence in favor of specific lipids. This is due to substantial correlation between lipids.

Table 3 shows the top ten lipids selected to the models (left column), in decreasing order of posterior probability of inclusion (right column). The table also gives the lipid identity in the second column for this selection (LIPID MAPS). To simplify, we will refer to lipids just by the lipid number (first column) in the text.

In addition, the table also shows effect estimates (B) for a between cases-and-controls group comparison with associated standard errors (SE), p-values (P), and (Bonferroni) corrected p-values from a classical mixed effect models fitted for each lipid separately and using the lipid levels as outcome, correcting for the same random (batch and family) and fixed effects (age and Framingham score). The top-three lipids, 36, 107, and 108 would appear most interesting. Lipid selection and ranking is identical for the top-three lipids between both analyses shown in this table, though the probability calibrations are incomparable. Following the top-three lipids, association between lipid ranking is much reduced between the "multivariate" Bayesian and univariate mixed-model analyses and thus results in a very different rankings between both.

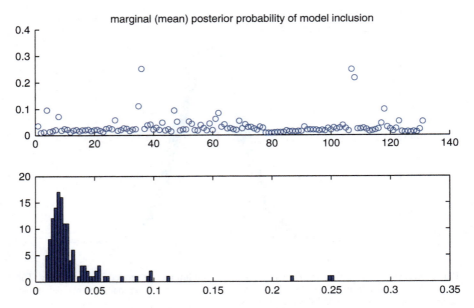

Fig. 3 Posterior probabilities of inclusion into the model for individual lipids. The *top plot* shows posterior probabilities versus lipid number. The *bottom plot* is a histogram of these probabilities across all 131 lipids in the analysis

Table 3 Table showing the top-ten selected lipids to the Bayesian model by lipid number and lipid identification (left two columns)

Lipid nr.	Lipid	B	SE	P	Bonferroni	Pprob
36	GPCho (O-36:3)	0.07	0.02	8.7e−05	0.01	0.25
107	TG (54:7)	−0.14	0.04	4.6e−05	0.006	0.25
108	TG (54:6)	−0.15	0.04	8.2e−05	0.01	0.22
35	GPCho (O-36:4)	0.06	0.02	0.0006	0.08	0.11
118	TG (56:6)	−0.10	0.03	0.0002	0.04	0.10
4	ChoE (22:6)	−0.09	0.03	0.002	0.3	0.10
47	GPCho (38:6)	−0.08	0.02	0.0007	0.1	0.09
62	SM (d18:1/15:0)	0.04	0.02	0.004	0.5	0.08
8	GPCho (O-16:0)	0.03	0.01	0.005	0.7	0.07
61	SM (d18:1/14:0)	0.03	0.02	0.02	2.7	0.06

Also shown are results from mixed models calculations for each lipid separately, with "B" the between-group effect, "SE" the associated standard error, and "P" and "Bonferroni" the standard p-values both with and without Bonferroni correction. The right column shows the inclusion probabilities from the Bayesian multivariate logistic regression model for each lipid

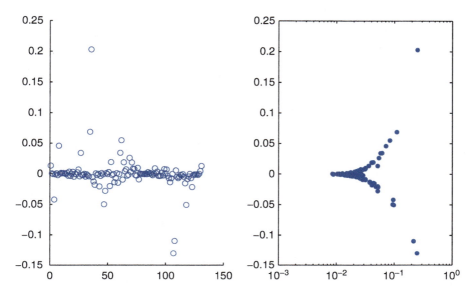

Fig. 4 The *left plot* shows marginal mean regression coefficients from the Bayesian model versus lipid number. The *right plot* shows the marginal mean regression coefficients versus the corresponding probability of selection for that lipid

Lipid Regression Effect

We can compute mean marginal regression effects across models as shown in the left-side plot of Fig. 4 for each lipid versus the lipid number. The right-side plot in this figure plots the same mean marginal regression effects versus the posterior probabilities of inclusion of the corresponding lipid into the model. Obviously, higher absolute mean marginal effects sizes are associated with lipids which have a higher posterior probability of selection, as one would expect. The top-three most-selected lipids (36, 107, and 108) also have the highest absolute average marginal effects. It must be noted that the averaging of the effects across all models, irrespective of dimensionality, hides a dependence of the calibrated effect on the dimensionality and hence also variance. This is shown in Fig. 5 which plots averaged effects for the top-seven lipids conditional on the dimensionality of the model. The plot shows another consequence of the association between variance and dimension. Higher-dimensional models are associated with smaller effects, as they can spread the effect across several (correlated) lipids. We can make a similar observation when investigating lipid selection conditional on model size. As models become larger, any individual lipid will have a smaller probability of being selected to the model (Fig. 6). Nevertheless, lipids selected to the models tend to remain the same across dimensions. Considering models with dimensions of 1 up to 5, the lipids 36, 107, 108, 47, and 35 are consistently among the top-ten selected lipids. Among these, lipids 36, 107, and 108 are always the top-three most-selected lipids within models.

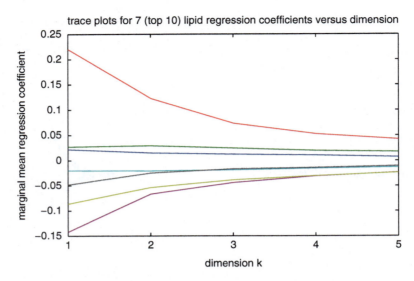

Fig. 5 Marginal mean regression parameters of the top-seven most selected lipids, computed conditional on fixed model dimensionalities (1–5) and plotted versus that model dimensionality

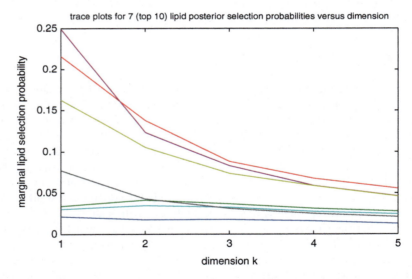

Fig. 6 Probability of selection to the model for the top-seven most selected lipids, computed conditional on fixed model dimensionalities (1–5) and plotted versus that model dimensionality

Lipid Co-selection

It is a particular feature of the Bayesian approach that we may not only investigate the extent to which individual lipids are selected to the model, but also the extent

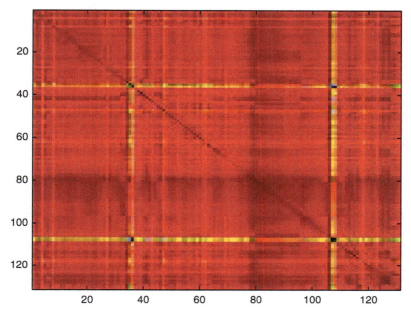

Fig. 7 Image showing the probability of co-selection of each pair of lipids jointly to the model (for models with dimensionality at least 2). *Deeper red* means higher frequency of selection of that pair, while *lighter colors* shows lower probability. The *horizontal and vertical axis* are the lipid numbers (1 to 131)

to which these are co-selected, or reversely, whether specific lipids tend to mutually exclude one another within the same model. Figure 7 shows an image representation of the matrix which stores the counts of the number of times two lipids are co-selected to the model. Some lipids seem to be mutually exclusive, such as 35 and 36, or 107 and 108, or 61 and 62, for example. Other pairs, on the other hand, seem to be more common. Examples are combinations of either 107 or 108 with 35 or 36. Another is 61 or 62 combined with either 107 or 108. Other interesting likely pairs are 36 or 62 together with 4. Yet another example is 61 or 62 with 36. The between-lipids correlations are an obvious driver behind these model selections. An interesting graphical representation of such co-selection is shown in the undirected graph Fig. 8. It represents frequently selected lipid pairs as connected through an edge, after application of a minimal threshold on connectivity. Besides the often occurring pairings, the image is also interesting for the connections which are absent such as between 107 and 108 which tend to be mutually exclusive. Lipids 36, 107, and 108 appear as obvious "hubs" in a network of co-selected lipids.

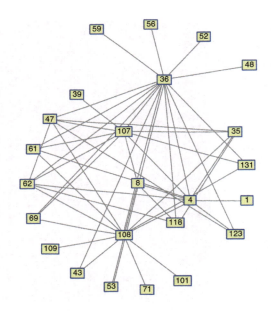

Fig. 8 Lipid pairs with high probability of co-selection to the model shown in an undirected graph, after application of a minimum threshold on connectivity. High probability of co-selection is shown through the presence of an edge between the corresponding lipids

5   Discussion

We have used a variable-dimension model to investigate added-value of lipids data, in addition to age and Framingham risk score in a case-control study. The model allows for assessment of optimal dimensionality of the lipid predictor set. We may also investigate lipid selection and co-selection, which is more difficult to achieve with classical regression approaches. The mutual exclusiveness of the lipids 35 and 36, 61 and 62, and 107 and 108 which was identified seems to have a biological basis. The paired lipids have almost identical structures: lipid 35 (GPCho(O-36:3)) is a glycerophosphocholine that is an alkyl, acyl-sn-glycero-3-phosphocholine in which the alkyl or acyl groups at positions 1 and 2 contain a total of 36 carbons and 3 double bonds, while lipid 36 (GPCho(O-36:2)) is exactly the same structure with only 2 double bonds. Similarly, lipid 61 (SM(d18:1/14:0)) is a sphingomyelin d18:1 in which the amino group of sphingosine is substituted as tetradecanoyl (myristoyl)), while lipid 62 (SM(d18:1/15:0)) is exactly the same structure in which the amino group of sphingosine is substituted by a pentadecanoyl group. Finally, lipid 107 (TG(54:7)) is a triglyceride in which the three acyl groups contain a total of 54 carbons and 7 double bonds, while lipid 108 (TG(54:7)) is a triglyceride with the same amount of 54 carbons but then contains 6 double bonds. Furthermore, the co-selection of lipid 35/36 (glycerophosphocholines) and 107/108 (triglycerides) in addition to the Framingham risk score seems to reflect the additional information in the lipid measures. The fact that glycerophosphocholines and triglycerides are clearly different lipid species, and lipids 61/62 (sphingomyelins) may be somewhat related to lipid 35/36, may explain why lipid 36, 107, and 108 appear as obvious "hubs" in the network of 17 co-selected lipids.

It should be noted that only a single imputation was used in the analysis of the Leiden Longevity study data. This was partly to reduce the computational overhead. It should, however, be noted that within the Bayesian context, multiple imputation can also be included as part of the model estimation procedure within the same MCMC run. Rather than spend time to amend to software to implement this, but certainly also for consistency with the original publication on this data, we decided to only use the original single-imputed dataset used in the first publication [7].

We included a rather extensive discussion on use of multiple imputation for (lipid)omic data in this paper. MI is useful for omic data which tends to contain large numbers of non-detects or censored observations. It must, however, be remarked that for left-censored observations due to detection limits, imputation of the detection limit itself, half the detection limit, or zero, is sometimes considered. When the lipids themselves are of interest (which lipids are associated with the outcome and with which effect), as in this paper, then MI can be applied to reduce bias.

Finally we want to point out that the approach using variable-dimension modeling discussed in this paper could be viewed as a Bayesian alternative to the frequentist cross-validation-based methods discussed by Rodríguez-Girondo (this volume [22]) to assess added-value of the lipids expression, after correcting for the known Framingham risk score and age.

Acknowledgements This work was supported by funding from the European Community's Seventh Framework Programme FP7/2011: Marie Curie Initial Training Network MEDIASRES ("Novel Statistical Methodology for Diagnostic/Prognostic and Therapeutic Studies and Systematic Reviews," www.mediasres-itn.eu) with the Grant Agreement Number 290025 and by funding from the European Union's Seventh Framework Programme FP7/ Health/F5/2012: MIMOmics ("Methods for Integrated Analysis of Multiple Omics Datasets," http://www.mimomics.eu) under the Grant Agreement Number 305280. The Leiden Longevity study was supported by the Innovation-Oriented Research Program on Genomics (SenterNovem IGE05007), the Centre for Medical Systems Biology, the Netherlands Consortium for Healthy Ageing (Grant 050-060-810), the Netherlands Metabolomics Centre (NMC) and the Biobanking and Biomolecular Resources Research Infrastructure (BBMRI) all in the framework of the Netherlands Genomics Initiative, the Netherlands Organization for Scientific Research (NWO), and the European Union's Seventh Framework Program (FP7/2007-2011) under grant agreement No. 259679. This research work was also partially supported by National Institutes of Health grant CA 170091-01A1 (Susmita Datta).

References

1. Azur, M. J., Stuart, E. A., Frangakis, C., & Leaf, P. J. (2011). Multiple imputation by chained equations: What is it and how does it work? *International Journal of Methods in Psychiatric Research, 20*(1), 40–49. doi:10.1002/mpr.329 PMID: 21499542.
2. Barker-Collo, S., Bennett, D. A., Krishnamurthi, R. V., Parmar, P., Feigin, V. L., Naghavi, M., et al. (2015). Sex differences in stroke incidence, prevalence, mortality and disability-adjusted life years: Results from the global burden of disease study 2013. *Neuroepidemiology, 45*, 203–214. doi:10.1159/000441103.

3. Brewer, L. C., Svatikova, A., & Mulvagh, S. L. (2015). The challenges of prevention, diagnosis and treatment of ischemic heart disease in women. *Cardiovascular Drugs and Therapy, 29,* 355–368. doi:10.1007/s10557-015-6607-4.

4. Carpenter, J., & Kenward, M. (2013). *Multiple imputation and its application.* New York: Wiley.

5. Cotter, D., Maer, A., Guda, C., Saunders, B., & Subramaniam, S. (2006). LMPD: LIPID MAPS proteome database. *Nucleic Acids Research, 34,* D507–D510.

6. Fahy, E., Subramaniam, S., Brown, H. A., Glass, C. K., Merrill, A. H. Jr., Murphy, R. C., et al. (2005). A comprehensive classification system for lipids. *Journal of Lipid Research, 46,* 839–862.

7. Gonzalez-Covarrubias, V., Beekman, M., Uh, H. W., Dane, A., Troost, J., Paliukhovich, I., et al. (2013). Lipidomics of familial longevity. *Aging Cell, 12*(3), 426–34. doi:10.1111/acel.12064. Epub 2013 Apr 2.

8. Hu, C., van Dommelen, J., van der Heijden, R., Spijksma, G., Reijmers, T. H., Wang, M., et al. (2008). RPLC-ion-trap-FTMS method for lipid profiling of plasma: Method validation and application to p53 mutant mouse model. *Journal of Proteome Research, 7,* 4982–4991.

9. Jiang, Z., Huang, X., Hang, S., Guo, H., Wang, L., Li, X., et al. (2016). Sex-related differences of lipid metabolism induced by triptolide: The possible role of the LXRa/SREBP-1 signaling pathway. *Frontiers in Pharmacology, 7,* 87. http://dx.doi.org/10.3389/fphar.2016.00087

10. Kakourou, A., Vach, W., & Mertens, B. (2016). Adapting censored regression methods to adjust for the limit of detection in the calibration of diagnostic rules for mass spectrometry proteomic data. arXiv:1606.09123 [stat.ME]

11. Kakourou, A., Vach, W., Nicolardi, S., van der Burgt, Y. & Mertens, B. (2016). Accounting for isotopic clustering in Fourier transform mass spectrometry data analysis for clinical diagnostic studies. *Statistical Applications in Genetics and Molecular Biology, 15*(5), 415–430. doi:10.1515/sagmb-2016-0005

12. Kujala, M., Nevalainen, J., März, W., Laaksonen, R., & Datta, S. (2015). Differential network analysis with multiply imputed lipidomic data. *PLoS ONE, 10*(3), e0121449. doi:http://dx.doi.org/10.1371/journal.pone.0121449.

13. Lagarde, M., Geloen, A., Record, M., Vance, D., & Spener, F. (2003). Lipidomics is emerging. *Biochimica et Biophysica Acta (BBA) - Molecular and Cell Biology of Lipids, 1634,* 61.

14. LIPID MAPS (2006). [http://www.lipidmaps.org].

15. Mertens, B. J. A. (2016). Logistic regression modeling on mass spectrometry data in proteomics case-control discriminant studies. In S. Datta & B. J. A. Mertens (Eds.), *Statistical analysis of proteomics, metabolomics, and lipidomics data using mass spectrometry.* New York: Springer. doi:http://dx.doi.org/10.1007/978-3-319-45809-0.

16. Molenberghs, G., Fitzmaurice, G., Kenward, M. G., Tsiatis, A., & Verbeke, G. (2014). *Handbook of missing data methodology.* London: CRC Press.

17. Molenberghs, G., & Kenward, M. (2007). *Missing data in clinical studies.* New York: Wiley.

18. Pradhan, A. D. (2014). Sex differences in the metabolic syndrome: Implications for cardiovascular health in women. *Clinical Chemistry, 60,* 44–52. doi:http://dx.doi.org/10.1373/clinchem.2013.202549.

19. Quehenberger, O., Armando, A. M., Brown, A. H., Milne, S. B., Myers, D. S., Merrill, A. H., et al. (2010). Lipidomics reveals a remarkable diversity of lipids in human plasma. *Journal of Lipid Research, 51*(11), 3299–3305. doi:http://dx.doi.org/10.1194/jlr.M009449. Epub 2010 Jul 29.

20. Quehenberger, O., & Dennis, E. A. (2011). The human plasma lipidome. *The New England Journal of Medicine, 365*(19), 1812–1823. doi:http://dx.doi.org/10.1056/NEJMra1104901.

21. Raghunathan, T. E., Lepkowksi, J. M., Hoewyk, J. V., & Solenbeger, P. (2001). A multivariate technique for multiply imputing missing values using a sequence of regression models. *Survey Methodology, 27,* 85–95.

22. Rodríguez-Girondo, M., Kakourou, A., Salo, P., Perola, M., Mesker, W. E., Tollenaar, R. A. E. M., et al. (2016). On the combination of omics data for prediction of binary outcomes. In S. Datta & B. J. A. Mertens (Eds.), *Statistical analysis of proteomics, metabolomics, and lipidomics data using mass spectrometry*. Springer, New York. doi:http://dx.doi.org/10.1007/978-3-319-45809-0.
23. Rubin, D. B. (2004). *Multiple imputation for nonresponse in surveys*. New York: Wiley.
24. Sud, M., Fahy, E., Cotter, D., Brown, A., Dennis, E. A., Glass, C. K., et al. (2007). LMSD: LIPID MAPS structure database. *Nucleic Acids Research, 35*, D527–D532. doi:http://dx.doi.org/10.1093/nar/gkl838. PMID: 17098933.
25. Sugiyama, M. G., & Agellon, L. B. (2012). Sex differences in lipid metabolism and metabolic disease risk. *Biochemistry and Cell Biology, 90*(2), 124–141. doi:http://dx.doi.org/10.1139/o11-067. Epub 2012 Jan 5.
26. van Buuren, S. (2007). Multiple imputation of discrete and continuous data by fully conditional specification. *Statistical Methods in Medical Research, 16*, 219–242. doi:http://dx.doi.org/10.1177/0962280206074463.
27. van Buuren, S., & Groothuis-Oudshoorn, K. (2011). mice: Multivariate imputation by chained equations in R. *Journal of Statistical Software, 45*(3), 1–67.
28. Watson, A. D. (2006). Thematic review series: Systems biology approaches to metabolic and cardiovascular disorders. Lipidomics: A global approach to lipid analysis in biological systems. *Journal of Lipid Research, 47*, 2101–2111.
29. WHO. (2009). *World health statistics 2009*. Geneva: WHO Press.
30. Wood, A. M., White, I. R., & Royston, P. (2008). How should variable selection be performed with multiply imputed data? *Statistics in Medicine, 27*, 3227–3246. doi:http://dx.doi.org/10.1002/sim.3717. PMID: 18203127.